INORGANIC SYNTHESES

Volume 29

Editor-in-Chief
RUSSELL N. GRIMES

●●●●●●●●●●●●●●●●●●●●●●●●●●●●●●●●

INORGANIC SYNTHESES

Volume 29

A Wiley-Interscience Publication
JOHN WILEY & SONS
New York   Chichester   Brisbane   Toronto   Singapore

Published by John Wiley & Sons, Inc.

Library of Congress Catalog Number: 39-23015

ISBN 0-471-54470-1

Printed in the United States of America

10 9 8 7 6 5 4 3 2 1

PREFACE

In its more than half-century of existence, the *Inorganic Syntheses* series has reflected, in the contents of its volumes, profound changes in the field of inorganic chemistry. The modern scope of this discipline would have been almost unimaginable to our counterparts of the 1930s and 1940s, who could not have anticipated huge subfields centered on transition metal organometallic complexes, electronically active solid state materials, or bioinorganic systems, not to mention metal clusters, carboranes, or rare-gas compounds. Whatever its formal definition, the fact is that in practice, inorganic synthesis today extends well beyond its traditional confines into the realms of biochemistry, materials science, physics, and organic chemistry (witness the inclusion of procedures for C_5HMe_5 in both this volume and its predecessor). In truth, inorganic chemistry has been enriched by the absorption of much that organic chemists have abdicated as outside their immediate concern. The discovery of ferrocene in 1951 might well have launched a novel area of organic chemistry, if the imaginative insight of the late Robert B. Woodward, who (with Geoffrey Wilkinson) correctly guessed the structure of $Fe(C_5H_5)_2$ and gave it its name, had been shared at that time by organic chemists generally. As it happened, the currently vast area of organotransition metal chemistry developed as an inorganic subdiscipline, much to the benefit of that field (the Inorganic Division of the American Chemical Society routinely sponsors the largest divisional programs at ACS national meetings).

The current volume of *Inorganic Syntheses* reflects this broad scope. In addition to many procedures for preparing metal coordination compounds and other substances within the more traditional inorganic areas, the book contains numerous syntheses of organometallics, boron and other main-group compounds, metal clusters, and transition metal–main-group clusters. One contribution that is particularly illustrative of the catholicity of modern inorganic chemistry is the synthesis of organic superconductors (synthetic metals) by a leading group in that area. Why "organic" metals in an *Inorganic Syntheses* volume? See the preceding paragraph!

Many of the procedures described herein will be important to a broad spectrum of inorganic and organometallic chemists, while others deal with more specialized areas but will be valuable to investigators in those fields. In making the inevitable judgments about the significance of each contribution and its appropriateness for *Inorganic Syntheses*, I have kept in mind that certain areas of research have intrinsic significance that belies small numbers of investigators—the fields of rare-gas compounds, boron clusters, and metal–nonmetal clusters, for example. Today's exotica may spark tomorrow's hot fields—when ferrocene was first prepared, there were only a half-dozen metallocene chemists in the world.

As in the case with my predecessors who edited earlier volumes in this distinguished series, I am indebted to many people including the authors, the checkers (the true heroes of *Inorganic Syntheses*, especially those who sent in their reports on time), my colleagues on the Editorial Board who reviewed countless manuscripts, and Tom Sloan of *Chemical Abstracts* for painstakingly checking the nomenclature on every submission and for preparing the index. Thanks also are due my secretary, Shirley Fuller, the Department of Chemistry at the University of Virginia for a variety of services, not least the absorption of postage costs, and my research group for their understanding and tolerance.

RUSSELL N. GRIMES

Charlottesville, Virginia
February 1992

CORRECTION TO VOLUME 24

TITANIUM(II) CHLORIDE

Submitted by JEAN'NE M. SHREEVE

The synthesis described in *Inorg. Synth.*, **1986**, *24*, 181, entitled "Titanium(II) Chloride", does not produce the title compound.

ERRATA: VOLUME 26, p. 372.

The second checker for procedure 68 should be Dominique Matt (not Watt as printed). Index of Contributors: change Watt, D., 26:372 to read Matt, D., 26:372.

NOTICE TO CONTRIBUTORS
AND CHECKERS

The *Inorganic Syntheses* series is published to provide all users of inorganic substances with detailed and foolproof procedures for the preparation of important and timely compounds. Thus the series is the concern of the entire scientific community. The Editorial Board hopes that all chemists will share in the responsibility of producing *Inorganic Syntheses* by offering their advice and assistance in both the formulation of and the laboratory evaluation of outstanding syntheses. Help of this kind will be invaluable in achieving excellence and pertinence to current scientific interests.

There is no rigid definition of what constitutes a suitable synthesis. The major criterion by which syntheses are judged is the potential value to the scientific community. For example, starting materials or intermediates that are useful for synthetic chemistry are appropriate. The synthesis also should represent the best available procedure, and new or improved syntheses are particularly appropriate. Syntheses of compounds that are available commercially at reasonable prices are not acceptable. We do not encourage the submission of compounds that are unreasonably hazardous, and in this connection, authors are requested to avoid procedures involving perchlorate salts due to the high risk of explosion in combination with organic or organometallic substances. Authors are also requested to avoid the use of solvents known to be carcinogenic.

The Editorial Board lists the following criteria of content for submitted manuscripts. Style should conform with that of previous volumes of *Inorganic Syntheses*. The introductory section should include a concise and critical summary of the available procedures for synthesis of the product in question. It should also include an estimate of the time required for the synthesis, an indication of the importance and utility of the product, and an admonition if any potential hazards are associated with the procedure. The procedure should present detailed and unambiguous laboratory directions and be written so that it anticipates possible mistakes and misunderstandings on the part of the person who attempts to duplicate the procedure. Any unusual equipment or procedure should be clearly described. Line drawings should be included when they can be helpful. All safety measures should be stated clearly. Sources of unusual starting materials must be given, and, if possible, minimal standards of purity of reagents and solvents should be stated. The scale should be reasonable for normal laboratory operation, and any problems involved in scaling the procedure either up or down should be discussed. The criteria for judging the purity of the final product should be delineated clearly. The section on Properties should supply and discuss those physical and chemical characteristics that are relevant to judging the purity of the product and to permitting its handling and use in an intelligent manner. Under References, all pertinent literature citations should be listed in order. A style sheet is available from the Secretary of the Editorial Board.

The Editorial Board determines whether submitted syntheses meet the general specifications outlined above, and the Editor-in-Chief sends the manuscript to an independent laboratory where the procedure must be satisfactorily reproduced.

Each manuscript should be submitted in duplicate to the Secretary of the Editorial Board, Professor Jay H. Worrell, Department of Chemistry, University of South Florida, Tampa, FL 33620. The manuscript should be typewritten in English. Nomenclature should be consistent and should follow the recommendations presented in *Nomenclature of Inorganic Chemistry*, 2nd ed., Butterworths & Co., London, 1970, and in *Pure and Applied Chemistry*, Volume 28, No. 1 (1971). Abbreviations should conform to those used in publications of the American Chemical Society, particularly *Inorganic Chemistry*.

Chemists willing to check syntheses should contact the editor of a future volume or make this information known to Professor Worrell.

TOXIC SUBSTANCES AND LABORATORY HAZARDS

Chemicals and chemistry are by their very nature hazardous. Chemical reactivity implies that reagents have the ability to combine. This process can be sufficiently vigorous as to cause flame, an explosion, or, often less immediately obvious, a toxic reaction.

The obvious hazards in the syntheses reported in this volume are delineated, where appropriate, in the experimental procedure. It is impossible, however, to forsee every eventuality, such as a new biological effect of a common laboratory reagent. As a consequence, *all* chemicals used and *all* reactions described in this volume should be viewed as potentially hazardous. Care should be taken to avoid inhalation or other physical contact with all reagents and solvents used in procedures described in this volume. In addition, particular attention should be paid to avoiding sparks, open flames, or other potential sources that could set fire to combustible vapors or gases.

A list of 400 toxic substances may be found in the *Federal Register*, Vol. 40, No. 23072, May 28, 1975. An abbreviated list may be obtained from *Inorganic Syntheses*, Volume 18, p. xv, 1978. A current assessment of the hazards associated with a particular chemical is available in the most recent edition of *Threshold Limit Values for Chemical Substances and Physical Agents in the Workroom Environment* published by the American Conference of Governmental Industrial Hygienists.

The drying of impure ethers can produce a violent explosion. Further information about this hazard may be found in *Inorganic Syntheses*, Volume 12, p. 317. A hazard associated with the synthesis of tetramethyldiphosphine disulfide [*Inorg. Synth.*, **15**, 186 (1974)] is cited in *Inorganic Syntheses*, Volume 23, p. 199.

CONTENTS

Chapter One   MAIN-GROUP COMPOUNDS—GENERAL

**Chapter Four   TRANSITION METAL ORGANOMETALLICS
                  AND LIGANDS**

**Chapter Five   CLUSTER AND CAGE COMPOUNDS
CONTAINING TRANSITION METALS**

INORGANIC SYNTHESES

Volume 29

Chapter One

MAIN-GROUP COMPOUNDS—GENERAL

1. XENON DIFLUORIDE (MODIFICATION)

Submitted by ANDREJ ŠMALC* and KAREL LUTAR*
Checked by SCOTT A. KINKEAD†

In the preparation of xenon difluoride by photosynthesis from the elements, described by S. M. Williamson [*Inorg. Synth.*, **11**, 147 (1968)], commercial fluorine was employed, which prior to use was purified from hydrogen fluoride by passing through a U trap cooled with dry ice. However, we have found[1] that hydrogen fluoride in amounts of up to about 1 mol % catalyses the photochemical reaction between xenon and fluorine and increases the rate of reaction by a factor of ~4 without noticeable damage to the Pyrex photochemical reactor. It is, therefore, pointless to purify commercial fluorine, which normally contains up to 0.5% hydrogen fluoride.

The reaction rate can be increased using mole ratios of F_2 : Xe higher than 1 : 1. However, in this case the reaction should be stopped in time (i.e., before all the xenon is consumed), in order to prevent the formation of xenon tetrafluoride.

The triple point of the product obtained is 129.10 ± 0.05°C (published value 129.03 ± 0.05°C).[2] Infrared spectra of the vapor show only the bands characteristic of XeF_2.[3]

* "J. Stefan" Institute, University of Ljubljana, 61000 Ljubljana, Slovenia.
† Los Alamos Scientific Laboratory, Los Alamos, NM 87545.

Procedure

The reaction is preferably carried out in a 1–1.5-L Pyrex glass photochemical reactor with a water-cooled well for the light source, shown in Fig. 1. The reactor is equipped with a Rotaflo* or similar Teflon-needle glass valve. The reactor is connected to an all-metal vacuum line schematically shown in

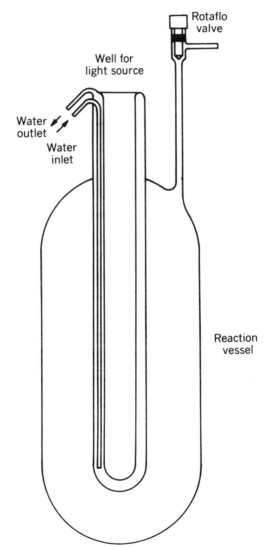

Fig. 1. Photochemical reactor for synthesis of xenon difluoride.

* J. Bibby Science Products Limited, Stone, Staffordshire ST 15 OSA, United Kingdom.

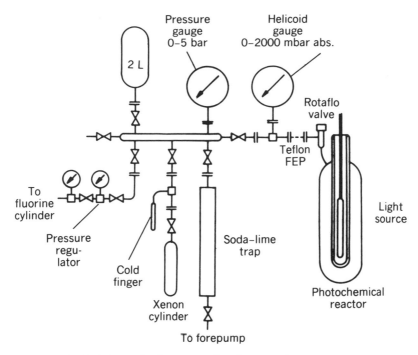

Fig. 2. Schematic diagram of fluorine metering system.

Fig. 2. The connection is best made by a piece of Teflon FEP tubing (i.d. 6–8 mm) additionally sealed with halocarbon wax (see preparation of KrF_2). The reactor is then evacuated to high vacuum for several days. It is advisable to heat it during evacuation in a hot-water bath. Finally the reactor is cooled down and seasoned once or twice by irradiation of about 500 mbar of fluorine for about an hour.

About 300 mbar of xenon followed by about 600 mbar of fluorine are introduced into the reactor and irradiated until the pressure in the reactor remains constant. If a 400-W medium pressure mercury lamp* is used as a light source, the reaction is finished in about 80 min with an XeF_2 yield of practically 100%. Approximately 2.5 g of XeF_2 is isolated.

Xenon difluoride is removed from the reactor by pumping the gases through a trap (preferably from Teflon FEP or Kel-F) cooled to $-78°C$. Unreacted fluorine as well as silicon tetrafluoride and other possible contaminants will pass through. Xenon difluoride will transfer more easily if a warm-water bath is placed around the reactor (after unreacted fluorine is removed) and siphoned into the water jacket of the lamp well.

---

* E.g., Applied Photophysics Ltd., 20 Albemarle Street, London W1X 3HA, United Kingdom.

■ **Caution.** *Fluorine is a potent oxidizing agent and must be handled with great care, preferably in nickel or copper equipment. The apparatus used should be thoroughly cleaned and degreased, and then prefluorinated prior to use. Organic materials other than fluoropolymers such as Teflon, FEP or PFA, and Kel-F should be absent. The use of fluorine requires that this procedure be carried out in an efficiently ventilated area. Only equipment specially designed for use with fluorine such as pressure regulators* for use with fluorine cylinders, pressure gauges, † and valves§ should be used. For more exhaustive precautions see* Inorganic Syntheses, Fluorine Compounds.[4] *In addition, safety glasses or a face shield, neoprene gloves, and protective shielding should be used in front of the reactor during operation. The reduced fluorine pressure in the fluorine metering system should not exceed a few bars.*

References

1. K. Lutar, A. Šmalc, and J. Slivnik, *Vestn. Slov. Kem. Drus.*, **26**, 435 (1979).
2. F. Schreiner, G. N. McDonald, and C. L. Chernick, *J. Phys. Chem.*, **72** 1162 (1968).
3. J. G. Malm, H. Selig, J. Jortner, and S. A. Rice, *Chem. Rev.*, **65**, 199 (1965); S. Reichman and F. Schreiner, *J. Chem. Phys.*, **51**, 2355 (1969)
4. Fluorine Compounds, *Inorg. Synth.*, **11**, 131 (1968).

2. XENON TETRAFLUORIDE

Submitted by KAREL LUTAR,* ANDREJ ŠMALC,* and BORIS ŽEMVA*
Checked by SCOTT A. KINKEAD†

Xenon tetrafluoride was first prepared by heating a gaseous mixture of xenon and fluorine under moderate pressure.[1,2] However, the product obtained by this method is always contaminated with xenon difluoride and hexafluoride, which are in chemical equilibrium with xenon tetrafluoride.[3] Xenon tetrafluoride obtained in this way can be purified by subsequent chemical purification with arsenic pentafluoride.[4]

However, pure xenon tetrafluoride can be obtained by thermal dissociation of xenon hexafluoride.[5]

* E.g., Matheson Co., Inc., East Rutherford, NJ or Matheson Gas Products, 2431 Oevel, Belgium.
† E.g., Helicoid Pressure Instruments, Helicoid Division, Bristol Babcock, Buckingham St., Watertown, CT 06795.
‡ E.g., Hoke Inc., 1 Tenakill Park, Creskill, NJ 07626, or Autoclave Engineers, Erie, PA 16501.
§ "J. Stefan" Institute, University of Ljubljana, 61000 Ljubljana, Slovenia.
¶ Los Alamos Scientific Laboratory, Los Alamos, NM 87545.

Xenon reacts with excess fluorine at 300°C under pressure, yielding xenon hexafluoride, which is combined with sodium fluoride to form a nonvolatile complex $Na_2[XeF_8]$

$$Xe + 3F_2 + 2NaF \rightarrow Na_2[XeF_8]$$

Xenon difluoride and tetrafluoride, which are also present in the reaction mixture, form no complex with sodium fluoride and can be pumped off at room temperature. Finally, $Na_2[XeF_8]$ is decomposed at 350°C:

$$Na_2[XeF_8] \rightarrow XeF_4 + F_2 + 2NaF$$

This method could also be used for the preparation of pure fluorine.

Procedure

The reaction is carried out in a 500-mL pressure vessel made of nickel or Monel and rated to 350 bar, equipped with a Monel high-pressure valve and NPT adapter.* Prior to first use about 15 g of sodium fluoride is added to the vessel. The reaction vessel is then heated in high vacuum at 450°C for 8 h. After cooling down it is pretreated once or twice with about 500 mbar of fluorine at 300°C for an hour and finally with 5 bar of fluorine at 350°C for 8 h. Fluorine is metered into the reaction vessel by repeated condensation from a storage vessel on the all-metal vacuum line shown schematically in Fig. 2 (see XeF_2 synthesis above, Section 1). It is advisable but not absolutely necessary to pretreat the vessel additionally with xenon hexafluoride at 150°C for 2 h. After seasoning is finished, the reaction vessel is evacuated at 350°C to high vacuum.

Approximately 3 bar of xenon (60 mmol) and 60 bar of fluorine are introduced into the reaction vessel which is then heated for 2 days at 350°C and one more day at 50°C. The last step ensures that all xenon hexafluoride formed combines with sodium fluoride. After excess fluorine is distilled off at − 183°C into another pressure vessel held at − 196°C, xenon difluoride and tetrafluoride are pumped off at 50°C. The reaction vessel is then closed and heated again to 350°C for 2.5 h in order to decompose the complex and dissociate the liberated xenon hexafluoride. The reaction vessel is then cooled down to − 196°C and the fluorine liberated is pumped off. To get a higher yield of xenon tetrafluoride, it is advisable to repeat the decomposition several times. In order to remove any possibly remaining xenon hexafluoride,

* E.g., Hoke, Inc., Tenakill Park, Cresskill, NJ 07626.

the reaction vessel is again heated at 50°C overnight. Xenon tetrafluoride is removed from the reaction vessel by pumping through a series of traps (preferably made of Teflon or Kel-F) cooled to -196°C. Subsequently the trap with the product is slowly warmed up to 0°C in order to remove unreacted xenon hexafluoride, which should be later disposed of with care (see **Caution**). The yield (based on xenon) is more than 70%. Infrared spectra of the product show only the bands associated with XeF_4. Approximately 9 g of XeF_4 is isolated.

■   **Caution.** *Fluorine must be handled with great care; see preparation of xenon difluoride (above) for descriptions of the apparatus and precautions required when withdrawing fluorine from a cylinder.*

*Xenon tetrafluoride and hexafluoride are very powerful oxidizing agents also. They are particularly dangerous because they react with water, giving explosive xenon trioxide.[6] (See **Cautions** under the syntheses of XeF_4[7] and XeO_3 solution.[8] The best way to dispose of xenon tetrafluoride or hexafluoride is to rinse the vessel (e.g., a protective trap) first with carbon tetrachloride and then with plenty of water, otherwise explosions may occur.)*

Properties

Xenon tetrafluoride is a colorless crystalline solid with a vapor pressure of 3.3 mbar at 25°C. Its triple point is 117.10 ± 0.05°C.[9] Xenon tetrafluoride may be identified by its strong absorption band at 586 cm^{-1}.[10] It can be kept in thoroughly dried glass or quartz and can be stored an indefinitely long time in Teflon, FEP, KelF, nickel, or Monel containers. XeF_4 hydrolyses rapidly to Xe, O_2, HF, and XeO_3. (**Caution:** *See above!*)

References

1. H. H. Claassen, H. Selig, and J. G. Malm, *J. Am. Chem. Soc.*, **84**, 3593 (1962).
2. J. Slivnik, B. S. Brčić, B. Volavšek, A. Šmalc, B. Frlec, A. Zemljič, and A. Anžur, *Croat. Chem. Acta*, **34**, 187 (1962).
3. B. Weinstock, E. E. Weaver, and C. P. Knop, *Inorg. Chem.*, **5**, 2189 (1966).
4. N. Bartlett and F. O. Sladky, *J. Am. Chem. Soc.*, **90**, 5316 (1968).
5. M. Bohinc and J. Slivnik, *Vestn. Slov. Kem. Drus.*, **20**, 9 (1973).
6. S. M. Williamson and C. W. Koch, *Science*, **139**, 1046 (1963); S. M. Williamson and C. W. Koch, in *Noble-Gas Compounds*, H. H. Hyman (ed.), University of Chicago Press, Chicago, 1963, p. 158.
7. J. G. Malm and C. L. Chernick, *Inorg. Synth.*, **8**, 254 (1966).
8. E. H. Appelman, *Inorg. Synth.*, **11**, 205 (1968).
9. F. Schreiner, G. N. McDonald, and C. L. Chernick, *J. Phys. Chem.*, **72**, 1162 (1968).
10. H. H. Claassen, C. L. Chernick, and J. G. Malm, *J. Am. Chem. Soc.*, **85**, 1927 (1963).

3. CHLORINE PENTAFLUORIDE

Submitted by KAREL LUTAR,* ANDREJ ŠMALC,* and BORIS ŽEMVA*
Checked by SCOTT A. KINKEAD†

$$ClF_3 + F_2 \rightarrow ClF_5$$

Chlorine pentafluoride was first synthesized by fluorination of chlorine trifluoride with fluorine under pressure of 250 bar at 360°C.[1] Later on this method was modified, for example, by use of alkali metal tetrafluorochlorates instead of chlorine trifluoride.[2] It has also been found that nickel(II) fluoride catalyzes the reaction between chlorine trifluoride and fluorine.[3]

Procedure

The reaction is carried out in a 150-mL heavy-wall Monel pressure vessel rated to 350 bar, equipped with an 0.25-in. NPT opening and corresponding adapter‡ with a Monel high-pressure valve with PTFE stem packing, rated to 700 bar.§ About 2 g of pulverized anhydrous nickel(II) fluoride is put into the vessel. The vessel is prefluorinated once or twice with about 500 mbar of fluorine and subsequently once or twice with about 500 mbar of chlorine trifluoride for an hour under occasional heating up to 300–400°C. Into the seasoned reaction vessel, cooled to − 196°C, about 40 mmol of chlorine trifluoride and about 400 mmol of fluorine are condensed.

■ **Caution.** *Fluorine should be metered into the reaction vessel correctly so that the pressure in the reaction vessel at room temperature does not exceed about 100 bar. For metering fluorine and precautions, see preparation of xenon difluoride. Safety shielding (e.g., a 8–10-mm-thick steel plate) should be placed in front of the furnace with the reaction vessel during the reaction. Chlorine trifluoride and pentafluoride are corrosive, highly oxidizing substances and should be handled under the same precautions as fluorine.*

The reaction vessel is then removed from the liquid-nitrogen bath and placed into a cold electric furnace with an aluminum block. The aluminum block is then heated to 200°C for 3 h. Subsequently, the reaction vessel is cooled with liquid oxygen to − 183°C and excess fluorine is distilled off into another container, cooled to − 196°C. The rest of the fluorine is pumped off through a soda-lime trap. Chlorine pentafluoride is removed from the reactor

---

* "J. Stefan" Institute, University of Ljubljana, 61000 Ljubljana, Slovenia.
† Los Alamos Scientific Laboratory, Los Alamos, NM 87545.
‡ E.g., Hoke, Inc., Tenakill Park, Cresskill, NJ 07626.
§ E.g., Autoclave Engineers, Erie, PA 16501.

by pumping off through a trap (preferably made of Teflon, FEP, or Kel-F) cooled to $-100°C$. The product is stored in a prefluorinated nickel, Monel, or stainless steel container. Under the conditions described, about 5 g of ClF_5 is synthesized in a single run. The yield of the synthesis is nearly 100%. Infrared spectra of the product show only the bands associated with ClF_5.

Properties

Chlorine pentafluoride at room temperature is a colorless gas that condenses to a colorless liquid (bp $-14°C$) and freezes to a white solid (mp $-103°C$). It appears to be less corrosive to metals than chlorine trifluoride and can be stored in nickel, Monel, or stainless steel containers. It is extremely reactive toward water.

References

1. D. F. Smith, *Science*, **141**, 1039 (1963).
2. D. Pilipovich, W. Maya, E. A. Lawton, H. F. Bauer, D. F. Sheenan, N. N. Ogimachi, R. D. Wilson, F. C. Gunderloy, Jr., and V. E. Bedwell, *Inorg. Chem.*, **6**, 1918 (1967).
3. A. Šmalc, B. Žemva, J. Slivnik, and K. Lutar, *J. Fluorine Chem.*, **17**, 381 (1981).

4. DIOXYGENYL HEXAFLUOROARSENATE

Submitted by ANDREJ ŠMALC* and KAREL LUTAR*
Checked by SCOTT A. KINKEAD†

$$O_2 + \tfrac{1}{2}F_2 + AsF_5 \xrightarrow{hv} [O_2][AsF_6]$$

Dioxygenyl hexafluoroarsenate can be prepared by the reaction between arsenic pentafluoride and dioxygen difluoride[1] or by heating a mixture of arsenic pentafluoride, fluorine, and oxygen.[2] However, photosynthesis,[3,4] using this latter mixture, appears to be the most convenient method for the preparation of dioxygenyl hexafluoroarsenate.

Procedure

The same photochemical reactor and apparatus is used for this synthesis as described previously (see procedure for xenon difluoride). About 280 mbar of

* "J. Stefan" Institute, University of Ljubljana, 61000 Ljubljana, Slovenia.
† Los Alamos Scientific Laboratory, Los Alamos, NM 87545.

arsenic pentafluoride, 280 mbar of oxygen, and 420 mbar of fluorine are introduced into the reactor and irradiated until the pressure in the reactor remains unchanged (if a 400-W lamp is used,* the reaction is finished in about 15 min).

■ **Caution.** *For a description of the apparatus and precautions required when withdrawing fluorine from a cylinder, see procedure for xenon difluoride. Arsenic pentafluoride is very toxic. It hydrolyzes easily and therefore is a blistering agent as is hydrogen fluoride. Care must be taken to avoid breathing in and contact with the skin.*

Subsequently, volatile reactants remaining in the photochemical reactor are pumped off. The product adhering to the wall of the reactor is loosened by warming the reactor with hot air or hot water. The product is transferred from the reactor into a container in a dry box. Emptying the reactor is easier if the reactor is fitted with an extra neck with a ground cone and socket fitting. Otherwise, the product is removed from the reactor through the valve from which the needle is removed. The yield of the synthesis is nearly 100%. Under the conditions described about 3.8 g of $[O_2]$ $[AsF_6]$ is prepared in a single run. If more product is needed, the procedure can be repeated several times without removing the product from the reactor.

Anal. Calcd. for $[O_2]$ $[AsF_6]$: F, 51.60%; As, 33.91%; O, 14.49%. Found: F, 50.8%; As, 33.7%. Infrared spectra show only the band associated with AsF_6^-.

Properties

Dioxygenyl hexafluoroarsenate is a white, loose, nonvolatile powder. It is stable at room temperature, but above 100°C it decomposes rapidly. It hydrolyzes readily, and therefore should be handled in a dry box.

References

1. A. R. Young, T. Hirata, and S. I. Morrow, *J. Am. Chem. Soc.*, **86**, 20 (1964).
2. A. J. Edwards, W. E. Falconer, J. E. Griffiths, W. A. Sunder, and M. J. Vasile, *J. Chem. Soc. Dalton Trans.*, **1974**, 1129.
3. S. Shamir and J. Binenboym, *Inorg. Chim. Acta*, **2**, 37 (1968).
4. A. Šmalc and K. Lutar, *J. Fluorine Chem.*, **9**, 399 (1977).

* E.g., Applied Photophysics Ltd., 20 Albemarle Street, London W1X 3HA, United Kingdom.

5. PEROXYDISULFURYL DIFLUORIDE (MODIFICATION)

Submitted by ANDREJ ŠMALC*
Checked by STEVEN G. MAYORGA† and NEIL BARTLETT†

$$2HSO_3F + 2[O_2] [AsF_6] \rightarrow S_2O_6F_2 + 2O_2 + 2AsF_5 + 2HF$$

Peroxydisulfuryl fluoride is usually synthesized by catalytic fluorination of sulfur trioxide.[1,2] However, pure peroxydisulfuryl fluoride can be obtained in a relatively simple way by the reaction between fluorosulfuric acid and dioxygenyl hexafluoroarsenate.[3] The method lends itself to preparation of smaller quantities (a few grams) of peroxydisulfuryl fluoride. The reaction can be carried out in an openable nickel or Monel pressure vessel, into which dioxygenyl hexafluoroarsenate is transferred from a photochemical reactor (see the preceding synthesis of dioxygenyl hexafluoroarsenate) or, more simply, in the same Pyrex-glass flask in which dioxygenyl hexafluoroarsenate has previously been synthesized.

Procedure

The reaction is carried out in a 100-mL round-bottomed flask fitted with a ground joint and glass vacuum stopcock lubricated with Kel-F grease. The flask is evacuated to high vacuum for one day and baked out several times. About 300 mbar of arsenic pentafluoride, then about 300 mbar of oxygen and about 400 mbar of fluorine are introduced into the flask using the same apparatus as described previously (see procedure for xenon difluoride). The flask is then exposed to daylight for a day or two. Unreacted gases are then pumped off briefly and, if more product needed, the procedure is repeated once or twice without removing the product from the reactor.

■ **Caution.** *For a description of the apparatus and precautions required when withdrawing fluorine from a cylinder, see procedure for xenon difluoride. Arsenic pentafluoride is very toxic. It hydrolyzes easily, and therefore is a blistering agent as is hydrogen fluoride. Care must be taken to avoid breathing it in and contact with the skin.*

About 10 g of fluorosulfuric acid is condensed into an evacuated flask containing dioxygenyl hexafluoroarsenate cooled to $-196°C$ and warmed upto room temperature. The products are pumped through three traps, cooled to 0, -78, and $-196°C$, respectively. In the first trap unreacted

* "J. Stefan" Institute, University of Ljubljana, 61000 Ljubljana, Slovenia.
† University of California, Berkeley, CA 94702.

fluorosulfuric acid is condensed; in the second, peroxydisulfuryl fluoride; and in the third, hydrogen fluoride and silicon tetrafluoride are collected. The yield of the synthesis (based on $[O_2]$ $[AsF_6]$) is nearly 100%. About 0.25 g of peroxydisulfuryl fluoride is obtained. Peroxydisulfuryl fluoride can be additionally purified by trap-to-trap distillation and condensed at $-78°C$. Infrared spectra of the product show only the bands characteristic for $S_2O_6F_2$.

Properties

Peroxydisulfuryl fluoride is a colorless liquid (bp 67.1°C). It reacts with water to give fluorosulfate ions and oxygen. It can be stored in a glass container equipped with a Teflon-needle glass valve. See also ref. 1.

References

1. J. M. Shreeve and G. H. Cady, *Inorg. Synth.*, **7**, 124 (1963).
2. G. H. Cady, *Inorg. Synth.*, **11**, 155 (1968).
3. A. Šmalc, *Vestn. Slov. Kem. Drus.*, **21**, 5 (1974).

6. KRYPTON DIFLUORIDE

Submitted by ANDREJ ŠMALC,* KAREL LUTAR,* and BORIS ŽEMVA*
Checked by SCOTT A. KINKEAD†

$$Kr + F_2 \xrightarrow{hv} KrF_2$$

The commonly used method for the preparation of krypton difluoride, which is based on passing an electric discharge through a gaseous mixture of krypton and fluorine at low temperature and reduced pressure,[1,2] is rather tedious because of the relatively low yield. However, the irradiation of a liquid krypton–fluorine mixture at $-196°C$[3] proved to be a very successful method for the preparation of krypton difluoride in quantities up to 10 g in a single run.

Procedure

The synthesis of krypton difluoride is carried out in a 150-mL low-temperature photochemical reactor (Fig. 1). The reactor consists of a ring-shaped

* "J. Stefan" Institute, University of Ljubljana, 61000 Ljubljana, Slovenia.
† Los Alamos Scientific Laboratory, Los Alamos, NM 87545.

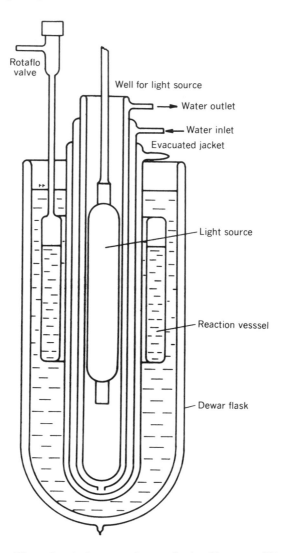

Fig. 1.   Photochemical reactor for synthesis of krypton difluoride.

reaction vessel in the middle of which an immersion well for the light source can be placed. The well is thermally insulated by an evacuated jacket which prevents the cooling water from freezing. The reaction vessel is equipped with a Teflon-needle glass valve.* The reactor is attached to an all-metal vacuum

* E.g., Rotaflo valve, supplied by J. Bibby Science Products Ltd., Stone, Staffordshire ST 15 OSA, United Kingdom.

line (see preparation of xenon difluoride; Section 1, Fig. 2). A piece of Teflon FEP tubing (i.d. 6 mm) serves as a metal-to-glass connection. If needed, the connection can be additionally sealed with halocarbon wax.* Prior to use the photochemical reactor is evacuated to high vacuum for a couple of days and baked out several times. Finally, about 150 mbar of fluorine is introduced into it and irradiated for an hour. Then the fluorine is pumped off and the reactor is evacuated to high vacuum.

About 2 mol of krypton is condensed into the lower part of the reactor. (During the condensation of krypton only the bottom of the reactor is cooled with liquid nitrogen.) Subsequently, the reactor is completely immersed in liquid nitrogen and about 0.5 mol of fluorine is condensed into it. After placing the well with the light source in the middle of the reaction vessel, the reaction mixture is irradiated for about 20 h. Using a 400-W medium-pressure mercury lamp,† 0.7–1 g of KrF_2 per hour is synthesized. The well with the light source is then removed and the liquid nitrogen bath is replaced with liquid oxygen. Excess fluorine is distilled off and condensed in a nickel or Monel pressure vessel, cooled to $-196°C$. The vessel is then connected to another pressure vessel for krypton, also cooled to $-196°C$. The liquid oxygen bath is then carefully removed by lowering it until the pressure in the reaction vessel reaches about 600 mbar. If needed, the reaction vessel is occasionally shortly immersed into liquid oxygen (or nitrogen) bath in order to maintain the pressure below 600 mbar (see **Caution**) until all krypton has sublimed off (i.e., until the pressure in the reaction vessel when it is removed from liquid oxygen–nitrogen bath remains below 400 mbar). After that the reaction vessel is placed in an acetone–dry ice bath ($-78°C$) in order to remove the remaining krypton from the product. Krypton difluoride is removed from the reaction vessel by pumping it off at room temperature through a Teflon FEP trap cooled to $-78°C$. About 17 g KrF_2 is obtained in 20 h. Infrared and Raman spectra of the product show only the bands characteristic for KrF_2.

■ **Caution.** *Fluorine must be handled with great care; see procedure for synthesis of xenon difluoride (above) for the description of the apparatus and precautions required when withdrawing fluorine from a cylinder. Special care must be taken in work with liquid fluorine. It should be done in an efficient hood and protective shielding should be used in front of the Dewar flask with the reactor. During the whole operation the pressure in the reactor should be measured to be sure that the temperature is low enough to prevent any*

---

* E.g., halocarbon wax series 12-00, supplied by Halocarbon Products Corp., 82 Burlews Court, Hackensack, NJ 076011.

† E.g., 400-W medium pressure mercury lamp, supplied by Applied Photophysics Ltd., 20 Albemarle Street, London WIX 3HA, United Kingdom.

overpressure. Therefore the photochemical reactor with liquid fluorine should always be completely immersed in liquid nitrogen. Application of an automatic device for maintaining the level of liquid nitrogen in the Dewar flask containing the reactor is recommended.

To avoid the possible overpressure in the system due to coolant loss that could consequently cause the bursting of the reactor, it is highly recommended to attach a large buffer reservoir to the system (e.g., instead of the 2-L vessel in Fig. 2 in Section I). The volume of this vessel should be sufficient (about 40 L to contain all of the gas at room temperature and at a pressure of < 2 bar. The vessel should be pressure-tested and resistant to fluorine (e.g., an empty fluorine gas cylinder would be the most appropriate) filled with fluorine to 400 mbar (vapor pressure of fluorine at − 196°C). The buffer reservoir should be opened to the reactor immediately after condensation of fluorine in the reactor. After the synthesis is finished and excess fluorine distilled off, the buffer reservoir should be closed again.

Instead of the buffer reservoir a proper vacuumtight pressure-relief valve could be installed onto a vacuum line, which opens automatically at just above atmospheric pressure and is, of course, vented to a proper hood.

As the light source a medium-pressure mercury lamp should be used only (in no case a high-pressure one).

When removing unreacted krypton by distillation, care should be taken that the pressure during this operation does not exceed 600 mbar, otherwise melting of solid krypton can occur, which may cause a sudden raising of the pressure due to better contact between krypton and the wall of the reaction vessel. The unreacted krypton may contain some fluorine, and therefore the container for unreacted krypton should be made of nickel or Monel and should be pretreated in the same way as usual for the fluorine containers (once or twice with about 500 mbar of fluorine for an hour under occasional heating up to 300–400°C).

Like xenon hexafluoride or xenon tetrafluoride, krypton difluoride reacts with water, giving highly explosive hydrolysis products. The best way for disposing of krypton difluoride is to allow it to react with carbon tetrachloride (see caution note under xenon tetrafluoride procedure).

Properties

Krypton difluoride is a white crystalline solid with a vapor pressure of about 40 mbar at 0°C. It is thermally unstable and decomposes slowly at room temperature. Therefore, it should be stored preferably in nickel or Monel containers held at low temperature (below − 60°C). Krypton difluoride is stronger oxidizing agent than elemental fluorine, dioxygen difluoride, or xenon hexafluoride.

References

1. F. Schreiner, J. G. Malm, and J. C. Hindman, *J. Am. Chem. Soc.*, **87**, 25 (1965).
2. I. F. Isaev, K. A. Kazanski, and V. F. Malcev, cited in V. N. Prusakov and V. B. Sokolov, *Atomnaya Energiya*, **31**, 259 (1971).
3. J. Slivnik, A. Šmalc, K. Lutar, B. Žemva, and B. Frlec, *J. Fluorine Chem.*, **5**, 273 (1975).

7. SIMPLE PREPARATION OF BENZENETHIOLATO AND BENZENESELENOLATO DERIVATIVES OF INDIUM AND TIN

Submitted by RAJESH KUMAR,* HASSAN E. MABROUK,* and DENNIS G. TUCK*
Checked by EDWARD GANJA† and THOMAS B. RAUCHFUSS†

Thiolato derivatives of main-group metals provide an interesting contrast to those of the transition metals, since despite the industrial application of compounds such as organotin sulfides, much remains to be done in the study of the chemistry of these interesting compounds. Methods for the preparation of organotin thiolates[1,2] and selenolates[2] have been reviewed, as have their physical and chemical properties. The preparation of tris(benzenethiolato)-indium by the reaction of $InCl_3$ with $NaSC_6H_5$ in methanol has recently been described, as have the properties of this compound.[3]

The syntheses described below for $In(EC_6H_5)_3$ (E = S, Se) and $Sn(EC_6H_5)_4$ all involve one-step preparation from the metallic elements and diphenyl disulfide or diselenide. The presence of iodine leads to the formation of $InI(EC_6H_5)_2$ in the indium/$E_2(C_6H_5)_2$ system. These methods appear to have many advantages over the metatheses used previously,[1-3] and in fact the indium–diphenyl diselenide reactions represent the first synthesis of $In(SeC_6H_5)_3$ and $InI(SeC_6H_5)_2$.

A. TRIS(BENZENETHIOLATO)INDIUM(III), $In(SC_6H_5)_3$

$$In(s) + \tfrac{3}{2}(C_6H_5)_2S_2(sol) \rightarrow In(SC_6H_5)_3$$

Procedure

■ **Caution.** *Diphenyl disulfide is a foul-smelling and irritant substance, and all operations should be carried out in a well-ventilated hood, with due*

*Department of Chemistry and Biochemistry, University of Windsor, Windsor, Ontario Canada, N9B 3P4.
† Department of Chemistry, University of Illinois, Urbana IL, 61801.

regard to the persistent and unpleasant nature of the compound, even at low concentrations.

A 250-mL round-bottomed flask is equipped with a magnetic stirrer–heater and a reflux condenser. The reaction mixture placed in the flask consists of toluene (100 mL; freshly distilled from calcium hydride), small chips of indium metal (1.5 g, 13.1 mmol), and diphenyl disulfide (4.3 g, 19.6 mmol). The chips of metal can be conveniently cut from a rod available commercially (99.9% purity). The reaction mixture is heated with stirring until the disulfide dissolves, and the temperature then raised to produce vigorous refluxing; no mechanical stirring is necessary at this stage. In order to achieve an appropriate rate of refluxing, it may be necessary to heat the flask sufficiently that local superheating will cause the indium metal to melt. The reaction mixture is refluxed for 6 h, or until all the metal has reacted. A microcrystalline white solid precipitates as the reaction proceeds. At the end of the refluxing, the flask and its contents are allowed to cool to room temperature, when more precipitate is thrown down. This solid is collected by filtration on a glass sinter, washed with *n*-pentane (2×10 mL), and dried *in vacuo*. The yield is 5.3 g, 11.9 mmol (91%).

Properties

Tris(benzenethiolato)indium(III) is an air-stable colorless microcrystalline solid, mp 245°C. It is soluble in donor solvents at room temperature, and in aromatic solvents at boiling point. The ^{1}H NMR spectrum in di(methyl-d_3) sulfoxide consists of a typical aromatic multiplet, 6.8–7.4 ppm from Me$_4$Si. The chemical properties are those of a typical InX$_3$ Lewis acid.[4]

B. TRIS(BENZENESELENOLATO)INDIUM(III), In(SeC$_6$H$_5$)$_3$

$$\text{In(s)} + \tfrac{3}{2}(C_6H_5)_2Se_2(\text{sol}) \rightarrow \text{In(SeC}_6\text{H}_5)_3$$

Procedure

■ **Caution.** *Diphenyl diselenide is an evil-smelling and highly toxic substance. All operations should be carried out in a well-ventilated fume hood, and rubber gloves should be worn when handling this material. Extreme care is necessary at all times.*

The procedure essentially mirrors that described in Section A. The reaction mixture consists of indium metal (1.15 g, 10.0 mmol) and diphenyl diselenide (4.73 g, 15.2 mmol) in dry toluene (100 mL). After 3 h of vigorous refluxing, the metal dissolves and the solution becomes light yellow in color. The mixture is filtered hot under dry nitrogen through a sinter to remove residual cloudiness; on cooling to room temperature, the filtrate deposits a yellow microcrystalline solid, which is collected by filtration, washed with n-pentane (2×10 mL), and dried *in vacuo*. Yield 5.4 g, 93%.

Properties

Tris(benzeneselenolato)indium(III) is an air-stable yellow microcrystalline solid, mp 150°, soluble in donor solvents and in hot aromatic solvents. It is a Lewis acid.[4] The ^{1}H NMR spectrum di(methyl-d_3 sulfoxide) consists of a complex multiplet, 6.6–7.2 ppm from Me$_4$Si.

C. IODOBIS(BENZENETHIOLATO)INDIUM(III), $InI(SC_6H_5)_2$

$$In(s) + (C_6H_5)_2S_2(sol) + \tfrac{1}{2} I_2(sol) \rightarrow InI(SC_6H_5)_2$$

Procedure

■ **Caution.**   *See note at the beginning of Section A above.*

The experiment uses the equipment and materials identified in Section A. The flask contains toluene (100 mL), indium chips (0.6 g, 5.2 mmol) and diphenyl disulfide (1.25 g, 5.7 mmol). The mixture is warmed gently until the disulfide dissolves, after which solid iodine (0.66 g, 2.6 mmol I_2) is added and the solvent brought to vigorous reflux. The color of the iodine is completely discharged after 1 h, and the indium completely reacted after 6 h. The final yellow solution is filtered hot under nitrogen to remove traces of unreacted metal, and after cooling the resultant filtrate slowly evaporated *in vacuo* to reduce the volume to 10 mL, when a yellow solid is thrown down. This product, $InI(SC_6H_5)_2$, is collected by filtration, washed with n-pentane (2×10 mL) and dried *in vacuo* at room temperature. Yield 2.2 g, 90%.

The selenium analog $InI(SeC_6H_5)_2$ can be prepared by a similar procedure; in a typical experiment, using quantities similar to those noted above, the color of iodine disappeared after refluxing for 15 min, and the metal was consumed in 5 h.

Properties

Iodobis(benzenethiolato)indium(III) is a yellow crystalline solid, mp 190°, soluble in donor solvents and in aromatic solvents. The solid is air-stable over several weeks, but in solution it is susceptible to aerial oxidation, and hence the need to filter the solution above under dry nitrogen. The ^{1}H NMR spectrum in di(methyl-d_3) sulfoxide consists of a complex multiplet in the region 7.06–7.57 ppm from Me$_4$Si. The chemical properties are apparently those of a Lewis acid,[4] since treatment with tetra-*n*-butylammonium iodide in acetonitrile gives a precipitate of $(C_4H_9)_4N[I_2In(SC_6H_5)_2]$ (see ref. 5 for another route to this anion).

Iodobis(benzeneselenolato)indium(III) is a yellow crystalline solid, mp 155°, soluble in donor solvents and hot aromatic solvents. The ^1H NMR spectrum di(methyl-d_3) sulfoxide consists of a complex aromatic multiplet at 7.06 ppm 7.56 from Me_4Si. The stability in air is similar to that of $InI(SC_6H_5)_2$.

D.   TETRAKIS(BENZENETHIOLATO)TIN(IV), $Sn(SC_6H_5)_4$

$$Sn(s) + 2(C_6H_5)_2S_2(sol) \rightarrow Sn(SC_6H_5)_4$$

Procedure

■ **Caution.**   *See note in Section A above.*

In this experiment, 0.76 g (6.40 mmol) of tin metal in the form of small chips shaved from a rod is added to a solution of diphenyl disulfide (2.79 g, 12.9 mmol) in freshly distilled toluene (40 mL) previously dried over calcium hydride. The mixture is refluxed vigorously for 5 h, by which time the tin is completely consumed and the original yellow color of the solution has given way to a yellow-green. The mixture is filtered hot under nitrogen and the filtrate evaporated *in vacuo* to yield a yellow-green solid that is triturated with *n*-pentane (10 mL). The pentane washings are discarded and the residual yellow solid collected and dried *in vacuo*. Yield 3.2 g 90%.

The substitution of diphenyl diselenide in the above procedure yields $Sn(SeC_6H_5)_4$ in 87% yield, after refluxing for 3 h.

Properties

The physical properties of $Sn(SC_6H_5)_4$ have been described in the literature,[2] and the material prepared in the way described is identical in all respects: mp 67°; cf. lit. value[6] 67°. The ^1H NMR spectrum in di(methyl-d_3) sulfoxide is a complex multiplet 7.19–7.47 ppm from Me_4Si. The compound is air-stable and is a Lewis acid.[7]

Tetrakis(benzeneselenolato)tin(IV) is a yellow crystalline solid; the properties of the material prepared by the preceding method are essentially identical with those given in the literature[2] (mp 80–83°; cf. lit. value[8] 83.5°). The ^1H NMR spectrum in di(methyl-d_3) sulfoxide is a multiplet 7.27–7.48 ppm from Me_4Si.

References

1. E. W. Abel and D. A. Armitage, *Adv. Organomet. Chem.*, **5**, 1 (1967).
2. H. Schumann, I. Schumann-Rudisch, and M. Schmidt, in *Organotin Compounds*, Vol. 2, A. K. Sawyer (ed.), Marcel Dekker, New York, 1971, p. 297.
3. R. K. Chadha, P. C. Hayes, H. E. Mabrouk, and D. G. Tuck, *Can. J. Chem.*, **65**, 804 (1987).
4. R. Kumar, H. E. Mabrouk, and D. G. Tuck, *J. Chem. Soc. Dalton Trans.*, **1988**, 1045.
5. C. Peppe and D. G. Tuck, *Can. J. Chem.*, **62**, 2798 (1984).
6. H. J. Backer and J. Kramer, *Rec. trav. chim. Pays-Bas*, **53**, 1101 (1934).
7. J. L. Hencher, M. A. Khan, F. F. Said, R. Sieler, and D. G. Tuck, *Inorg. Chem.*, **21**, 2787 (1982).
8. H. J. Backer and J. C. B. Hurenkamp, *Rec. trav. chim. Pays-Bas*, **61**, 802 (1942).

8. TETRAKIS (2,4,6-TRIMETHYLPHENYL)DISILENE (TETRAMESITYLDISILENE)

Submitted by ROBIN P. TAN,* GREGORY R. GILLETTE,* HOWARD B. YOKELSON,* and ROBERT WEST*
Checked by PHILIP BOUDJOUK†

As silicon is the element most similar to carbon, compounds containing multiple bonds to Si have long been sought.[1] Tetramesityldisilene, the first stable compound containing a silicon–silicon double bond, was initially prepared in 1981 by ultraviolet irradiation of 2,2-dimesityl-1,1,1,3,3,3-hexamethyltrisilane.[2a, b] Photolysis of the trisilane produces dimesitylsilylene, which dimerizes to form the disilene.

The synthesis detailed here is a variation of the original photolytic method, requiring the synthesis of the trisilane as a precursor. The trisilane is prepared by coupling dichlorodimesitylsilane with chlorotrimethylsilane, using lithium metal in tetrahydrofuran (THF). The synthesis of dichloro-dimesitylsilane has been described previously.[2b] It is also available commercially.‡

Similar photolysis of trisilanes may be used to generate other stable disilenes.[3] Stable disilenes can also be synthesized by photoinduced fragmentation of cyclotrisilanes,[4] and, in some cases, by dehalogenation of dihalo-diorganosilanes.[5]

General Procedure

All reactions are performed under an atmosphere of argon or dry nitrogen in oven-dried glassware (120°C). Because of the moisture-sensitive nature of

* Department of Chemistry, University of Wisconsin, Madison, WI 53706.
† Department of Chemistry, North Dakota State University, Fargo, ND 58105.
‡ Available from Petrarch Systems, Inc., Bristol, PA 19007.

these silicon compounds, all manipulations, unless otherwise specified, are carried out using standard inert atmosphere techniques.[6] Tetrahydrofuran is dried and freshly distilled from sodium–benzophenone. Gas-chromatographic analyses were performed on a Hewlett-Packard 5890A gas chromatograph with HP series 530μ fused-silica column and HP 3390A integrating recorder. Other column packing materials such as SE-30 silicone are also quite satisfactory.

A.   2,2-DIMESITYL-1,1,1,3,3,3-HEXAMETHYLTRISILANE

$$Mes_2SiCl_2 + 2Me_3SiCl + 4Li \rightarrow Mes_2Si(SiMe_3)_2 + 4LiCl$$

$$(Mes = 2,4,6\text{-trimethylphenyl})$$

Procedure

A three-necked 100-mL flask is equipped with a reflux condenser, gas inlet, 25-mL pressure-equalizing addition funnel, and a magnetic stirring bar. The system is assembled hot and flushed with dry argon. The flask is charged with 15 mL of dry THF, and 0.9 g (130 mmol) of high sodium content ($\sim 1\%$) lithium wire* is cut into small pieces directly into the reaction vessel under positive argon flow. The flask is then placed in an ice bath, and 4.5 mL (36 mmol) of chlorotrimethylsilane,† freshly distilled from K_2CO_3, is added via syringe. Next, the addition funnel is charged with 5.4 g (16 mmol) of Mes_2SiCl_2 dissolved in 15 mL of dry THF. The Mes_2SiCl_2 solution is added dropwise over 1 h with stirring. The mixture is stirred for an additional 2 h; then the ice bath is removed and the reaction mixture is allowed to warm to room temperature. Progress of the reaction is evidenced by discoloration of the metal surface. After 4–6 h at room temperature any unreacted lithium metal is removed from the reaction mixture by vacuum filtration through a sintered-glass frit. At this point, exclusion of moisture is no longer critical. The filtrate is evaporated to a solid and 100 mL of hexane is added. The lithium salts are insoluble in hexane and a white suspension results. On addition of 50 mL of water these salts dissolve in the resultant aqueous layer. The two-phase mixture is then transferred to a 250-mL separatory funnel and the aqueous layer is removed. The organic layer is washed three times with 50-mL portions of water and once with brine. The clear pale yellow solution is then dried over Na_2SO_4 and evaporated to a solid. This is redissolved in hot hexane and cooled to 0° to yield colorless crystals of $Mes_2Si(SiMe_3)_2$. The crystals are filtered, rinsed with cold hexane, and dried at room

* Available from Aldrich Chemical Co., Milwaukee, WI 53233.
† Available from Aldrich Chemical Co., Milwaukee, WI 53233.

temperature *in vacuo* (5 h, 0.01 torr). The isolated yield is 5.27 g (80%), mp 169–171°.

Anal. Calcd. for $C_{24}H_{40}Si_3$: C, 69.82; H, 9.77; Si, 20.41. Found: C, 69.61; H, 9.73; Si, 20.23.

Properties

The trisilane is inert to both air and moisture. Purity may be determined by gas chromatography and ^{1}H NMR. The trisilane should be kept dry as traces of moisture would destroy the disilene formed in the photolysis step.

^{1}H NMR (C_6D_6) δ 0.25 (s, 18H), 2.11 (s, 6H), 2.28 (br, 12H), 6.74 (s, 4H). ^{29}Si NMR (C_6D_6) δ − 12.7, − 47.8.

B. TETRAMESITYLDISILENE

$$2Mes_2Si\,(SiMe_3)_2 \xrightarrow{hv} Mes_2Si{=}SiMes_2 + 2Me_3SiSiMe_3$$

Procedure

The compound $Mes_2Si(SiMe_3)_2$ (2.0 g, 4.9 mmol) is placed in the dry quartz photolysis tube (A) that has a specially designed frit attachment (Fig. 1). The system is evacuated for about 10 min to remove any trace of moisture from the trisilane, and then backfilled with dry argon. Next, the system is opened briefly at the 24/40 joint. The frit attachment is held a little above the photolysis tube (A) and the argon flow is continued while 80 mL of dry deolefinated pentane* and 40 mL of dry THF are syringed into tube A. The system is closed, using plastic clamps on both joints, and the trisilane is dissolved using the magnetic stirring bar. Next, the system is purged by five freeze–pump–thaw degassing cycles. (Exclusion of oxygen is critical for a maximum yield of disilene.) The cell is then cooled to − 50°C in a quartz Dewar vessel equipped with a liquid-nitrogen blowoff system, a proportional controller,† and a magnetic stirrer (Fig. 2). The stirred, colorless solution is irradiated at 254 nm in a Rayonet Model RPR-100 photoreactor‡ equipped with 254 nm lamps (No. RPR 2537A) for 48 h. The appearance of a deep

---

* Pentane (350 mL) is washed two times with 1 : 1 H_2SO_4 : HNO_3 (100 mL), two times with H_2SO_4 (100 mL), two times with distilled water (100 mL), and then dried over $MgSO_4$ and distilled from $LiAlH_4$.

† Model 4202-F-02, available from Omega Engineering Inc., Stamford, CT 06514.

‡ Available from Southern New England Ultraviolet Co., Hamden, CT 06514.

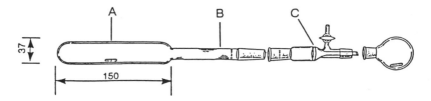

Fig. 1.   Glassware for photolysis. A, quartz tube (dimensions in millimeters); B, 24/40 quartz inner member joint (available from Ace Glass); C, fritted glass filter (Ace Glass: No. 7205-18; porosity D (10–20 μ).

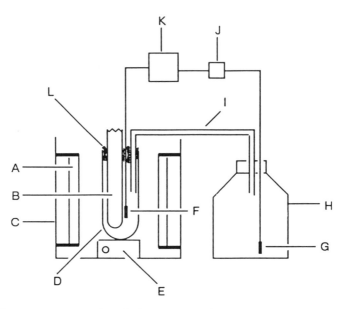

Fig. 2.   Low-temperature photolysis apparatus. A, 254-nm lamp; B, quartz photolysis tube (see Fig. 1); C, Rayonet photoreactor; D, double-walled quartz dewar; E, magnetic stirrer; F, Platinum RTD (available from Omega Engineering, Inc., No. KGC-0107); G, resistance heater (available from Watlow Electric Mfg. Co., St. Louis, Missouri 63146: No. G2J46, Firerod Cartridge Heater, 120 V, 250 W); H, double-walled reservoir Dewar for liquid nitrogen; I, insulated stainless steel tubing; J, variable transformer; K, proportional controller, L, glass wool and cheese cloth.

yellow color in solution at a short irradiation time is evidence for successful generation of the disilene species. As the photoreaction proceeds, tetramesityldisilene begins to precipitate from the photomixture. After the photolysis is completed, the solvent is removed via vacuum until about 10 mL of the mixture remains.

The reaction mixture is cooled for 5 min in a solid CO_2/acetone bath, and then the tube is inverted to filter the cooled mixture. The round bottom flask is then immersed in the cooling bath to pull the solvent through the frit. The tetramesityldisilene that remains on the side of the photolysis tube is washed into the frit by condensing solvent onto the side of the tube. This is best done by soaking a pair of cotton gloves in liquid nitrogen and applying it to the side of the tube. This process is repeated until all the tetramesityldisilene is washed into the filter frit.

The disilene is dried in a vacuum for 2 h. The whole photolysis setup is then transferred to a dry box with an argon or nitrogen atmosphere and the tetramesityldisilene is removed from the frit and weighed, yielding 1.08 g (67%) of tetramesityldisilene · 2THF. If tetramesityldisilene without any THF is desired, 120 mL of deolefinated pentane can be used as solvent instead of a pentane/THF mixture. However, because tetramesityldisilene is more soluble in pentane than in the pentane THF mixture, the isolated yield of tetramesityldisilene is somewhat lower. Mp 178–181°.

Anal. Calcd. for $C_{36}H_{44}Si_2$: C, 81.12; H, 8.33; Si, 10.55. Found: C, 80.20; H, 8.42; Si, 10.67.

When commercial glassware was used in place of the apparatus shown in Fig. 1, the yield of tetramesityldisilene was reduced by about 50%. The checkers carried out the photolysis without cooling, at 35°C, and obtained a yield of 44%.

Properties

Tetramesityldisilene is an air- and moisture-sensitive material and must be handled under an inert atmosphere. It reacts with oxygen and decolorizes, forming 2,2,4,4-tetramesityl-1,3-dioxetane,[7] a colorless crystalline solid, mp 215°. The chemical properties of tetramesityldisilene are described in reviews.[8] The disilene is characterized by [1]-NMR (C_6D_6) δ 2.03 (s, 12H), 2.49 (s, 24H), 6.70 (s, 8H), and ^{29}Si NMR (C_6D_6) δ 63.7.

References

1. For a historical discussion, see G. Raabe and J. Michl, *Chem. Rev.*, **85**, 419–509 (1985).
2. (a) R. West, M. J. Fink, and J. Michl, *Science* (Washington, DC) **214**, 1343 (1981); (b) M. J. Fink, M. J. Michalczyk, K. J. Haller, R. West, and J. Michl, *Organometallics*, **3**, 793 (1984).
3. G. R. Gillette, H. B. Yokelson, and R. West, *Organomet. Synth.*, **4**, 529–533 (1988).
4. S. Masamune, Y. Hanzawa, S. Murakami, T. Bally, and J. F. Blount, *J. Am. Chem. Soc.*, **104**, 4992 (1982); S. Murakami, S. Collins, and S. Masamune, *Tetrahedron Lett.*, **25**, 2131 (1984).
5. P. Boudjouk, B.-H. Han, and K. R. Anderson, *J. Am. Chem. Soc.*, **104**, 4992 (1984); S. Masamune, Y. Eriyama, and T. Kawase, *Angew. Chem. Int. Ed. Engl.*, **26**, 584 (1987);

H. Watanabe, K. Takeuchi, N. Fukawa, M. Kato, M. Goto, and Y. Nagai, *Chem. Lett.*, **1989**, 1131.
6. D. F. Shriver, *The Manipulation of Air Sensitive Compounds*, McGraw-Hill, New York, 1969.
7. M. J. Fink, K. J. Haller, R. West, and J. Michl, *J. Am. Chem. Soc.*, **106**, 822 (1984).
8. R. West, *Angew. Chem. Int. Ed. Engl.*, **26**, 1201–1211 (1987); R. West, G. R. Gillette, H. B. Yokelson, and A. J. Millevolte, *Phosphorus, Sulfur, Silicon & Related Elements*, **41**, 3–14 (1989).

9. MONOORGANOTIN PHOSPHINATE CLUSTERS: THE OXYGEN-CAPPED CLUSTER [(n-BuSn(OH)O$_2$PPh$_2$)$_3$O][Ph$_2$PO$_2$] (1) AND ITS CONVERSION TO THE CUBIC CLUSTER [n-BuSn(O)O$_2$PPh$_2$]$_4$(2)

$$3n\text{-BuSn(O)OH} + 4R_2PO_2H \longrightarrow$$

$$[(n\text{-BuSn(OH)O}_2PR_2)_3O][R_2PO_2] + 2H_2O \tag{1}$$

(1)

$$[(n\text{-BuSn(OH)O}_2PR_2)_3O][R_2PO_2] + n\text{-BuSn(O)OH} \xrightarrow[0.4-0.5\,\text{torr}]{200-205°}$$

$$[n\text{-BuSn(O)O}_2PR_2]_4 + 2H_2O \tag{2}$$

(2)

Submitted by K. C. KUMARA SWAMY* and ROBERT R. HOLMES*
Checked by KENNETH C. CARTER† and CHRISTOPHER W. ALLEN†

Formation of cubic clusters (2) from the reaction of n-butylstannonic acid with phosphinic acids[1,2] may be viewed as occurring via an intermediate stage involving an oxygen-capped cluster of 1.[3] The stannoxane skeleton for the oxygen-capped cluster (1) resembles the cube (2) with a corner missing. This is shown here in schematic form where the curved line represents R$_2$PO$_2$:

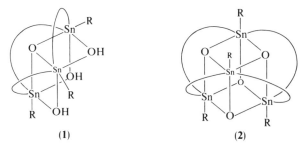

(1)   (2)

* Department of Chemistry, University of Massachusetts at Amherst, Amherst, MA 01003.
† Department of Chemistry, University of Vermont, Burlington, VT 05405.

Interconversion between the two forms has been observed in several cases.[4] This takes place via elimination or addition of water. The interrelationship is demonstrated in the procedure described here for preparing the cubic diphenylphosphinate derivative (2) from the corresponding O-capped structure (1).

A. OXYGEN-CAPPED CLUSTER [(*n*-BuSn(OH)O₂PPh₂)₃O] [Ph₂PO₂] (1)

In a 250-mL round-bottomed flask equipped with a magnetic stirring bar and a reflux condenser connected to the flask through a Dean–Stark apparatus (25-mL capacity), a mixture of *n*-butylstannonic acid (*n*-butyltin hydroxide oxide, Alfa, 95%, 2.09 g, 10.0 mmol) and diphenyl phosphinic acid (Aldrich, 2.20 g, 10.0 mmol) in toluene (Fisher, spectral grade, 120 mL) is heated with stirring under reflux for $\frac{1}{2}$ h. A heating mantle is used, and azeotropic removal of water occurs during this time. The reaction mixture is cooled and filtered, the solvent evaporated *in vacuo* with a rotary evaporator, and the residue dissolved in diethyl ether (200 mL). Filtration of this turbid solution, followed by concentrating it to a volume of 70 mL on a water bath and then evaporating the solvent slowly at 20° over an 18-h period to a volume of 20 mL, affords a white crystalline precipitate of 1. This is collected by filtration followed by washing with 10 mL of hexane: mp 188–200° (lit. 198–208°)[2] (yield 1.7 g, 35–45%). ³¹P{¹H}NMR (CDCl₃, ppm)[5]: 29.40 (s with¹¹⁹,¹¹⁷Sn satellites, 3P, P$_{bridging}$, ²P (¹¹⁹,¹¹⁷Sn–O–P) = 131.8, 125.4 Hz), 18.16 (s, 1P, P in Ph₂PO₂⁻). ¹¹⁹Sn{¹H}NMR (CDCl₃, ppm)[6]: − 497.03 ppm (t, ²*J* (Sn–O–P) = 132.2 Hz).

Anal. Calcd. for C₆₀H₇₀O₁₂P₄Sn₃; C, 49.25; H, 4.82. Found: C, 49.01; H, 4.97.

B. CUBIC CLUSTER [*n*-BuSn(O)O₂PPh₂]₄ (2)

The oxygen-capped cluster (1) as prepared above (0.509 g, 0.35 mmol) and *n*-butylstannonic acid (Alfa, 0.0726 g, 0.35 mmol) are powdered, thoroughly mixed, and transferred to a Shlenk tube (10-cm length) fitted with a 14/20 stopper. This mixture is heated *in vacuo* (0.4–0.5 mm Hg) on an oil bath (or in a sand bath) maintained at 200–205° for $\frac{1}{2}$ h, until the bubbling stopped completely. Then, with a continuous flow of dry nitrogen, the glassy material is transferred, using a spatula to separate the solid adhering to the surface, to a modified Schlenk-type 200-mL round-bottomed flask and dissolved in 70 mL of dry toluene (distilled over P₄O₁₀ directly into the flask). The turbid solution is filtered using Schlenk techniques[7] and the solvent evaporated *in vacuo* to dryness to yield the cubic cluster (2) as a white powdery material (yield 0.3 g, 51%). This material does not melt sharply but becomes a syrupy liquid at about 215°. It is best characterized by its ^{119}Sn and ^{31}P NMR spectra. ^{31}P{^{1}H}NMR (toluene-d_8 + toluene, 1 : 1, ppm): 31.74 (s with Sn satellites, ^2J(Sn–O–P) = 109.0 Hz). ^{119}Sn{^{1}H}NMR (toluene-d_8 + toluene, 1 : 1, ppm): $-$ 460.55(^2J(Sn–O–P) = 111.5 Hz).

Anal. Calcd. for $C_{64}H_{76}O_{12}P_4Sn_4$: C, 46.98; H, 4.65. Found: C, 47.07; H, 4.62.[8]

Properties

The oxygen-capped cluster (1) is an air-stable, crystalline solid and can be preserved in ordinary vials for a prolonged period of time without decomposition. It is also stable in solution ($CDCl_3$, C_6D_6). The cubic cluster (2) can be momentarily handled in air but in solution, if traces of moisture are present, it rapidly converts to the oxygen-capped cluster (1). This process can be conveniently monitored in $CDCl_3$ solutions by ^{119}Sn NMR spectroscopy.[4]

References and Notes

1. K. C. Kumara Swamy, R. O. Day, and R. R. Holmes, *J. Am. Chem. Soc.*, **109**, 5546 (1987).
2. R. R. Holmes, K. C. Kumara Swamy, C. G. Schmid, and R. O. Day, *J. Am. Chem. Soc.*, **110**, 7060 (1988).
3. R. O. Day, J. M. Holmes, V. Chandrasekhar, and R. R. Holmes, *J. Am. Chem. Soc.* **109**, 940 (1987).
4. K. C. Kumara Swamy, C. G. Schmid, H. Nadim, S. D. Burton, R. O. Day, and R. R. Holmes, Third Chemical Congress of North America, Toronto, Ontario, Canada, June 5–10, 1988, Abstract No. INOR 508; R. R. Holmes, K. C. Kumara Swamy, S. D. Burton, H. Nadim, and R. O. Day, unpublished work; R. R. Holmes, *Acc. Chem. Res.*, **22**, 190 (1989).
5. ^{31}P and ^{119}Sn NMR spectra were recorded on a Varian XL 300 FT-NMR spectrometer equipped with a multinuclear broadband probe and operated at 121.42 and 111.86 MHz, respectively. ^{31}P NMR spectra were recorded by setting triphenyl phosphate (in $CDCl_3$) at (P) = $-$ 18 ppm relative to 85% H_3PO_4; see J. Emsley and D. Hall, *The Chemistry of Phosphorus*, Wiley, New York, 1976, p. 82. Upfield shifts are negative.
6. The (^{119}Sn) shift for tetramethyltin in $CDCl_3$ is 0 ppm.
7. D. F. Shriver and M. A. Drezdzon, *The Manipulation of Air-Sensitive Compounds*, 2nd ed., Wiley, New York, 1986.
8. An analytical service familiar with handling moisture-sensitive compounds is recommended.

10. (TRIPHENYLPHOSPHORANYLIDENE)SULFAMOYL CHLORIDE

$$Ph_3P + Cl_2 \rightarrow Ph_3PCl_2$$

$$2Ph_3PCl_2 + H_2NSO_2OH \rightarrow Ph_3PNSO_2Cl + Ph_3PO + 3HCl$$

Submitted by DALE E. ARRINGTON*
Checked by ARLAN D. NORMAN†

(Triphenylphosphoranylidene)sulfamoyl chloride has been prepared by the reaction of triphenylphosphine imine with sulfuryl chloride,[1] the reaction of dichlorotriphenylphosphorane with sulfamic acid in acetonitrile,[2] and the reaction of triphenylphosphine with sulfuryl azide chloride.[3] Of the three methods, the second is the best for the large-scale preparation of this compound because of the ready availability of high-purity starting materials and ease of isolation of the product.

The synthesis reported here is an improvement over that originally reported[2] in that the use of benzene as the solvent enables the product to be isolated in excellent yield, and in an analytically pure state, by simple filtration.

Procedure

■ **Caution.** *In view of the toxic nature of chlorine gas and the suspected carcinogenicity of benzene, all operations should be conducted in a good hood.*

Triphenylphosphine (Aldrich; 267.5 g, 1.02 mol of 99% pure material) is placed in a three-necked, 2-L, round-bottomed flask equipped with a mechanical stirrer, a gas inlet tube, and a thermometer–gas outlet adapter fitted with a thermometer and a drying tube ($CaSO_4$). The flask is flushed with dry nitrogen and 600–700 mL of benzene, dried by refluxing over, and distillation from, calcium hydride is added. The gas inlet tube is connected by Tygon tubing to a gas washing bottle containing concentrated sulfuric acid, which, in turn, is connected to a cylinder of chlorine. The flask is surrounded by a cooling bath containing isopropyl alcohol. The stirrer is started (the triphenylphosphine dissolves) and the position of the gas inlet tube is adjusted to be slightly *above* the surface of the stirred solution. It is essential

* Department of Chemistry, South Dakota School of Mines and Technology, Rapid City, SD 57701.
† Department of Chemistry, University of Colorado, Boulder, CO 80309.

that the tip be above the surface of the solution, otherwise it will become plugged with dichlorotriphenylphosphorane during the reaction.

Chunks of Dry Ice are added to the bath to maintain a bath temperature of -20 to $-30°C$. When the reaction flask contents reach $0°C$, the addition of chlorine, at a rate of 30–50 bubbles per minute passing through the wash bottle, is begun. The reaction is exothermic and the internal temperature is kept between 0 and $10°$ by maintaining the bath temperature between $-20°$ and $-30°C$ and by regulating the rate of addition of chlorine. Toward the end of the chlorine addition, dichlorotriphenylphosphorane suddenly precipitates from the solution as a white solid. At this point, the bath temperature is allowed to rise to $-10°$ and chlorination is continued until the benzene acquires a distinct yellow-green color due to excess chlorine.

When chlorination is complete, the cooling bath is removed and the gas inlet tube connected to the nitrogen source. Against a countercurrent of nitrogen, the thermometer–gas outlet adapter is removed and finely ground sulfamic acid (48.5 g, 0.500 mol) is added. (**Note**: *Chlorine will be swept out of the flask by the nitrogen flow and the operator should take particular care to avoid inhaling the fumes.*) A reflux condenser, connected through a drying tube ($CaSO_4$) to a gas absorption trap, is attached; the gas inlet tube is replaced by a glass stopper and the contents heated to reflux, with stirring, using a heating mantle. As the mixture warms to reflux, the reactants are converted to a yellow, oily material and evolution of HCl is noted. As refluxing continues, (triphenylphosphoranylidene)sulfamoyl chloride precipitates from solution as small, white crystals. Although the reaction is generally complete in 3–4 h, refluxing is continued overnight to ensure complete reaction and to remove most of the HCl from the solvent.

After cooling the golden-brown mixture to room temperature, the apparatus is dismantled and most of the solvent, which contains the very soluble triphenylphosphine oxide, is removed by careful decantation under a modest N_2 flow. The product is washed in the flask with two 200-mL portions of fresh benzene and the washings combined with the bulk of the solvent for recovery. Finally, the product is suction filtered on a large Büchner funnel, washed with benzene, and dried in vacuo. (**Note**: In humid climates, product filtration is best done under nitrogen). The yield of snow-white, crystalline product is 178 g (95%).

Anal. Calcd. for $C_{18}H_{15}ClNO_2PS$: C, 57.54; H, 3.99; N, 3.86. Found: C, 57.55; H, 4.08; N, 3.86.

Most of the benzene used in this preparation can be recovered by distillation. The residue in the distillation flask should be poured from the flask while hot to prevent crystallization of triphenylphosphine oxide in the flask.

There is little difficulty in scaling the procedure up or down from the conditions reported here; runs using up to 4 mol of Ph_3P and 2.5 L of benzene in a 5-L flask have been safely made. In the larger-scale runs, however, it is critical to use a powerful stirrer and the requirement of larger cooling baths, etc., may be too cumbersome for some tastes.

Properties

(Triphenylphosphoranylidene)sulfamoyl chloride is a white, crystalline solid that melts at 213–216°C. It is insoluble in benzene, diethyl ether, and most hydrocarbons and moderately soluble in chloroform and dichloromethane. It reacts with dimethyl sulfoxide and *N,N*-dimethylformamide. If desired, it may be recrystallized from dry acetonitrile ($10 \, mL \, g^{-1}$; dissolution is slow), but this is generally unnecessary for further synthetic purposes if pure starting materials are used. It is slowly hydrolyzed by moisture and should be protected from the atmosphere in humid climates. The compound has been used to prepare a series of compounds containing the Ph_3PNSO_2— group; e.g., some substituted ureas and guanidines,[4] and some imino-1,3-dithianes and -dithiolanes.[5] The compound has the interesting property, previously unreported, of showing a green fluorescence under short-wave UV light and is weakly phosphorescent as well; its derivatives do not show these properties.

Infrared bands (KBr, 0.75%) appear at 3060 (w), 1590 (w), 1485 (m), 1435 (s), 1325 (s), 1195 (s), 1175 (s), 1155 (s), 1125 (s), 1110 (s), 990 (m), 770 (m), 750 (s), 735 (s), 715 (s), 680 (s), 555 (s), and 495 (s) cm^{-1}.

References

1. A. S. Shtepanek, V. A. Zasorina, and A. V. Kirsanov, *J. Gen. Chem. USSR* (Eng. transl.), **43**, 21 (1973).
2. D. E. Arrington, *Synth. React. Inorg. Met. Org. Chem.*, **4**, 107 (1974).
3. C. A. Obafemi, *Synth. React. Inorg. Met. Org. Chem.*, **12**, 41 (1982).
4. J. A. Moreshead and D. E. Arrington, *J. Chem. Eng. Data*, **23**, 353 (1978).
5. D. E. Westberg and D. E. Arrington, *Synth. React. Inorg. Met. Org. Chem.*, **18**, 307 (1988).

11. HEXAMETHYLDISILATHIANE

$$S + 2Na \xrightarrow[\text{THF}]{\text{naphthalene}} Na_2S \xrightarrow{2\,Me_3SiCl} Me_3Si\text{-}S\text{-}SiMe_3$$

$$I$$

Submitted by JEUNG-HO SO* and PHILIP BOUDJOUK*
Checked by HARRY H. HONG† and WILLIAM P. WEBER†

Hexamethyldisilathiane(I) has been widely used in synthesis,[1] particularly as a sulfur transfer agent or silylating reagent. While there are several routes[2] to **I**, the best procedures have been those of Harpp and Steliou,[3] which employs hexamethyldisilazane and hydrogen sulfide, and of Detty and Seidler,[4] which requires lithium triethylhydroborate (1 −). Both procedures, while efficient for the production of **I**, have inconvenient aspects. The first procedure utilizes hydrogen sulfide and generates ammonium sulfide. For large-scale synthesis neither compound is desirable: hydrogen sulfide is toxic and diammonium sulfide clogs the condenser during the reaction. The lithium triethylhydroborate route is an expensive one and generates flammable gases (hydrogen and triethylborane). The method that uses iodosilane and mercury(II) sulfide[5] requires special apparatus and is not suitable for large-scale synthesis.

The synthesis of hexamethyldisilathiane from sodium sulfide and chlorotrimethylsilane is described here. The present method is based on the convenient *in situ* syntheses of alkali metal selenides and diselenides.[6] Commercial sodium sulfide or lithium sulfide are reported to be poor substitutes for *in situ* generated sulfides in this reaction. For example, in 1961 Abel reported that disodium sulfide reacts with chlorotrimethylsilane in pressure vessels at 250°C for 20 h to produce **I**.[7] Our procedure is very convenient, utilizing readily available starting materials and apparatus under mild conditions. The yields are typically 80–88% at 0.3-mol scale. However, it can be improved to 90–95% on small scale (∼ 50-mmol) reactions. This procedure can be applied to the synthesis of various disilathianes.

Materials

Commercially available sulfur (Mallinckrodt, Inc.) and naphthalene (J. T. Baker Chemical Co.) were used without further purification. The amount of

* Department of Chemistry, North Dakota State University, Fargo, ND 58105.
† Loker Hydrocarbon Research Institute, University of Southern California, Los Angeles, CA 90089.

naphthalene can vary from 2 to 10 mol % of the sodium used. Higher naphthalene loading will shorten the time to reduce the sulfur but can lead to contamination problems during distillation. Commercial chlorotrimethylsilane (Aldrich Chemical Co.) is distilled from calcium hydride under nitrogen and stored under a nitrogen atmosphere. The submitter used sodium spheres (3–8 mm in mineral spirits, Aldrich Chemical Co.) washed with hexane and cut in half to increase reactivity. Tetrahydrofuran (THF) was distilled from sodium–benzophenone under a nitrogen atmosphere immediately before use.

Procedure

■ **Caution.** *Hexamethyldisilathiane should be regarded as toxic and vile-smelling. All operations should be conducted in a well-ventilated hood, and rubber gloves should be worn.*

All glassware and magnetic stirrer bar were dried at 140°C for several hours before use. All reaction procedures are conducted under a nitrogen atmosphere. A three-necked 500-mL flask is equipped with a pressure equalized funnel, reflux condenser, magnetic stirrer, and nitrogen inlet. The vessel is purged with nitrogen and charged with 9.6 g of sulfur (0.3 mol), 14.0 g of sodium (0.609 mol), 3.0 g (0.023 mol) of naphthalene, and 250 mL of THF. This mixture is refluxed for 12 h. The color of the mixture changes from yellow to orange to brown to light yellow to green to white. (Alternatively, disodium sulfide can be prepared by sonicating a mixture of sodium and sulfur in THF. For the quantities in the preparation of disodium sulfide, placing the reaction in an ultrasonic cleaner filled with water at a point that leads to maximum agitation in the flask from cavitation of the solvent[8] results in quantitative reduction of sulfur in 7–9 h. Reaction times of at least 48 h are required with room temperature stirring in the absence of ultrasonic waves.)

■ **Caution.** *It is important that chlorotrimethylsilane be distilled under nitrogen, preferably from calcium hydride.*[9]

The mixture was cooled to 0° with an ice-water bath followed by slow addition of 90 mL of chlorotrimethylsilane (0.71 mol) through a pressure equalized addition funnel. The reaction is exothermic. This reaction was stirred for 2 h at 0°C and 8 h at room temperature. The reaction was followed by ¹H NMR and GC and was completed at this point. To maximize yield, 40 mL of hexadecane was added as a kicker. Using a 9-in. Vigreux column, THF and excess chlorotrimethylsilane were distilled slowly at atmospheric pressure, followed by vacuum distillation to give 42.7–47.0 g of hexamethyldisilathiane (bp 91–95°C at 100 torr, 80–88% yields based on sulfur used). Distillation of the product at atmospheric pressure gave slightly lower yields.

When the reaction was carried out on a smaller scale (e.g., 0.05-mol sulfur) the reduction of sulfur required only 4 h, the reaction of sodium sulfide with chlorotrimethylsilane took 6 h, and 90–95% yields of product were obtained after distillation.

Properties

Hexamethyldisilathiane is a colorless liquid (bp 162°C at 760 torr)[10] and vile-smelling. It is stable for a long period of time at room temperature under oxygen- and moisture-free conditions. It displays the following spectroscopic properties: ^{1}H NMR (CCl$_4$, 90 MHz): 0.31 (s); IR (neat) cm^{-1}: 2950 (s), 2890 (s), 1450, 1402, 1250 (s), 850 (vs), 750 (s), 690; MS: m/z 178 (M$^+$, 50.41%), 163 (M$^+$–Me, 100%).

Acknowledgement

Financial support from the Air Force Office of Scientific Research through Grant No. 88-0060 is gratefully acknowledged.

References

1. (a) H. Berwe and A. Haas, *Chem. Ber.*, **120**, 1175 (1987); (b) K. Steliou, P, Salama, and J. Corriveau, *J. Org. Chem.*, **50**, 4969 (1985); (c) W. Ando, T. Furuhata, H. Tsumaki, and A. Sekiguchi, *Chem. Lett.*, **1982**, 885; (d) W. Ando, T. Furuhata, H. Tsumaki, and A. Sekiguchi, *Synth. Comm.*, **12**, 627 (1982); (e) T. Aida, T. H. Chan, and D. N. Harpp, *Tetrahedron Lett.*, **22**, 1089 (1981); (f) W. Haubold and U. Kraatz, *Chem. Ber.*, **112**, 1083 (1979); (g) H. S. D. Soysa and W. P. Weber, *Tetrahedron Lett.*, **1978**, 235; (h) E. W. Abel, D. A. Armitage, and R. P. Bush, *J. Chem. Soc.*, **1964**, 2455.
2. (a) I. Kuwajima and T. Abe, *Bull. Chem. Soc. Chem. Jpn.*, **51**, 2183 (1978); (b) E. J. Louis and G. Urry, *Inorg. Chem.*, **7**, 1253 (1968); (c) K. A. Hooten and A. L. Allred, *Inorg. Chem.*, **4**, 671 (1965); (d) M. Fild, W. Sundermeyer, and O. Glemser, *Chem. Ber.*, **97**, 620 (1964).
3. D. N. Harpp and K. Steliou, *Synthesis*, **1976**, 721.
4. M. R. Detty and M. D. Seidler, *J. Org. Chem.*, **47**, 1354 (1982).
5. J. E. Drake, B. M. Glavincevski, R. T. Hemmings, and H. E. Henderson, *Inorg. Synth.*, **19**, 274 (1979).
6. D. P. Thompson and P. Boudjouk, *J. Org. Chem.*, **53**, 2109 (1988).
7. E. W. Abel, *J. Chem. Soc.*, **1961**, 4933.
8. For recent reviews on the effect of ultrasound on heterogenous reactions, see (a) P. Boudjouk, in *Ultrasound: its Chemical, Physical and Biological Effects*; K. S. Suslick (ed.), VCH Publishers, New York, 1988; (b) P. Boudjouk, in *High Energy Processes in Organometallic Chemistry*; ACS Symposium Series No. 333, K. S. Suslick (ed.), American Chemical Society, Washington, DC, 1987; (c) P. Boudjouk, *J. Chem Educ.*, **63**, 427 (1986).
9. J. J. Bloomfield and J. M. Nelke, *Org. Synth.*, **57**, 1 (1977).
10. C. Eaborn, *J. Chem. Soc.*, **1950**, 3077.

12. ADDITION OF SULFUR TRIOXIDE TO PENTAFLUOROTHIO (SF$_5$) (PENTAFLUORO-λ^6-SULFANYL) CONTAINING FLUOROOLEFINS

$$SF_5Br + CHX{=}CF_2 \rightarrow SF_5CHXCF_2Br \tag{1}$$

$$SF_5CHXCF_2Br + xs\ KOH \rightarrow SF_5CX{=}CF_2 + KBr \tag{2}$$

$$SF_5CX{=}CF_2 + SO_3\ (\text{trimer}) \xrightarrow[1\text{-}3\,\text{days}]{100\,\pm\,10^\circ} SF_5\overline{CXCF_2O\dot{S}O_2} \tag{3}$$

$$(X = F,H)$$

Submitted by JAVID MOHTASHAM,* ROBIN J. TERJESON,*
and GARY L. GARD*
Checked by ROBERT A. SCOTT,† KRISHNAN V. MADAPPAT,†
and JOSEPH S. THRASHER †

It is known that the incorporation of pentafluorothio groups (SF$_5$) (pentafluoro-λ^6-sulfanyl) into molecular systems can bring about significant changes in their physical, chemical, and biological properties. These properties are manifested by various applications, such as solvents for polymers, perfluorinated blood substitutes, surface-active agents (surfactants), and fumigants and as thermally and chemically stable systems.[1]

We have found a facile synthesis for preparing SF$_5$ containing sultones in high yields. These unique sultones are capable of incorporating not only SF$_5$ groups but also sulfonyl fluoride groupings (SO$_2$F) into molecular systems; sulfonyl fluoride groups serve as precursors to compounds useful as ion exchange resins, surface active agents and strong sulfonic acids.[2-5]

Procedure

■ **Caution.** *Sulfur trioxide is a strong oxidizer and is hazardous. Pentafluorosulfur bromide is an extremely toxic and moisture-sensitive compound. The toxicity of fluorinated β-sultones are unknown, but they should also be regarded as hazardous. All operations should be carried out in a well-constructed vacuum line or in a well-ventilated hood. They must not be allowed to touch the skin or organic materials, such as grease, other than halocarbon types. Gloves, face shields, and other protective devices must be utilized where contacts with these compounds are possible. Sulfur trioxide (MCB) is available in sealed borosilicate glass ampules in 2-lb quantities. The ampule should be opened and*

* Department of Chemistry, Portland State University, Portland, OR 97207.
† Department of Chemistry, University of Alabama, Tuscaloosa, AL 35487.

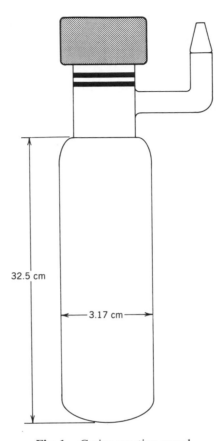

Fig. 1. Carius reaction vessel.

handled only in a hood in accordance with manufacturer's data sheet. In handling SO₃, place the ampule in a pan, scratch the neck with a file, and, wearing rubber gloves and a face shield, snap the top of the neck. Transfer and distill the SO₃ (bp 43± 1°C). The distilled SO₃ is then stored in 100–200-mL Pyrex-glass vessels equipped with Kontes Teflon valves.

The Pyrex-glass Carius reaction vessel (see Fig. 1) was crafted from Corning's heavy-wall tubing (3.17 cm o.d., 2.37 cm i.d.) and attached to a Kontes Teflon stopcock (0–12-mm bore); the sidearm is tipped with a 10/30 ground joint.

A.   SF₅CFHCF₂Br

Into an evacuated 150-mL Hoke stainless steel vessel equipped with a Whitey stainless steel valve, SF₅Br (34.0 g, 164.3 mmol; prepared according to the literature[6]) and $CFH=CF_2$ (13.5 g, 164.6 mmol; PCR, technical grade, used as received) are condensed at $-196°C$. The mixture is slowly warmed to room temperature and then heated at 70°C for 3 days. The pure product, SF_5CFHCF_2Br (30.7 g, 106.2 mmol), is obtained by distillation in a Kontes (19/22) all glass apparatus at atmospheric pressure (bp 75 ± 1°C). This modified procedure gave a yield of 64.6%.[7]

B. $SF_5CF=CF_2$

In a 250-mL, three-necked, round-bottomed Pyrex-glass vessel, equipped with a Teflon stirring bar, an addition funnel (125 mL), and a West (Kontes) reflux condenser to which a 100-mL Pyrex glass vacuum trap (cooled to $-80°C$) is attached, petroleum ether (75 mL, 90–120°C fraction; EM Science) is heated to reflux and KOH (4.8 g, 85.7 mmol; Baker) is added. The compound, SF_5CFHCF_2Br (12.6 g, 43.6 mmol), is added slowly over 1.0 h, and additional KOH (5.2 g, 92.9 mmol) is added during this period (the KOH turns brown and sludge-like). The mixture is allowed to heat at reflux for an additional 0.5 h. The pure product, $SF_5CF=CF_2$ (7.5 g, 36.1 mmol) is collected in the $-80°C$ trap in a 82.8% yield. The IR spectrum agrees with that reported previously.[7,8]

C. $SF_5CH_2CF_2Br$

Into an evacuated 150-mL Hoke stainless steel vessel equipped with a Whitey stainless steel valve, SF_5Br (17.5 g, 84.5 mmol; prepared according to the literature[6]) and $CH_2=CF_2$ (5.4 g, 84.4 mmol; PCR, technical grade, used as received) are condensed at $-196°C$. The mixture is slowly warmed to room temperature and then heated at 70°C for 2 days. The pure product, $SF_5CH_2CF_2Br$ (10.0 g, 36.9 mmol), is obtained by distillation in a Kontes (19/22) all glass apparatus at atmospheric pressure (bp 87 ± 1°C). This modified procedure gave a yield of 43.7%.[7]

D. $SF_5CH=CF_2$

In a three-necked 250-mL round-bottomed Pyrex-glass vessel, equipped with a Teflon stirring bar, an addition funnel (125 mL), and a West (Kontes) reflux condenser to which a 100-mL Pyrex glass vacuum trap (cooled to $-80°C$) is attached, petroleum ether (75 mL, 90–120°C fraction; EM Science) is heated to reflux and KOH (4.1 g, 73.2 mmol; Baker) is added. The compound, $SF_5CH_2CF_2Br$ (10.0 g, 36.9 mmol), is added slowly over 1.0 h, and additional KOH (4.5 g, 80.4 mmol) is added during this period. (The KOH turns brown and sludge-like.) The mixture is allowed to heat at reflux for an additional 0.5 h. The pure product, $SF_5CH=CF_2$ (4.2 g, 22.1 mmol) is collected in the

$- 80°C$ trap in a 59.9% yield. The IR spectrum agrees with that reported previously.[9,10]

E. $SF_5CFCF_2OSO_2$

Into an evacuated 143-mL Pyrex-glass Carius vessel equipped with a Kontes Teflon valve (see Fig. 1), trimer SO_3 (6.16 g, 77.0 mmol) and $SF_5CF=CF_2$ (18.7 g, 89.9 mmol; prepared according to the literature[7]) are condensed at $- 196°C$.* The reaction mixture is heated at 105–110°C for 72 h while the upper portion of the Carius tube is cooled with pressurized air. The pure product, 3,4,4-trifluoro-2,2-dioxo-3 (pentafluoro-λ^6-sulfanyl)-1,2λ^6-oxathietane, $SF_5CFCF_2OSO_2$ (13.7 g, 47.6 mmol), is obtained, after vacuum transfer, by distillation in a Kontes (14/20), all-glass apparatus, at atmospheric pressure (bp 88°C), in 61.8% yield. The IR spectrum agrees with that reported previously.[2]

F. $SF_5CHCF_2OSO_2$

Into an evacuated 143-mL Pyrex-glass Carius vessel equipped with a Kontes Teflon valve (see Fig. 1) trimer SO_3 (4.18 g, 52.2 mmol) and $SF_5CH=CF_2$ (10.8 g, 56.8 mmol; prepared according to the literature[9]) are condensed at $- 196°C$.†

The reaction mixture is heated at $95 \pm 5°C$ for 24 h while the upper portion of the Carius tube is cooled with pressurized air. The pure solid product, 4,4-difluoro-2,2-dioxo-3-(pentafluoro-λ^6-sulfanyl)-1,2λ^6-oxathietane, $SF_5CHCF_2OSO_2$ (8.19 g, 30.3 mmol), is obtained after vacuum transfer by distillation in a Kontes (14/20), all-glass apparatus at 600 mm (bp 108–111°C; mp 47–48 °C), in 58% yield. The infrared spectrum agrees with that reported previously.[3]

G. $SF_5CFCF_2OSO_2$ AND $SF_5CHCF_2OSO_2$

The β-sultone, 3,4,4-trifluoro-2,2-dioxo-3-(pentafluoro-λ^6-sulfanyl)-1,2λ^6-oxathietane $SF_5CFCF_2OSO_2$, is a fuming colorless stable liquid with a boiling point of 88°C.

---

* It is important to condense the $(SO_3)_3$ near the bottom of the Carius tube; during heating monomeric SO_3 is formed and reacts with the fluoroolefins.

† The checkers found that pretreatment of the SF_5 olefins with trimeric SO_3 at *room temperature* was beneficial; the treated SF_5 olefins were then used as described above.

The IR spectrum contains the following bands (cm^{-1}): 1448 (s, with sh at 1460 and 1438), 1372 (w), 1346 (w), 1304 (s), 1240 (vs), 1214 (vs with sh at 1158), 1096 (s), 1020 (m), 896 (vs), 854 (vs), 788 (s), 692 (mw), 654 (ms), 632 (w), 610 (ms), 596 (ms), 570 (m), 532 (ms), 476 (w), 460 (w), 386 (vw).

The ^{19}F NMR spectrum of $F_{(A)}SF_{4(B)}\overline{CF_{(C)}CF}_{D(E)}F_{E(D)}OSO_2$ are as follows: ϕ_A, 62.7; ϕ_B, 57.0; ϕ_C, − 116.9; ϕ_D, − 81.1; ϕ_E, − 83.3 ppm. The coupling constants are J_{AB} = 146.6, J_{BD} = 22.9, and J_{DE} = 102.8 Hz.*

A molecular ion was not observed, but other appropriate fragment ions were found. Mass spectrum (m/e): 231 (M–3F)$^+$, 219 (M–CF$_3$)$^+$, 208 (M–SO$_3$)$^+$, 181 [M–(CF$_3$ + 2F)]$^+$, 161 (M–SF$_5$ or C$_2$F$_3$OS$_2$)$^+$, 129 (C$_2$F$_3$OS)$^+$, 127 (SF$_5$)$^+$, 113 (C$_2$F$_3$S)$^+$, 108 (SF$_4$)$^+$, 101 (SF$_3$C)$^+$, 100 (C$_2$F$_4$)$^+$, 97 (C$_2$F$_3$O)$^+$, 89 (SF$_3$)$^+$, 81 (C$_2$F$_3$)$^+$, 80 (SO$_3$)$^+$, 70 (SF$_2$)$^+$, 66 (COF$_2$)$^+$,64 (SO$_2$)$^+$, 60 (SOC)$^+$, 51 (SF)$^+$, 50 (CF$_2$)$^+$, 48 (SO)$^+$, 47 (COF)$^+$, 32 (S)$^+$, 31 (CF)$^+$.

Anal. Calcd. for C$_2$F$_8$O$_3$S$_2$: C, 8.34; F, 52.7. Found: C, 8.44; F, 52.4.

The β-sultone, 4,4-difluoro-2,2-dioxo-3-(pentafluoro-λ^6-sulfanyl)-1,2λ^6- oxathietane $SF_5\overline{CHCF_2O}SO_2$ is a stable crystalline solid with a vapor pressure of 9 torr at 22°C.

The IR spectrum contains the following bands (cm^{-1}): 3002 (wm), 1419 (s), 1315 (s), 1271 (s), 1203 (vs), 1106 (s), 1078 (s), 965 (m), 916 (vw), 878, 845, 819 (vs, b), 750 (vs), 684 (s), 669 (w), 656 (ms), 612 (ms), 575 (m, sh at 565), 525 (s), 444 (m), 403 (m).

The ^{19}F NMR spectrum of $F_{(A)}SF_{4(B)}\overline{CH_{(C)}CF}_{D(E)}F_{E(D)}OSO_2$ is as follows: ϕ_A, 68.3; ϕ_B, 69.3; ϕ_D, − 72.4; ϕ_E, − 79.3 ppm. The ^{1}H NMR spectrum is δ 6.58 ppm. The coupling constants are J_{AB} = 150.8, J_{DE} = 103 Hz.†

The negative ion (CI) mass spectrum (m/e, species): 269 (M–H)$^-$, 142 (C$_2$F$_2$SO$_3$)$^-$, 138 (SC$_2$SO$_3$H$_2$)$^-$, 136 (SC$_2$SO$_3$)$^-$, 127 (SF$_5^-$ or FSCSO$_2^-$), 123 (C$_2$FSO$_3$)$^-$, 83 (SF$_2$CH$^-$ or SO$_2$F$^-$), 79 (CHCF$_2$O)$^-$.

Anal. Calcd. for C$_2$HF$_7$O$_3$S$_2$: C, 8.89; H, 0.37; S, 23.74; F, 49.3. Found: C, 9.05; H, 0.51; S, 23.63; F, 48.9.

References

1. See, for example, G. L. Gard and C. W. Woolf, U.S. Patent 3,448,121, (1969); G. L. Gard, J. Bach, C. W. Woolf, Br. Patent 1,167,112, (1969); E. E. Gilbert and G. L. Gard, U.S. Patent 3,475,453, (1969); R. E. Banks and R. N. Haszeldine, Br. Patent 1,145,263; Y. Michimasa, *Chem. Abstr.*, **82**, 175255g (1975); W. A. Sheppard, U.S. Patent 3,219,690, (1965).

* The checkers found some additional couplings, and these are included.
† The checkers found some additional couplings, and these are included.

2. J. M. Canich, M. M. Ludvig, G. L. Gard, and J. M. Shreeve, *Inorg. Chem.*, **23**, 4403 (1984).
3. R. J. Terjeson, J. Mohtasham, and G. L. Gard, *Inorg. Chem.*, **27**, 2916 (1988).
4. R. N. Haszeldine and J. M. Kidd, *J. Chem. Soc.*, **1954**, 4228.
5. T. Gramstad and R. N. Haszeldine, *J. Chem. Soc.*, **1957**, 2640.
6. R. J. Terjeson, J. M. Canich, and G. L. Gard, *Inorg. Synth.*, **27**, 329 (1990)
7. J. Steward, L. Kegley, H. F. White, and G. L. Gard, *J. Org. Chem.*, **34**, 760 (1969).
8. J. R. Case, N. H. Ray, and H. C. Roberts, *J. Chem. soc.*, **1961**, 2070.
9. G. L. Gard, "Sulfur–Fluorine Chemistry," research report, Allied-Signal Corp., 1968.
10. R. A. Demarco and W. B. Fox, *J. Fluorine Chem.*, **12**, 137 (1978).

13. PENTAFLUORO(ISOCYANATO)-λ^6-SULFANE (SF$_5$N=C=O)

$$N{\equiv}SF_3 + COF_2 + HF \rightarrow SF_5NHC(O)F$$

$$SF_5NHC(O)F + NaF \rightarrow SF_5N{=}C{=}O + NaHF_2$$

Submitted by JOSEPH S. THRASHER,* MATTHEW CLARK,* JON B. NIELSEN,* CARLOS ALVARADO,* and MARK T. ANDERSON†
Checked by GÜNTHER STEINKE,‡ THOMAS MEIER,‡ and RÜDIGER MEWS‡

The reactive isocyanate SF$_5$NCO was first prepared in 1964.[1] Some time later a new, more convenient route was found using the reaction of NSF$_3$, COF$_2$, and HF.[2] It has been determined that SF$_5$NCO of high purity can be prepared through the isolation of the intermediate SF$_5$NHC(O)F. The isolation of this nonvolatile compound greatly simplifies the separation of NSF$_3$ and other volatile by-products from the desired SF$_5$NCO.[3] Pure SF$_5$NCO is obtained in high yield when hydrogen fluoride is eliminated in the presence of NaF.

Pentafluoro(isocyanato)-λ^6-sulfane is a convenient starting material for the syntheses of compounds containing the SF$_5$N<group. Thus far, the reactions of SF$_5$NCO with carboxylic acids,[3] alcohols,[1,4] aldehydes,[4] amines,[4] and compounds with acidic C–H bonds[4] have been studied. In these cases, the chemistry of the isocyanate group follows expected reaction paths. The reactions of SF$_5$NCO with silylated nucleophiles have produced a more diverse chemistry.[5] Recently, the isocyanate has been used to incorporate SF$_5$

* Department of Chemistry, The University of Alabama, Tuscaloosa, AL 35487.
† Summer Undergraduate Research Fellow (NIH-BRSG), Department of Chemistry, Gustavus Adolphus College, St. Peter, MN 56082.
‡ Institute of Inorganic and Physical Chemistry, University of Bremen, Leobenerstrasse, NW-2, W-2800 Bremen, Germany.

groups into polynitroaliphatic explosives with the expressed purpose of improving the properties of these explosive materials.[6]

Procedure

■ **Caution.** *All compounds used in this procedure are extremely toxic and moisture sensitive. Gloves, face shields, and other protective devices must be utilized whenever handling anhydrous hydrogen fluoride so as to avoid contact of both the vapor and the liquid with the skin. The toxicity of SF$_5$NCO is unknown, but the compound should be regarded as hazardous inasmuch as it hydrolyzes to form hydrogen fluoride. All operations should be carried out in a well-constructed vacuum line within a well-ventilated hood.*

Thiazyl trifluoride[7] (63.8 g; 0.619 mol), carbonyl fluoride[7] (51.0 g; 0.772 mol) and anhydrous hydrogen fluoride (12.6 g; 0.630 mol) are measured out by PVT (pressure, volume, temperature) methods and condensed into a 500-mL stainless steel Hoke cylinder [1800 psig (pounds per square inch, gauge)] attached to a stainless steel vacuum line. The cylinder is allowed to warm to room temperature. Then it is placed in a rocker and allowed to shake.

After one week, the vessel is attached to a stainless steel tee. The tee is connected to the vacuum line, and the reaction cylinder is chilled to $-78°$ in a Dry Ice–acetone slush bath. A second 500-mL stainless steel Hoke vessel is attached to the other leg of the tee and cooled to $-196°$. The volatile contents are then stripped from the cold reaction vessel for several hours, under static vacuum, into the second cylinder, which is held at $-196°$. The slush bath around the reaction cylinder is allowed to warm slowly to $-10°$ during this procedure. It is crucial that all residual NSF$_3$ be removed at this point since separation of NSF$_3$ and SF$_5$NCO by trap-to-trap distillation is difficult. At the conclusion of the stripping, the reaction cylinder contains pure SF$_5$NHC(O)F which is nonvolatile at $-10°$. The cylinder containing the volatile materials is removed from the tee, and the contents are discarded. A third 500-mL stainless steel Hoke cylinder is filled half way with dried NaF pellets, attached to the tee, evacuated, and chilled to $-196°$. The contents of the reaction vessel are then transferred onto the NaF pellets under vacuum. The tee and the cylinder containing the SF$_5$NHC(O)F are heated gently with a heating tape to ensure complete transfer.

After 24 h at room temperature, the product is distilled from the NaF pellets through a series of traps at -125 and $-196°$. The trap at $-125°$ collects pure SF$_5$NCO (60.0 g; 57% yield). The purity of the SF$_5$NCO can be checked by IR and ^{19}F NMR spectroscopy. The most likely contaminant is a small amount of NSF$_3$, which exhibits a characteristic IR absorption at

$1523 \ cm^{-1}$ and a triplet at 70.0 ppm (J_{N-F} 27 Hz) in the ^{19}F NMR spectrum.[7] The amount of NSF_3 can be minimized by carrying out the reaction in as close a 1 : 1 : 1 molar ratio of NSF_3, COF_2, and HF as possible. Large excesses of HF should be avoided because the formation of SF_5NCO is known to be reversible; that is, large amounts of COF_2 and SF_5NH_2 are formed when SF_5NCO is reacted with HF.[3] Any SF_5NH_2 formed would end up as NSF_3 following treatment with NaF.

Properties

Pentafluoro(isocyanato)-λ^6-sulfane is a colorless gas, bp 5–5.5°.[1] IR Spectrum (gas): 2270 (vs) (ν_{asym} N=C=O), 1378 (m) (ν_{sym} N=C=O), 909 (vs) (ν S—F), 870 (vs) (ν S—F), 610 (s) (δ F—S—F) cm^{-1}. ^{19}F NMR (CCl_3F): δ_A 70.0, δ_B 87.5, J_{AB} 156.4 Hz. ^{33}S NMR (CS_2): δS − 191.3, J_{S-F} 264 Hz, linewidth 85 Hz.[8] ^{14}N NMR ($MeNO_2$): δNCO − 271.8, linewidth 120 Hz.[8] ^{13}C NMR (TMS): δNCO 130.7 (bm).[4] The structure of SF_5NCO has been investigated by electron diffraction,[9] microwave,[10] and computational methods.[10,11]

References and Note

1. C. W. Tullock, D. D. Coffmann, and E. L. Muetterties, *J. Am. Chem. Soc.*, **86**, 357 (1964); C. W. Tullock, U.S. Patent 3,347,644; *Chem. Abstr.*, **67**, 110203j (1967); *Chem Abstr.* registry [2375-30-6].

2. L. C. Duncan, T. C. Rhyne, A. F. Clifford, R. E. Shaddix, and J. W. Thompson, *J. Inorg. Nucl. Chem. Suppl.*, **33** (1976); A. F. Clifford, T. C. Rhyne, and J. W. Thompson, U.S. Patent 3,666,784; *Chem. Abstr.*, **77**, 100776r (1972).

3. J. S. Thrasher, J. L. Howell, and A. F. Clifford, *Inorg. Chem.*, **21**, 1616 (1982).

4. J. S. Thrasher, J. L. Howell, and A. F. Clifford, *Chem. Ber.*, **117**, 1707 (1984).

5. A. Clifford and J. Howell, *J. Fluorine Chem.*, **10**, 431 (1977); J. S. Thrasher, J. L. Howell, M. Clark, and A. F. Clifford, *J. Am. Chem. Soc.*, **108**, 3526 (1986); M. Clark, J. S. Thrasher, and P. L. Huston, unpublished results.

6. M. E. Sitzmann, W. H. Gilligan, D. L. Ornellas, and J. S. Thrasher, *J. Energ. Mat.*, **8**, 352 (1990); M. E. Sitzmann and W. H. Gilligan, U.S. Patent Appl., NTIS Order No. PAT-APPL-6-213038; *Chem. Abstr.*, **111**, 60565h (1989).

7. R. Mews, K. Keller, and O. Glemser, *Inorg. Synth.*, **24**, 12 (1986). The COF_2 produced in this reaction is saved and distilled through − 150° and − 196° traps. The trap at − 196° collects COF_2 and a small amount of SF_6.

8. K. Seppelt and H. Oberhammer, *Inorg. Chem.*, **24**, 1227 (1985).

9. H. Oberhammer, K. Seppelt, and R. Mews, *J. Mol. Struct.*, **101**, 32 (1983).

10. L. L. Tho, J. D. Graybeal, M. Clark, J. S. Thrasher, and M. H. Palmer, unpublished results.

11. P. Zylka, H.-G. Mack, A. Schmuck, K. Seppelt, and H. Oberhammer, *Inorg. Chem.*, **30**, 59 (1991).

14. AMBIENT-PRESSURE SUPERCONDUCTING SYNTHETIC METALS β-(BEDT–TTF)$_2$X, X = I$_3^-$, IBr$_2^-$, AND AuI$_2^-$

Submitted by H. HAU WANG* and JACK M. WILLIAMS*

Organic superconductors with critical temperatures (T_c) beyond 1–2 K have recently been realized in the (BEDT–TTF)$_2$X system where BEDT–TTF is 3,3′,4,4′-bis(ethylenedithio)2,2′,5,5′-tetrathiafulvalene, or 2-(5,6-dihydro-1,3-dithiolo [4,5-*b*][1,4]dithiin-2-ylidene)-5,6-dihydro-1,3-dithiolo[4,5-*b*] [1,4]-dithiin, C$_{10}$H$_8$S$_8$, and X is a monovalent anion. While there is only one ambient-pressure superconductor in the (TMTSF)$_2$X family (X = ClO$_4^-$, T_c, 1.2 K, TMTSF is tetramethyltetraselenafulvalene or 4,4′,5,5′-tetramethyl-2,2′-bi-1,3-diselenolylidene),[1] there are 13 ambient-pressure superconductors reported among the BEDT–TTF containing charge transfer complexes: β-(BEDT–TTF)$_2$X with X = I$_3^-$, (T_c, 1.4 K),[2] IBr$_2^-$ (2.8 K),[3] and AuI$_2^-$ (4.98 K);[4] α_t-(BEDT–TTF)$_2$I$_3$ (7 K);[5] γ-(BEDT–TTF)$_3$(I$_3$) (2.5 K);[6] θ-(BEDT–TTF)$_2$I$_3$ (3.6 K);[7] α-(BEDT–TTF)$_2$[(NH$_4$)Hg(SCN)$_4$] (1.15 K);[8] and κ-(BEDT–TTF)$_2$X, with X = I$_3^-$ (3.6 K),[9] [Hg$_{2.89}$Br$_8^{2-}$]$_{0.5}$ (4.3 K),[10] [Hg$_{1.41}$Br$_4^-$] (2.0 K),[11] [Cu(NCS)$_2$]$^-$ (10.4 K),[12] Ag(CN)$_2^-$·H$_2$O (5.0 K),[13] and Cu[N(CN)$_2$]Br$^-$ (11.6 K).[14] Single-crystal structural analyses indicate that the BEDT–TTF molecules form a two-dimensional network with loosely stacked BEDT–TTF columns and short (< 3.6 Å, the van der Waals radius of S) side by side interstack S $\cdots$ S contacts.[15] The two-dimensional physical properties of the (BEDT–TTF)$_2$X salts were confirmed by electrical, magnetic, and optical measurements. Because of the great flexibility of the BEDT–TTF molecule in terms of packing in the crystalline state, multiple phases with the same stoichiometry, but with totally different physical properties, are commonly found. Under a hydrostatic pressure of 0.5 kbar, β-(BEDT–TTF)$_2$I$_3$ reaches a high T_c state (β*) with the onset temperature as high as 8 K.[16–18] The novel physical properties, interesting pressure effects, large variety of crystal packing motifs, and promise of even higher T_c values in the (BEDT–TTF)$_2$X compounds have attracted many chemists as well as physicists to this expanding area of research. Herein, we present the detailed synthetic procedures for the ambient-pressure superconductors β-(BEDT–TTF)$_2$X with X = I$_3^-$, IBr$_2^-$, and AuI$_2^-$.

*Chemistry and Materials Science Divisions, Argonne National Laboratory, Argonne, IL 60439. Work performed under the auspices of the Office of Basic Energy Sciences, Division of Materials Sciences, of the U.S. Department of Energy under Contract W-31-109-ENG-38.

Materials

The following reagents are used as received: tetrabutylammonium iodide, tetrabutylammonium bromide, iodine, iodine monobromide, and potassium iodide (all from Aldrich, ACS reagent grade), potassium tetrabromoaurate (Alfa), and bromine (Fluka, puriss). Organic solvents are purified according to the literature.[19, 20] Tetrahydrofuran (Aldrich, gold label) is distilled from sodium-benzophenone. 1,1,2-Trichloroethane (Aldrich, reagent) is distilled from phosphorus pentoxide. Benzonitrile (Fluka, puriss) is stirred over calcium chloride and vacuum-distilled over P_2O_5. BEDT–TTF is prepared by following the literature porcedure[21] or purchased from Strem Chemicals, Inc. It is best stored under argon in a freezer.

A. TETRABUTYLAMMONIUM TRIIODIDE AND SUPERCONDUCTING BIS(BISETHYLENEDITHIO-TETRATHIAFULVALENIUM)TRIIODIDE (2 : 1)

Chemical Synthesis

$$n\text{-Bu}_4\text{NI} + \text{I}_2 \xrightarrow[\text{KI(aq)}]{} n\text{-Bu}_4\text{N}[\text{I}_3]$$

Electrocrystallization

$$2\text{ BEDT–TTF} + \text{excess } n\text{-Bu}_4\text{N}[\text{I}_3] \xrightarrow[\text{THF}]{1\ \mu\text{A cm}^{-2}}$$

$$\beta\text{-(BEDT–TTF)}_2\text{I}_3 + n\text{-Bu}_4\text{N}[\text{I}_3]$$

**Submitted by H. HAU WANG,* JAN D. COOK,† PATRICIA L. JACKSON,†
SCOTT E. PERSCHKE,† MILLICENT A. FIRESTONE,†
and JACK M. WILLIAMS***
Checked by LEON J. TILLEY‡ and LAWRENCE K. MONTGOMERY‡

Procedure

Chemical Synthesis

■ **Caution.** *Iodine is intensely irritating to eyes, skin, and mucous membranes. Preparation should be carried out in a hood.*

* Correspondent, Chemistry and Materials Science Divisions, Argonne National Laboratory, Argonne, IL 60439.
† Student Research Participants sponsored by the Argonne Division of Educational Programs: from Carnegie-Mellon University, Pittsburg, PA; Cumberland College, Williamsburg, KY; Liberty Baptist College, Lynchburg, VA; and Indiana University of Pennsylvania, Indiana, PA, respectively.
‡ Department of Chemistry, Indiana University, Bloomington, IN 47405.

The preparation of n-Bu$_4$N[I$_3$] has been reported in the literature.[22] A comparable but different route is presented as follows. A 1.4 M aqueous KI solution is prepared by adding 119 g of KI to 500-mL boiled distilled water in a 1-L beaker. Then 20.0 g of n-Bu$_4$NI (54.1 mmol) is suspended in the KI solution. The mixture is kept at 60°C while 13.8 g of I$_2$ (54.4 mmol) is slowly added. A deep brown oil forms immediately, which solidifies on cooling to room temperature. The brown solid mass is recrystallized twice from absolute methanol to yield 28.7 g of n-Bu$_4$N[I$_3$] (4.61×10^{-2} mol, 85% yield) as black needles.

Anal. Calcd. for C$_{16}$H$_{36}$NI$_3$: C, 30.84, H, 5.82; N, 2.25; I, 61.09. Found:[23] C, 30.57; H, 6.21, N, 2.21; I, 60.80.

Properties

Tetrabutylammonium triiodide crystallizes as black needles that melt at 70–71°C. It is very soluble in most polar organic solvents, slightly soluble in diethyl ether, and insoluble in water. n-Bu$_4$N[I$_3$] is best kept in a desiccator and separated from any other polyhalides to avoid cross-contamination.

Electrocrystallization

■ **Caution.** *1,1,2-Trichloroethane is a suspected carcinogen and benzonitrile causes skin irritation. Preparation should be carried out in a hood.*

The following electrocrystallization is carried out in a 45-mL-capacity H cell with a commercially available constant current source.[24] The detailed procedures to prepare the platinum electrodes and H cells were described previously.[25, 26] Then 34.3 mg of ET (8.92×10^{-5} mol) is dissolved in 15.0 mL of dry THF and is added to the anode compartment of the H cell. Similarly, 30.0 mL of THF solution containing 1.78 g of n-Bu$_4$N[I$_3$] (2.86×10^{-3} mol) is prepared and used to fill the cathode compartment and to equalize fluid levels on both sides. The above solutions are purged with argon before the electrodes (cleaned and dried) are inserted. Crystals are grown with a constant current density of 1.0 μA cm^{-2} at 22°C and the first few crystals can be observed to form within 1–3 days. The yield depends on the growth conditions; however, slow growth (1 μA cm^{-2}) usually gives high-quality crystals. In the above experiment, 35.7 mg of shiny black distorted-hexagon shaped crystals of β-(BEDT–TTF)$_2$I$_3$ (3.10×10^{-5} mol) were harvested after six weeks and washed with THF to give a 70% yield. The β-phase assignment was confirmed by using ESR linewidth techniques. Mixtures of α and β phases of (BEDT–TTF)$_2$I$_3$ have been obtained when 1,1,2 trichloroethane or benzonitrile are used as solvents.

Properties

Black distorted-hexagon-shaped crystals of β-(BEDT–TTF)$_2$I$_3$ are insoluble in most organic solvents at room temperature. The best way to identify the superconducting β phase is by use of an ESR instrument such as a Varian E-9 spectrometer operated at 9.14 GHz with 100-kHz field modulation. The α crystals give a peak-to-peak linewidth of about 70–110 G while the β crystals show a much narrower linewidth of about 18.5–24 G. The triclinic β-(BEDT–TTF)$_2$I$_3$ crystal (space group $P\bar{1}$, $Z = 1$) has cell dimensions (298 K) of: $a = 6.615(1)$ Å, $b = 9.100(1)$ Å, $c = 15.286(2)$ Å, $\alpha = 94.38(1)°$, $\beta = 95.59(1)°$, $\gamma = 109.78(1)°$, $V = 855.9(2)$ Å^3.[15] The triclinic α-(BEDT–TTF)$_2$I$_3$ crystal, which undergoes a metal–insulator transition at 135 K, has cell dimensions (298 K, $P\bar{1}$, $Z = 2$) of: $a = 9.183(1)$ Å, $b = 10.804(2)$ Å, $c = 17.422(2)$ Å, $\alpha = 96.96(1)°$, $\beta = 97.93(1)°$, $\gamma = 90.85(1)°$, $V = 1698.4(4)$ Å^3.[15] The β-(BEDT–TTF)$_2$I$_3$ is an ambient-pressure superconductor with a T_c of 1.4 K, while the α-(BEDT–TTF)$_2$I$_3$ can be thermally converted (70°C, 4 days) to α_t-(BEDT–TTF)$_2$I$_3$ with T_c around 7 K.[5]

B. TETRABUTYLAMMONIUM DIBROMOIODIDE AND SUPERCONDUCTING BIS(BISETHYLENEDITHIOTETRATHIAFULVALENIUM) DIBROMOIODIDE

Chemical Synthesis

$$n\text{-Bu}_4\text{NI} + \text{Br}_2 \xrightarrow[\text{EtOH}]{} n\text{-Bu}_4\text{N}[\text{IBr}_2] \qquad (1)$$

$$n\text{-Bu}_4\text{NBr} + \text{IBr} \xrightarrow[\text{EtOH}]{} n\text{-Bu}_4\text{N}[\text{IBr}_2] \qquad (2)$$

Electrocrystallization

$$2\text{BEDT–TTF} + \text{excess } n\text{-Bu}_4\text{N}[\text{IBr}_2] \xrightarrow[\text{THF}]{1 \ \mu\text{A cm}^{-2}}$$

$$\alpha\text{- } + \beta\text{-(BEDT–TTF)}_2\text{IBr}_2 + n\text{-Bu}_4\text{N}[\text{IBr}_2]$$

Submitted by **H. HAU WANG,* PAUL R. RUST,† CLAUDE MERTZENICH,†
MILLICENT A. FIRESTONE,† KEVIN S. WEBB,† and JACK M. WILLIAMS***
Checked by **CHAD A. HUSTING‡ and LAWRENCE K. MONTGOMERY‡**

Procedure

Chemical Synthesis 1

■ **Caution.** *Bromine is extremely toxic. Contact with skin of concentrated solutions causes severe irritation. Serious irritation of the respiratory tract may follow inhalation of the vapor. Syntheses should be done in a well-ventilated hood and gloves should be worn at all times. Alcoholic $Na_2S_2O_3$ solution should be prepared beforehand for use in case of a bromine spill.*

In a 600-mL beaker, 30.0 g of n-Bu$_4$NI (81.2 mmol) is dissolved in 400 mL of absolute ethanol that is purged with Ar for 15 min. Then 4.4 mL of Br$_2$ (85.9 mmol) is added dropwise at room temperature and the color of the solution turns deep red. By the end of the bromine addition, an orange-colored microcrystalline product precipitates. The reaction mixture is brought to about 65°C for 15 min to remove excess bromine and to dissolve all precipitates, and is then slowly cooled to ambient temperature. The crude product seperates and is recrystallized twice from absolute ethanol to give 36.5 g of n-Bu$_4$N[IBr$_2$] (69.0 mmol, 85% yield).

Anal. Calcd. for $C_{16}H_{36}NIBr_2$: C, 36.31, H, 6.86; N, 2.65; I, 23.98. Found:[23] C, 36.38; H, 6.95; N, 2.72; I, 23.76.

Chemical Synthesis 2

■ **Caution.** *IBr vapors are corrosive to the eyes and mucous membranes. All preparations should be carried out in a hood and gloves should be worn at all times.*

In a 600-mL beaker equipped with a magnetic stirrer is added 400 mL of absolute ethanol, which is purged with Ar gas for 10 min. Then, 20.44 g of n-Bu$_4$NBr (63.4 mmol) is dissolved in the alcohol and 12.09 g of IBr (58.4 mmol) is quickly added to the solution. On dissolving, the solution

* Correspondent, Chemistry and Materials Science Divisions, Argonne National Laboratory, Argonne, IL 60439.
† Student Research Participants sponsored by the Argonne Division of Educational Programs: from the University of Wisconsin, Eau Claire, WI; Carthage College, Kenosha, WI; Indiana University of Pennsylvania, Indiana, PA; and Saint Michael's College, Winooski, VT, respectively.
‡ Department of Chemistry, Indiana University, Bloomington, IN 47405.

changes to a dark red color. After 30 min of constant stirring at room temperature, the solution is covered and stored in the freezer for 10 h, yielding 30.0 g of orange crystalline product. The crude product is re-crystallized twice from absolute ethanol to give 25.1 g of n-Bu$_4$N[IBr$_2$] (81% yield).

Properties

Tetrabutylammonium dibromoiodide crystallizes as orange rods that melt at 62–64 °C. It is very soluble in most polar organic solvents and insoluble in water. n-Bu$_4$N[IBr$_2$] should be kept in a desiccator separated from other polyhalides to avoid cross-contamination.

Electrocrystallization

Electrolytic cells are set up by following the previous procedure. The solutions needed for the cell consist of 24.8 mg of BEDT–TTF (6.44 ×10^{-5} mol) in 15.0 mL of dry THF and 1.46 g of n-Bu$_4$N[IBr$_2$] (2.57 ×10^{-3} mol) in 30.0 mL of dry THF. By use of this procedure, crystals of two different phases (α and β) are grown simultaneously on the anode at a current density of 1 μA cm^{-2}. After 9 days, the lustrous black crystals are harvested to give 13 mg of (BEDT–TTF)$_2$IBr$_2$ (1.23×10^{-5} mol, total yield of two phases, 38%). The superconducting β crystals are separated from the α phase by use of ESR linewidth measurements.

Properties

Both α- and β-(BEDT–TTF)$_2$IBr$_2$ are black shiny crystalline materials. They are insoluble in most organic solvents at room temperature. The α phase gives an ESR linewidth of around 50 G, while the superconducting β phase shows a narrower linewidth of about 20 G. The triclinic β-(BEDT–TTF)$_2$IBr$_2$ is isostructural to β-(BEDT–TTF)$_2$I$_3$ and the cell dimensions (298 K) are: $a = 6.593(1)$ Å, $b = 8.975(2)$ Å, $c = 15.093(5)$ Å, $\alpha = 93.79(2)°$, $\beta = 94.96(2)°$, $\gamma = 110.54(2)°$, $V = 828.7(5)$ Å^3.[15] The triclinic α-(BEDT–TTF)$_2$IBr$_2$ has cell dimensions (298 K, space group $P\bar{1}$, $Z = 2$) of: $a = 8.901(3)$ Å, $b = 12.023(4)$ Å, $c = 16.399(8)$ Å, $\alpha = 85.16(3)°$, $\beta = 88.74(3)°$, $\gamma = 70.88(2)°$, $V = 1652(1)$ Å^3.[15] The β-(BEDT–TTF)$_2$IBr$_2$ is an ambient-pressure super-conductor with T_c of 2.8 K,[3] while the semiconducting α-(BEDT–TTF)$_2$IBr$_2$ can be thermally converted (140 °C) to β-(BEDT–TTF)$_2$IBr$_2$.[27]

C. TETRABUTYLAMMONIUM DIBROMOAURATE(I), TETRABUTYLAMMONIUM DIIODOAURATE(I), AND BIS(BISETHYLENEDITHIOTETRATHIAFULVALENIUM) DIIODOAURATE(I)

Chemical Synthesis[28]

$$n\text{-}Bu_4NBr + K[AuBr_4] \xrightarrow[\text{abs. EtOH}]{} n\text{-}Bu_4N[AuBr_4] + KBr$$

$$n\text{-}Bu_4N[AuBr_4] + CH_3COCH_3 \xrightarrow[70°C/Ar]{\text{abs. EtOH}} BrCH_2COCH_3 + HBr$$

$$+ n\text{-}Bu_4N[AuBr_2]$$

$$n\text{-}Bu_4N[AuBr_2] + 2n\text{-}Bu_4NI \xrightarrow[70°C/Ar]{\text{abs. EtOH}} n\text{-}Bu_4N[AuI_2] + 2n\text{-}Bu_4NBr$$

Electrocrystallization

$$2BEDT\text{-}TTF + \text{excess } n\text{-}Bu_4N[AuI_2] \xrightarrow[\text{THF/Ar}]{0.35\,\mu A\ cm^{-2}} \beta\text{-}(BEDT\text{-}TTF)_2AuI_2$$

$$+ n\text{-}Bu_4N[AuI_2]$$

Submitted by KELVIN S. WEBB,* ANN B. DUNN,* H. HAU WANG,† and JACK M. WILLIAMS†
Checked by LAWRENCE K. MONTGOMERY‡

Procedure

Chemical Synthesis

The $n\text{-}Bu_4N[AuBr_4]$ was prepared from a reaction of $K[AuBr_4]$ (4.04 g, 7.3 mmol) and $n\text{-}Bu_4NBr$ (2.32 g, 7.2 mmol) in absolute ethanol at 70°C. It is recrystallized from absolute ethanol to give 72% yield, mp 179–183°C.

The following procedures were carried out under argon with standard Schlenk glassware. The acetone and absolute ethanol solvents are degassed

* Student Research Participants sponsored by the Argonne Division of Educational Programs: from the Saint Michael's College, Winooski, VT; Princeton University, Princeton, NJ, respectively.
† Correspondent, Chemistry and Materials Science Divisions, Argonne National Laboratory, Argonne, IL 60439.
‡ Department of Chemistry, Indiana University, Bloomington, IN 47405.

prior to being used. In a 200-mL Schlenk flask, 80 mL of degassed absolute ethanol is added to 9.06 g of n-Bu$_4$N[AuBr$_4$]. The suspension is then heated in a water bath to 70°C and stirred for 10 min. Next, 5.0 mL of acetone is slowly added via a syringe and the suspension is stirred for an additional 15 min at 60°C. After 10 min at 60°C, the maroon color of n-Bu$_4$N[AuBr$_4$] disappears, producing a very light yellow color. The solution is allowed to cool slowly to room temperature and then stored in the refrigerator (6°C), resulting in the formation of long white needle-like crystals. The solution is filtered and the crystals washed with cold degassed absolute ethanol in a nitrogen gas purged dry box. After vacuum drying overnight 6.30 g (88.1% yield) of tetrabutylammonium dibromoaurate(I) (mp 93–94°C) is harvested.

Anal. Calcd. for $C_{16}H_{36}NAuBr_2$: C, 32.05; H, 6.05; N, 2.34; Br, 26.67. Found:[23] C, 32.89; H, 6.34; N, 2.15; Br, 26.41.

To a 200-mL Schlenk flask with 4.64 g (7.7 mmol) of tetrabutylammonium dibromoaurate(I) is added 80 mL of degassed absolute ethanol. To a second 100-mL Schlenk flask with 5.73 g (15.5 mmol) of tetrabutylammonium iodide is added 60 mL of degassed absolute ethanol. The tetrabutylammonium iodide solution is transferred via Teflon tubing to the tetrabutylammonium dibromoaurate(I) suspension. A water bath is then used to heat the resulting mixture, and a constant temperature of 60°C is maintained for 25 min, during which time the mixture turns from clear to pale yellow and finally to yellow. The mixture is then filtered while hot via a Teflon tube and a Schlenk filter under argon. The solution is allowed to cool slowly to room temperature and then stored in the refrigerator. Then 3.13 g (58% yield) of tetrabutylammonium diiodoaurate(I) (light yellow needles, mp 77.5–79.0°C) are formed.

Anal. Calcd. for $C_{16}H_{36}NAuI_2$: C, 27.70; H, 5.23; N, 2.02; I, 36.62. Found:[23] C, 27.64; H, 5.61; N, 2.29; I, 36.39.

Electrocrystallization

Electrolytic cells are set up by following the previous procedure. Because the n-Bu$_4$N[AuI$_2$] is found to be air-sensitive, the cell is prepared inside a dry box and sealed with parafilm. The solutions needed for the cell consist of 330 mg (0.48 mmol) of n-Bu$_4$N[AuI$_2$] in 30 mL of dried tetrahydrofuran (THF) and 18 mg (0.047 mmol) of BEDT–TTF in 15 mL of THF. The crystals are grown at a current density of 0.35 μA cm^{-2} and harvested at the desired stage of crystal growth (approximately 2–3 weeks) to yield metallic black crystals of β-(BEDT–TTF)$_2$[AuI$_2$].

Properties

Crystals of β-(BEDT–TTF)$_2$AuI$_2$ are lustrous metallic-black in color and superconducting with a T_c of 4.98 K at ambient pressure.[4] The crystallographic lattice parameters for the triclinic unit cell (298 K) are: $a = 6.603(1)$ Å, $b = 9.015(2)$ Å, $c = 15.403(4)$ Å, $\alpha = 94.95(2)°$, $\beta = 96.19(2)°$, $\gamma = 110.66(1)°$, $V_c = 845.2$ Å^3 (triclinic, space group $P\bar{1}$, $Z = 1$).[15] The room temperature ESR peak-to-peak linewidth of β-(BEDT–TTF)$_2$[AuI$_2$] is 16–20 G.

References and Notes

1. K. Bechgaard, K. Carneiro, F. B. Rasmussen, M. Olsen, G. Rindorf, C. S. Jacobson, H. J. Pedersen, and J. C. Scott, *J. Am. Chem. Soc.*, **103**, 2440 (1981).
2. E. B. Yagubskii, I. F. Shchegolev, V. N. Laukhin, P. A. Kononovich, M. W. Karatsovnik, A. V. Zvarykina, and L. I. Buravov, *Pis'ma Zh. Eksp. Teor. Fiz.*, **39**, 12 (1984) [*JETP Lett.* (Engl. transl.), **39**, 12 (1984)].
3. J. M. Williams, H. H. Wang, M. A. Beno, T. J. Emge, L. M. Sowa, P. T. Copps, F. Behroozi, L. N. Hall, K. D. Carlson, and G. W. Crabtree, *Inorg. Chem.*, **23**, 3839 (1984).
4. H. H. Wang, M. A. Beno, U. Geiser, M. A. Firestone, K. S. Webb, L. Nuñez, G. W. Crabtree, K. D. Carlson, J. M. Williams, L. J. Azevedo, J. F. Kwak, and J. E. Schirber, *Inorg. Chem.*, **24**, 2465 (1985).
5. G. O. Baram, L. I. Buravov, L. S. Degtyarev, M. E. Kozlov, V. N. Laukhin, E. E. Laukhina, V. G. Onishchenko, K. I. Pokhodnya, M. K. Shienkman, R. P. Shibaeva, and E. B. Yagubskii, *JETP* (Engl. transl.), **44**, 376 (1986).
6. R. P. Shibaeva, V. F. Kaminskii, and E. B. Yagubskii, *Mol. Cryst. Liq. Cryst.*, **119**, 361 (1985).
7. H. Kobayashi, R. Kato, A. Kobayashi, Y. Nishio, K. Kajita, and W. Sasaki, *Chem. Lett.*, **1986**, 833.
8. H. H. Wang, K. D. Carlson, U. Geiser, W. K. Kwok, M. D. Vashon, J. E. Thompson, N. E. Larsen, G. D. McCabe, R. S. Hulscher, and J. M. Williams, *Physica C*, **57**, 166 (1990).
9. A. Kobayashi, R. Kato, H. Kobayashi, S. Moriyama, Y. Nishio, K. Kajita, and W. Sasaki, *Chem. Lett.*, **1987**, 459.
10. R. N. Lyubovskaya, E. I. Zhilyaeva, S. I. Pesotskii, R. B. Lyubovskii, L. O. Atovmyan, O. A. D'yachenko, and T. G. Takhirov, *JETP Lett.* (Engl. transl.), **46**, 188 (1987).
11. O. A. D'yachenko and R. N. Lyubovskaya, ICSM 90 (International Conference on Science and Technology of Synthetic Metals) (Tübingen) Abstracts, Cry 3.4, p. 254 (1990).
12. H. Urayama, H. Yamochi, G. Saito, K. Nozawa, T. Sugano, M. Kinoshita, S. Sato, K. Oshima, A. Kawamoto, and J. Tanaka, *Chem. Lett.*, **1988**, 55.
13. H. Mori, I. Hirabayashi, S. Tanaka, T. Mori, and H. Inokuchi, *Solid State Commun.*, **76**, 35 (1990).
14. A. M. Kini, U. Geiser, H. H. Wang, K. D. Carlson, J. M. Williams, W. K. Kwok, K. G. Vandervoort, J. E. Thompson, D. L. Stupka, D. Jung, and M.-H. Whangbo, *Inorg. Chem.*, **29**, 2555 (1990).
15. J. M. Williams, H. H. Wang, T. J. Emge, U. Geiser, M. A. Beno, P. C. W. Leung, K. D. Carlson, R. J. Thorn, and A. J. Schultz, *Progr. Inorg. Chem.*, **35**, 51 (1987).
16. V. N. Laukhin, E. E. Kostyuchenko, Yu. B. Sushko, I. F. Shchegolev, and E. B. Yagubskii, *Pis'ma Zh. Eksp. Teor. Fiz.*, **41**, 68 (1985).

17. K. Murata, M. Tokumoto, H. Anzai, H. Bando, G. Saito, K. Kajimura, and T. Ishiguro, *J. Phys. Soc. Jpn.*, **54**, 1236 (1985).
18. A. J. Schultz, H. H. Wang, J. M. Williams, and A. Filhol, *J. Am. Chem. Soc.*, **108**, 7853 (1986).
19. A. J. Gordon and R. A. Ford, *The Chemist's Companion*, Wiley-Interscience, New York, 1972.
20. D. D. Perrin, W. L. F. Armarego, and D. R. Perrin, *Purification of Laboratory Chemicals*, Pergamon Press, London, 1983.
21. P. E. Reed, J. M. Braam, L. M. Sowa, R. A. Barkhau, G. S. Blackman, D. D. Cox, G. A. Ball, H. H. Wang, and J. M. Williams, *Inorg. Synth.*, **26**, 386 (1989).
22. A. I. Popov and R. E. Buckles, *Inorg. Synth.*, **5**, 167 (1957).
23. Midwest Microlab, Indianapolis, IN.
24. Enquiries to: Sambrook Engineering, Department of Chemistry, University College of North Wales, Bangor Gwynedd LL57 2UW, United Kingdom.
25. T. J. Emge, H. H. Wang, M. A. Beno, J. M. Williams, M.-H. Whangbo, and M. Evain, *J. Am. Chem. Soc.*, **108**, 8215 (1986).
26. D. A. Stephens, A. E. Rehan, S. J. Compton, R. A. Barkhau, and J. M. Williams, *Inorg. Synth.*, **24**, 136 (1986).
27. H. H. Wang, K. D. Carlson, L. K. Montgomery, J. A. Schleuter, C. S. Cariss, W. K. Kwok, U. Geiser, G. W. Crabtree, and J. M. Williams, *Solid State Commun.*, **66**, 1113 (1988).
28. P. Braunstein and R. J. H. Clark, *J. Chem. Soc. Dalton Trans.*, **1973**, 1845.

Chapter Two

BORON COMPOUNDS

15. TRIMETHYLAMINE–TRIBROMOBORANE

$$(CH_3)_3N \cdot BH_3 + \tfrac{1}{2}Br_2 \rightarrow (CH_3)_3N \cdot BH_2Br + \tfrac{1}{2}H_2 \tag{1}$$

$$(CH_3)_3N \cdot BH_2Br + Br_2 \rightarrow (CH_3)_3N \cdot BHBr_2 + HBr \tag{2}$$

$$(CH_3)_3N \cdot BHBr_2 + Br_2 \rightarrow (CH_3)_3N \cdot BBr_3 + HBr \tag{3}$$

Submitted by MILAP A. MATHUR,* DENNIS A. MOORE,† RONALD E. POPHAM,† and HARRY H. SISLER‡
Checked by SHAWN DOLAN§ and SHELDON G. SHORE§

Various procedures for the synthesis of trimethylamine–tribromoborane have reported.[1-6] These involve either the direct combination of BBr_3 with $(CH_3)_3N$[1,4,5] or the reaction mixture of BBr_3 and $(CH)_3N \cdot BH_3$[2,3,6] at elevated temperatures. Because of the sensitivity of BBr_3 to hydrolysis, the requirement of complex equipment, or the necessity for disposing of diborane,[6] these procedures are inconvenient. A convenient alternate is the reaction of $(CH_3)_3N \cdot BH_3$ with elemental bromine.[7-9] A three-step procedure based on this reaction has been developed and is presented here. Elemental hydrogen and hydrogen bromide are produced that can be discharged safely using ordinary procedures.

* To whom the correspondence should be addressed. Department 68J/61-2 IBM-GPD, Tucson, AZ 85744.

† Department of Chemistry, Southeast Missouri State University, Cape Girardeau, MO 63701.

‡ Department of Chemistry, University of Florida, Gainesville, FL 32611.

§ Department of Chemistry, Ohio State University, Columbus, OH 43210.

Procedure

A solution of 3.65 g (50.0 mmol) of trimethylamine–borane (obtained from Callery Chemical Co.) in 80 mL of benzene (dried over CaH_2) is placed in a three-necked 500-mL flask, fitted with a N_2 inlet, a pressure-compensated calibrated separatory funnel, and a vent tube protected by a drying tube filled with $CaSO_4$. The flask contains a magnetic stirring bar. The apparatus and contents are then flushed with dry N_2 for 15 min.

■ **Cautions.** *The following cautions apply to chemicals and procedures employed in the synthesis of trimethylamine–tribromoborane*:

1. *This synthesis should be carried out in a hood away from flame and free of electrical spark hazards to avoid fire or explosion due to the evolution of hydrogen gas in step 1 and to avoid inhalation of hazardous vapors.*
2. *Bromine will burn and blister skin on contact and cause serious irritation of the respiratory tract on inhalation. When handling bromine, always keep ammonia water within reach.*
3. *Benzene is a carcinogen. Avoid skin contact and handle in hood to avoid inhalation. Xylene can be substituted for benzene in the synthesis. Xylene may be narcotic at high concentration, but is less toxic than benzene.*
4. *Hydrogen bromide is highly irritating to eyes, skin, and the respiratory tract. Avoid inhalation and do not discharge into the atmosphere.*

Step 1 (caution 1, above). Add 25.1 mmol of Br_2 [45.5 mL of 0.552 M of Br_2 (caution 2) in benzene (caution 3)] by means of the dropping funnel to the stirred solution of $(CH_3)_3N \cdot BH_3$ at the rate of 50 drops per minute. A rapid exothermic reaction ensues, and hydrogen gas (caution 1) is evolved, leaving an almost colorless solution that contains $(CH_3)_3N \cdot BH_2Br$.

Step 2. Add 50.2 mmol of Br_2 (91 mL of a 0.552 M solution of Br_2 in benzene) to the stirred solution obtained in step 1 at the rate of 100 drops per minute. The reaction is slow and its rate decreases as the reaction proceeds. At the end of this step, the reaction mixture has a persistent yellow color. The hydrogen bromide (see caution 4, caution list above) discharged during the reaction is passed through a trap containing sodium hydroxide solution and cooled externally with ice-cold water.

Step 3. Add 49.7 mmol Br_2 (90 mL of a 0.552 M solution of Br_2 in benzene) all at once to the stirred solution remaining from step 2. Allow the reaction to proceed with continuous stirring for 22 h at 25°C. During this period, a white crystalline solid precipitates, leaving a light yellow solution. The HBr produced is flushed out with dry N_2, flowing at 2 to 3 mL per minute, and the reaction mixture is heated or refluxed for 90 min at 40°C, or until there is no further change in the intensity of the color of the solution. The remaining Br_2 is removed by the addition of a few drops of cyclohexene. The solid product is removed by filtering, washed with hexane, and dried under vacuum (wt 9.18 g, mp 228–230°C; lit., 230°C).

Evaporation of the benzene in a rotary evaporator yields 4.75 g of additional product melting at 222–224°C. Sublimation at a pressure of 10^{-3} to 10^{-4} torr and between 40 and 78°C results in a loss of weight of only 0.1 g, and the melting point of the product is 228–230°C. The combined product yield of 13.83 g is 89.3% of the theoretical yield of 15.48 g. Analysis of the bromine content in the product yields 76.51% compared to the theoretical 77.42%. Recrystallization from a dichloromethane–hexane mixture does not change the melting point or bromine analysis results for the product.

Properties

Trimethylamine–tribromoborane is appreciably soluble in benzene, chloroform, dichloromethane, acetone, and nitromethane. It is insoluble in water, CCl_4, pentane, hexane, and heptane. The infrared spectrum (obtained in a KBr pellet) has absorptions at 2960 (w), 1485 (s), 1460 (s), 1450 (m, sh), 1410 (s), 1225 (m), 1110 (s), 980 (m, sh), 950 (s), 820 (s), 745 (m, sh), 725 (s), 700 (s), 665 (s) cm^{-1} (s = strong; m = medium; w = weak, sh = shoulder) which is in agreement with IR data reported in the literature.[3, 10-14]

The ^1H NMR spectrum in CH_2Cl_2 solution has a 1 : 1 : 1 : 1 quartet concentrated at 3.16 ppm downfield from internal $Si(CH_3)_4$. The boron–hydrogen coupling constant is 3.14 Hz. The ^1H NMR data are in agreement with literature reported values[7, 9, 15] of 3.13 ppm for the quartet and 3.1 for the boron–hydrogen coupling constant. The chemical shift is 0.25 ppm farther downfield than in $(CH_3)_3N \cdot BHBr_2$. The ^{11}B NMR spectrum of this compound in dichlormethane is reported[7] to have a singlet at 22.3 ppm upfield from external trimethyl borate.

Derivative

A 3.09-g (10-mmol) sample of $(CH_3)_3N \cdot BBr_3$ is treated with 11.5 mL (100 mmol) of 4-methylpyridine, according to a procedure described elsewhere.[6] The amount of tetrakis(4-methylpyridine)–boron(3+) bromide obtained is 0.91 g or 1.46 mmol, which is 14.6% of theory. The melting point of this bromide salt is 254–255°C with decomposition (lit. 253–254°C with decomposition). The ^1H NMR spectrum of this compound in D_2O has peaks at 8.61 (broad), 8.17 (doublet), and 2.86 (singlet) ppm, respectively, downfield from internal DDS NMR reference.

Acknowledgment

The authors gratefully acknowledge the support of the research by the Grant and Research funding Committee of the Southeast Missouri State University.

References

1. C. M. Bax, A. R. Katritzky, and L. E. Sutton, *J. Chem. Soc.,* **1958**, 1258.
2. H. Noth and H. Beyer, *Chem. Ber.,* **93**, 2251 (1960).
3. J. M. Miller and M. Onyszchuk, *Can. J. Chem.,* **41**(11), 2898 (1963).
4. J. M. Makosky, G. L. Galloway, and G. E. Ryschkewitsch, *Inorg. Chem.,* **6**, 1972 (1967).
5. T. Okuda, H. Ishara, K. Yamada, and H. Negita, *Bull, Chem. Soc. Jpn.,* **52**(2), 307 (1979).
6. G. E. Ryschkewitsch and G. L. Galloway, *Inorg. Synthe.,* **12**, 141 (1970).
7. M. A. Mathur, Ph.D. dissertation, University of Florida, 1970.
8. M. A. Mathur and G. E. Ryschkewitsch, *Inorg. Synth.,* **12**, 123 (1970).
9. W. H. Myers, G. E. Ryschkewitsch, M. A. Mathur, and R. W. King, *Inorg. Chem.,* **14**, 2874 (1975).
10. A. R. Katritzky, *J. Chem. Soc.,* **1959**, 2049.
11. R. C. Taylor, "Boron–Nitrogen Chemistry." Advances in Chemistry Series No. 42, American Chemical Society, Washington, DC, 1964.
12. W. Sawodny and J. Goubeau, *Z. Physik Chem.,* **44**, 227 (1965).
13. R. L. Amster and R. C. Taylor, *Spectrochim. Acta,* **20**(10), 1487 (1964).
14. K. C. Nainen, Ph.D. dissertation, University of Florida, 1969.
15. J. M. Miller and M. Onyszchuk, *Can. J. Chem.,* **42**, 1518 (1964).

16. 1,4-DI-*tert*-BUTYL-2,6-DIISOPROPYL-3,5-BIS(TRIMETHYLSILYL)-3,5-DIAZA-*nido*-HEXABORANE(6)

$$\text{Li}(C_4H_9) + \text{HN}[\text{Si}(CH_3)_3]_2 \rightarrow \text{Li}\{\text{N}[\text{Si}(CH_3)_3]_2\} + C_4H_{10}$$

$$\text{Li}\{\text{N}[\text{Si}(CH_3)_3]_2\} + (t\text{-}C_4H_9)\text{BCl}_2 \rightarrow \text{Cl}(t\text{-}C_4H_9)\text{B}{=}\text{N}[\text{Si}(CH_3)_3]_2 + \text{LiCl}$$

$$\text{Cl}(t\text{-}C_4H_9)\,\text{B}{=}\text{N}[\text{Si}(CH_3)_3]_2 \rightarrow (t\text{-}C_4H_9)\text{B}{\equiv}\text{NSi}(CH_3)_3 + \text{ClSi}(CH_3)_3$$

$$(t\text{-}C_4H_9)\text{B}{\equiv}\text{NSi}(CH_3)_3 + (i\text{-}C_3H_7)\text{BCl}_2 \rightarrow$$

$$\text{Cl}(t\text{-}C_4H_9)\text{B}{-}\text{N}[\text{Si}(CH_3)_3]{-}\text{B}(i\text{-}C_3H_7)\text{Cl}$$

$$2\,\text{Cl}(t\text{-}C_4H_9)\text{B}{-}\text{N}[\text{Si}(CH_3)_3]{-}\text{B}(i\text{-}C_3H_7)\text{Cl} + 4\,\text{Li} \rightarrow 2$$

$$4\,\text{LiCl} +$$

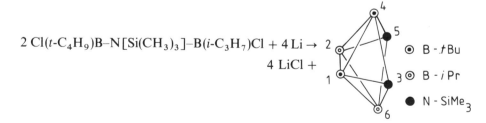

◉ B - *t* Bu

◉ B - *i* Pr

● N - SiMe₃

Submitted by P. PAETZOLD* and B. REDENZ-STORMANNS*
Checked by N. S. HOSMANE†

Hexaalkyl diazahexaboranes are derived from $nido$-$N_2B_4H_6$. $nido$-Clusters with six vertices normally correspond to the pentagonal bipyramid as the $closo$-structure by taking off one of the apices. In diazahexaboranes, however, one of the equatorial vertices is missing, thus leaving a trapezoid fragment of the corresponding pentagonal basis with the nitrogen atoms flanking the hole. By this pattern, these highly electronegative atoms remain four-coordinated, and a higher coordination number, typical for electron-deficient cluster molecules, is restricted to boron. A structure of this type was first reported for $[N(t\text{-}Bu)]_2(BMe)_4$ and was confirmed by the determination of its crystal structure.[1] A more complex derivative, $[(Nt\text{-}Bu)_2[B(i\text{-}Pr)]_2[B(t\text{-}Bu)]_2$, revealed an unsymmetric distribution of the two alkyl groups to apical and equatorial boron atoms. The synthetic route to these clusters started from the haloboration of iminoboranes,[2] $RB{\equiv}NR'$, with subsequent dehalogenation of the haloboration products, Hal—BR—NR'—BR''—Hal, with alkali metal. The originally expected dehalogenation product, a three-membered ring [—BR—NR'—BR''—], is formed only in the case of three huge ligands, e.g., $R = R' = R'' = t\text{-}Bu$.[1] The crucial point of the synthetic route seems to be the formation and haloboration of the iminoborane. A still unpublished product has been picked out here, which promises to be synthesized rather easily.

Procedure

A commercially available 1.6 M solution of n-butyl lithium in hexane (37.5 mL, 0.0600 mol) is dropped into a stirred solution of commercial bis(trimethylsilyl)amine (9.66 g, 0.0600 mol; Aldrich, cat. no. H 1,000-2) in diethyl ether (40 mL) at 0°. The formation of the corresponding lithium amide is completed by stirring the solution at 25°C for 1 h. The resulting solution is slowly dropped into a solution of $tert$-butyldichloroborane[3] (8.34 g, 0.0601 mol) in 90 mL hexane, cooled by Dry Ice. After stirring for 2 h at 25°, lithium chloride is filtered off and the solvents are removed under vacuum. The pure colorless [bis(trimethylsilyl)amino]-$tert$-butylchloroborane is gained by distillation at 36° at 0.005 mm (lit.[4]: 50–53° at 0.2 mm), with the distillate cooled by Dry Ice (12.02 g, 76% yield).

* RWTH Aachen University of Technology, Institute of Inorganic Chemistry, Templergraben 55, D-5100 Aachen, Germany.
† Department of Chemistry, Southern Methodist University, Dallas, TX 75275.

The elimination of chlorotrimethylsilane is achieved in a quartz tube (3 cm diameter, 60 cm length) in a vertical tube furnace (50 cm length). The tube is filled with bits of quartz glass in order to afford sufficient heat transport. The upper end is connected to a 100-mL flask via a U-shaped glass tube that is surrounded by a heating band. The lower end leads to a vacuum pump through a cooling trap, where pentane (5 mL) is stored. The aminoborane, $Cl(t\text{-}C_4H_9)B{=}N[Si(CH_3)_3]_2$, (10.54 g, 0.0400 mol) is filled into the flask, vacuum is established in the whole system (< 0.1 mm), the quartz tube and the U-shaped tube are heated to 570 and 100°, respectively, and the amino-borane is then distilled at 60°C (bath temperature) into the quartz tube within 2 h. Pentane and the products iminoborane, $(t\text{-}C_4H_9)B{\equiv}NSi(CH_3)_3$, and trichloromethylsilane are frozen out in the trap by liquid nitrogen. The system is filled with dry nitrogen, the trap, still cooled, is separated from the quartz tube in a rapid stream of nitrogen, and a dropping funnel is attached, equipped with a pressure balance. The liquid nitrogen is now replaced by Dry Ice cooling. Dichloroisopropylborane[3] (5.00 g, 0.0401 mol) is added dropwise to the cooled solution in the trap, and the mixture is brought to room temperature and stirred for 1 h. After removing pentane and chlorosilane *in vacuo* (0.005 mm), the remaining liquid material is taken into a syringe and added slowly to a mixture of tetrahydrofuran (10 mL), hexane (10 mL), and lithium (1.68 g, 0.242 mol) in an ice-cooled flask. After stirring 1 h at 0° and 1 h at room temperature, most of the solvent is removed *in vacuo*. Hexane (30 mL) is added again. The precipitated lithium chloride and excess lithium are filtered off. Remaining products are collected by washing the solids twice with hexane (10 mL). All volatile components are removed from the joint solutions at 40°C *in vacuo* (0.005 mm). Recrystallizing from diethyl ether gives pure cluster product (3.50 g, 42% yield).

Anal. Calcd. for $B_4N_2C_{20}H_{50}Si_2$: C, 57.5; H, 12.2; N, 6.70. Found: C, 57.7; H, 12.1; N, 6.69.

Properties

The colorless crystals of the diazahexaborane $[NSi(CH_3)_3]_2[B(i\text{-}C_3H_7)]_2[B(t\text{-}Bu)]_2$ decompose on heating above 250°. They are soluble in chlorinated hydrocarbons. The ^{13}C NMR spectrum (in $CDCl_3$, TMS as standard) exhibits eight quartets for CH_3 groups: two quartets (3.0, 4.3 ppm) represent two nonequivalent $NSiMe_3$ vertices, two quartets (31.9, 32.9 ppm) stand for an apical and an equatorial $B(t\text{-}Bu)$ vertex, respectively, and four quartets (22.5, 23.1, 23.3, 23.4 ppm) are shown by four nonequivalent CH_3 groups from two $B(i\text{-}Pr)$ vertices, an apical and an equatorial one. According to that, four ^{11}B NMR signals are found (-11.6, -13.6, 7.1, 8.6 ppm, $Et_2O \cdot BF_3$ as standard). From four hypothetical distributions of two pairs of alkyl groups (i-Pr,i-Pr/t-Bu,t-Bu) to two apical ($a1$ and $a2$) and two equatorial ($e1$ and $e2$) positions, the distributions $a1$, $e1/a2$, $e2$ and $a1$, $e2/a2$, $e1$ with different configurations are realized, representing a racemate, but not the distributions $a1$, $a2/e1$, $e2$ and $e1$, $e2/a1$, $a2$ with different constitutions.

References

1. R. Boese, B. Kröckert, and P. Paetzold, *Chem. Ber.*, **120**, 1913 (1987).
2. P. Paetzold, *Adv. Inorg. Chem.*, **31**, 123 (1987).
3. P. A. McCusker, E. C. Ashby, and H. S. Makowski, *J. Am. Chem. Soc.*, **79**, 5182 (1957).
4. B.-L. Li, M. A. Goodman, and R. H. Neilson, *Inorg. Chem.*, **23**, 1368 (1984).
5. K.-H. van Bonn, T. von Bennigsen-Mackiewicz, J. Kiesgen, C. von Plotho, and P. Paetzold, *Z. Naturforsch.*, **43b**, 61 (1988).

17. POLYCYCLIC BORAZINES

Submitted by D. T. HAWORTH* and G.-Y. LIN KIEL*
Checked by LEE J. TODD† and MARK W. BAIZE†

$$Pb(OCOCH_3)_2 + 2RSH \rightarrow Pb(SR)_2 + 2CH_3COOH$$

$$3Pb(SR)_2 + 2BCl_3 \rightarrow 2B(SR)_3 + 3PbCl_2$$

$$3B(SR)_3 + 3HX(CH_2)_n NH_2 \longrightarrow \left[B \begin{matrix} X \\ (CH_2)_n \\ N \end{matrix} \right]_3 + 9RSH$$

In earlier volumes of *Inorganic Syntheses* the preparation of borazine,[1] 2,4,6-trichloroborazine,[2,3] 1,3,5-trimethylborazine,[4]‡ and 1,3,5-trimethyl-2,4,6-trichloroborazine[5] are given. We report here the details of procedures for the preparation of several polycyclic borazines using the three-step synthesis shown above. The procedures described here are based on the original

* Department of Chemistry, Marquette University, Milwaukee, WI 53233.
† Department of Chemistry, Indiana University, Bloomington, IN 47405.
‡ Borazine: 1,3,5,2,4,6-triazatriborine.

syntheses;[6-8] however, we include additional NMR data of carbon resonances of the tris(alkylthio)boranes and the effect of the borazine ring on some of the carbon resonances of the exocyclic —X—$(CH_2)_n$— ring system where X is NH or NCH_3 and n is 2 or 3. The solid starting material, $Pb(SR)_2$, is easier to handle than the volatile thiols.

Procedure

Lead Thiolates

■ **Caution.** *Thiols have a strong, disagreeable odor. Work in a well-ventilated fume hood.*

Lead thiolates are obtained in near quantitative yield from the reaction of the thiol with a 95% ethanol solution of lead acetate. To a stirred solution of 50.0 g (0.13 mol) of $Pb(OCOCH_3)_3 \cdot 3H_2O$* in a mixture of 350 mL of 95% ethanol plus 35 mL H_2O, 18.3 g (0.30 mol) of ethanethiol is slowly added. Filtration after 0.5 h affords a yellow solid that is washed with water, then ethanol- and vacuum-dried, yielding 38.2 g (89% yield) of $Pb(SC_2H_5)_2$.[6]

The corresponding $Pb(SC_3H_7)_2$ (42.4 g, 91% yield) is prepared in an analogous manner from 50.0 g (0.13 mol) of $Pb(OCOCH_3)_2 \cdot 3H_2O$ and 23.0 g (0.30 mol) of 1-propanethiol.[6]

Tris(alkylthio)boranes. A two-necked 250-mL flask is charged with lead ethanethiolate (20.0 g, 0.061 mol), 50 mL of petroleum ether, and a magnetic stirring bar. The flask is fitted with a condenser and a Dry Ice charged cold finger that is attached to a tank of BCl_3†. After the addition of 7.0 g (0.06 mol) of BCl_3, the mixture is refluxed for 4 h. The solids collected by vacuum filtration are washed with two 15-mL portions of petroleum ether. Vacuum distillation of the filtrate affords 6.7 g (0.035 mol) of $B(SC_2H_5)_3$, bp 72–74°/0.07 torr.[6] Yield: 58%.

The tris(propylthio)borane is prepared in an analogous manner from the reaction of $Pb(SC_3H_7)_2$ (20.0 g, 0.056 mol) and BCl_3 (4.0 g, 0.043 mol) to afford 9.0 g (0.039 mol) of $B(SC_3H_7)_3$, bp 83.5° at 0.2 torr.[6] Yield: 89%.

Because of the extreme moisture sensitivity of the tris(alkylthio)boranes and the subsequent polycyclic borazines, all the manipulations made during their synthesis should be executed with the exclusion of water.

Polycyclic Borazines. All the following operations are performed under a pure dry nitrogen atmosphere. A double necked, round-bottomed, 120-mL

---

* Available from Aldrich Chemical Company, Milwaukee, WI 53233.
† Available from Matheson Gas Products, Joliet, IL 60434.

flask containing a magnetic stirring bar is fitted with a water-cooled reflux condenser. The condenser is topped with a T-tube that is connected to a nitrogen source and a pressure-release oil bubbler. The flask is charged with 0.8 g (10.8 mmol) of *N*-methyl-1,2-ethanediamine* in 50 mL of dry benzene. A 2.05-g (10.5-mmol) sample of $B(SC_2H_5)_3$ is introduced by syringe into the flask. The flask is immersed in an oil bath, and the solution is allowed to reflux for 3 h. After cooling and vacuum removal of the solvent, the solids are transferred (under N_2) to a vacuum sublimator. Vacuum sublimation at 105° and 0.3 torr gives the product hexahydro-1,6,11-trimethyl-1*H*,6*H*, 1*H*-tris[1,3,2]diazaborolo[1,2-*a*:1′,2′-*c*:1″,2″-*e*][1,3,5,2,4,6]triazatriborine [52813-38-4], $(\overline{BN(CH_3)C_2H_4N})_3$. Yield: 0.45 g (52%), mp 160–162°.[8]

Anal. Calcd. for $C_9H_{21}B_3N_6$: C, 43.90; H, 8.55. Found: C, 43.76; H, 8.49.

Using the same apparatus and workup procedure, 0.93 g (10.6 mmol) of *N*-methyl-1,3-propanediamine* was allowed to react with 2.5 g (7.4 mmol) of $B(SC_3H_7)_3$ in benzene to give the product dodecahydro-1,7-13-trimethyltris[1,3,2]diazaborino[1,2-*a*: 1′,2′-*c*: 1″,2″-*e*] [1,3,5,2,4,6]triazatriborine [57907-40-1], $(\overline{BN(CH_3)C_3H_6N})_3$. Yield: 0.45 g (44%), mp 145–148°.[8]

Anal. Calcd. for $C_{12}H_{27}B_3N_6$: C, 50.00; H, 9.38. Found: C, 49.75; H, 9.42.

1,3-Propanediamine* (0.31 g, 4.19 mmol) and $B(SC_3H_7)_3$ (1.1 g, 7.74 mmol) in benzene give the product dodecahydrotris[1,3,2]diazaborino[1,2-*a*: 1′,2′-*c*: 1″,2″-*e*][1,3,5,2,4,6]triazatriborine [6063-61-2], $(\overline{BN(H)C_3H_6N})_3$. Yield 0.30 g (85%), mp 150–152°.[8]

Anal. Calcd. for $C_9H_{21}B_3N_6$: C, 43.90; N, 8.55. Found: C, 43.69; H, 8.46.

Properties

^{13}C NMR spectra of $B(SC_2H_5)_3$ in $CDCl_3$ gives resonances at δ-values of 16.4 (CH_2) and 25.2 (CH_3) ppm (TMS standard). $B(SC_3H_7)_3$ gives resonances at δ values of 13.3 (S—CH_2—), 31.5 (CH_2—CH_2—CH_3), and 24.5 (—CH_3) ppm.‡

These thermally stable planar borazine compounds having exocyclic groups all give IR spectra (KBr pellet technique) for the B—N (ring out of plane), B—N (ring deformation), and B—N (cyclic) vibrations in the ranges 680–720, 900–950, and 1345–1375 cm^{-1}, respectively.[8,9] ^{13}C NMR taken in

---

‡ The checkers report that the ^{11}B NMR spectra recorded in CH_2Cl_2 gave δ values of + 59.4 ppm for $B(SC_3H_7)_3$, + 23.6 ppm for $(\overline{BN(H)C_3H_6N})_3$ and + 25.1 ppm for $(\overline{BN(CH_3)C_3H_6N})_3$ using $Et_2O \cdot BF_3$ as an external standard.

$CDCl_3$ gives N—CH_3 carbon resonances at 33.2 and 28.6 ppm for compounds $(\overline{BN(CH_3)C_2H_4N})_3$ and $(\overline{BN(CH_3)C_3H_6N})_3$, respectively. This compares to 36.3 and 33.8 ppm in the respective starting materials, N-methyl-1,2-ethanediamine and N-methyl-1,3-propanediamine. The compound $(\overline{BN(H)C_3H_6N})_3$ can also be prepared by the reaction of tris-(dimethylamino)borane and 1,3-propanediamine[10,11] and the compound $(\overline{BN(CH_3)C_2H_6N})_3$ by the reaction of tris(dimethylamino)borane or tris-(ethylamino)borane with N-methyl-1,2-ethanediamine.[12] Thus the interaction of a tris(dialkylamino)borane with a diamine or amine alcohol is an alternate method for the synthesis of polycyclic borazines.[13] Several macrocyclic and linear oligomeric condensates containing the borazine ring have also been reported in the literature.[14]

References

1. K. Niedenzu and J. W. Dawson, *Inorg. Synth.*, **10**, 142 (1967).
2. K. Niedenzu and J. W. Dawson, *Inorg. Synth.*, **10**, 139 (1967).
3. D. T. Hawroth, *Inorg. Synth.*, **13**, 41 (1972).
4. J. Bonham and R. S. Drago, *Inorg. Synth.*, **9**, 8 (1967).
5. D. T. Haworth, *Inorg. Synth.*, **13**, 43 (1972).
6. R. H. Cragg J. P. N. Husband, and A. F. Weston, *J. Inorg. Nucl. Chem.*, **35**, 3685 (1973).
7. R. H. Cragg and A. F. Weston, *J. Chem. Soc. Chem. Commun.*, **1972**, 79.
8. R. H. Cragg and A. F. Weston, *J. Chem. Soc. Dalton Trans.*, **1975**, 1761.
9. A. Meller, *Organomet. Chem. Rev.*, **2**, 1 (1967).
10. K. Niedenzu and P. Pritz, *Z. Inorg. Allgem. Chem.*, **340**, 329 (1965).
11. K. Niedenzu, P. J. Busse, and C. D. Miller, *Inorg. Chem.*, **9**, 977 (1970).
12. R. H. Cragg and M. Nazerv, *Inorg. Nucl. Chem. Lett.*, **10**, 481 (1974).
13. R. H. Cragg and M. Nazerv, *J. Organomet. Chem.*, **303**, 329 (1986).
14. A. Meller, H-J. Fullgrabe, and C. D. Habben, *Chem. Ber.*, **112**, 1252 (1979).

18.   1,5-CYCLOOCTANEDIYLBORYLSULFIDES

Submitted by ROLAND KÖSTER* and GÜNTER SEIDEL*
Checked by WALTER SIEBERT† and BERND GANGNUS†

A number of organoboron chalcogen compounds formed between the 1,5-cyclooctanediylboryl residue(s) and the chalcogenides sulfide and selenide,

* Max-Planck-Institut für Kohlenforschung, Kaiser-Wilhelm-Platz 1, D-4330 Mülheim an der Ruhr, Germany.
† Anorganisch-Chemisches Institut der Universität Heidelberg, Im Neuenheimer Feld 270, D-6900 Heidelberg, Germany.

respectively, are very simply prepared and are widely applicable in synthesis. Furthermore, the compounds can be easily purified if necessary. In two sections the syntheses and properties of six derivatives with the 1,5-cyclo-octanediylboryl residue bonded to chalcogenides are described. Three pure compounds having a 1,5-cyclooctanediylboryl-sulfide unit are known:[1] bis(1,5-cyclooctanediylboryl)monosulfide, 9-hydrothio-9-borabicyclo[3.3.1]-nonane, and bis(1,5-cyclooctanediylboryl)disulfide. All of these compounds may be conveniently prepared by reactions from simple adducts and are easy to manipulate with the provision that the strict exclusion of air and parti-cularly of moisture is maintained. Compared with R_2BS compounds having monofunctional organic residues[2] the three 1,5-cyclooctanediylboryl-sulfides are much more uniform because neither substituent exchanges nor redox reactions of the boron sulfur grouping to S_1/S_2 mixed ring compounds[2] take place.

The described reactions start from the compounds bis-9-bora-bicyclo[3.3.1]nonane $(9H\text{-}9\text{-}BBN)_2$,[3] elemental sulfur (S_8), and dihydrogen sulfide or from the easily preparable 9-iodo-9-borabicyclo[3.3.1]nonane.[4,5] The procedures are simple and practicable, and the yields are nearly quantitative.* The $(9H\text{-}9\text{-}BBN)_2$–sulfur reaction can be followed not only by NMR spectroscopy but also by volumetric measurements of the evolved gas (H_2). It is also remarkable that the BC bonds of the 1,5-cyclooctanediylboryl group are completely stable toward sulfidation[6,7] at temperatures up to about 150°. The three compounds described here can be prepared easily on a scale of some 100 g of high purity. If necessary, the purification is possible, involving distillation or sublimation under vacuum without any decom-position.

Procedure

■ **Caution.** *The starting material $(9H\text{-}9\text{-}BBN)_2$ is toxic and care must be exercised to avoid allowing the material or its solutions to contact the skin. Because of the extremely high reactivity of the boron–sulfur compounds to moisture, care must be taken in these syntheses to maintain strictly anhydrous conditions. Therefore, all parts of the apparatus used must be dried carefully and are best kept under an atmosphere of dry oxygen-free nitrogen or argon. Reaction with moisture causes change to boron–oxygen compounds with libera-tion of H_2S. All operations and manipulations are carried out in an efficiently ventilated hood, and provisions must be made for destroying effluent gas (e.g., traces of H_2S) either by absorption or by freezing out. Thus, the conditions of*

---

* Checkers obtained about 2% lower yields of the three organoboron sulfides using one-tenth of the procedure scale shown below.

the following preparations should be maintained strictly. Any liquid used for cooling (bath, reflux condenser) should be inert such as aliphatin.* Care must also be taken when cleaning the apparatus. Propanols admitted under inert gas (Ar, N₂) before dismantelling the apparatus will efficiently dissolve and destroy residual compounds.

A.   BIS(1,5-CYCLOOCTANEDIYLBORYL)MONOSULFIDE

Bis(1,5-cyclooctanediylboryl)monosulfide is an extremely efficient reagent for the exchange of the sulfide–sulfur toward the oxide–oxygen of organic and inorganic substrates. The monosulfide is very soluble in the common organic solvents, also in dioxane, with which the reagent does not react even on heating. All solvents with active hydrogen must be avoided for handling the borane. The monosulfide is a starting material for preparing the hydrothio- and the disulfide compounds (see below).

The pure monosulfide can be prepared nearly quantitatively from (9H-9-BBN)₂ with elemental sulfur. The yellow solutions of both reactants in mesitylene are mixed and heated to 130–140°. The liberation of gas (H₂) starts slowly at about 90°. The initially yellow solution is decolorized slowly. After more than 20 h at 130° a colorless clear solution is obtained. Only the monosulfide (^{11}B, δ 83.7) and the quantitative amount of gas (H₂) are obtained. The stoichiometric ratio of the two reactants according to the above equation must be realized exactly, because the disulfide (^{11}B, δ 78.4) is an intermediate of the S₈ degradation with the $>$BH₂B$<$ reagent. Despite the precipitation of the slightly soluble disulfide, one obtains solutions with

---

* Aliphatin is a mixture of saturated hydrocarbons with bp 190°, available from Esso AG, D-2000 Hamburg, Germany.

the monosulfide and a very small amount of the disulfide, which have an averaging ^{11}B NMR signal at $\delta < 82$. If the $>$BH$_2$B$<$ starting compound (^{11}B, δ 27) is present after the end of the H$_2$ evolution, it is possible to transfer this compound quantitatively into the monosulfide by adding the stoichiometric amount of sulfur. The $>$BH$_2$B$<$ reagent can also be destroyed quantitatively with ethene at about 80° to form the easily distillable 9-ethyl-9-BBN. The monosulfide is purified, if necessary, by sublimation under vacuum. The colorless pure borane melts at 104–105° without decomposition.

The pure bis(1,5-cyclooctanediylboryl)monosulfide is much easier to prepare than the other bis(diorganoboryl)monosulfides. The compound has a high thermal stability, allowing it to be prepared in a highly pure form. The preparation of (R$_2$B)$_2$S compounds with monofunctional organic residues from (R$_2$BH)$_2$ boranes and elemental sulfur is not possible. In analogy to the corresponding RBSe$_3$BR ring compounds[8] the five-membered RBS$_3$BR rings,[9] which have both a S$_1$ and a S$_2$ bridge between the two boron atoms, are formed very quickly. These mixed sulfidated organoboranes are also obtained from alkyl haloboranes with elemental sulfur.[10,11]

Procedure

A 1-L apparatus consisting of a three-necked flask with a magnetic stirrer is fitted with a jacket for an inside thermometer and a reflux condenser capped with a bypass for inert gas and a bubbler, connected with a gasometer (for measuring the amount of the evolved gas). The whole apparatus is evacuated, then filled with inert gas (Ar, N$_2$) and charged with oxygen-free, dry mesitylene (about 300 mL), elemental sulfur* (6.78 g, 212 mmol), and (9H-9-BBN)$_2$[3],† (52.5 g, 212 mmol).

A stirred light yellow suspension is slowly heated to 130–140°. At about 90° a slow gas evolution takes place and a yellow solution is formed. After 24–26 h at 130° the reaction is complete and 4.74 L [100%, STP (standard temperature and pressure)] pure hydrogen (mass spectrum) are collected in the gasometer. Ethene gas is bubbled through the colorless solution for nearly 30 min at about 60° (bath) to destroy quantitatively the last amounts of (9H-9-BBN)$_2$ (^{11}B, δ 27). The solvent is distilled off under reduced pressure (0.001 torr, bath $\leq$ 60°) and the solid residue is sublimed (bath: 85–110°/

---

* Elemental Sulfur (S$_8$) was purchased from Reininghaus, Chemische Fabrik, 4300 Essen, Germany.

† Highly pure (9H-9-BBN)$_2$ has mp 157° (DSC: 155°). It is commercially available from Fluka Chemie AG, Buchs, Switzerland or from Aldrich Chemical Co., Milwaukee, WI 53233.

0.001 torr) to give a pure, colorless monosulfide (48.5 g, 83%); mp 104–105° (DSC: 104.2°).

Anal. Calcd. for $C_{16}H_{28}B_2S$ (274.1): C, 70.12; H, 10.29; B, 7.88; S, 11.71. Found: C, 70.35; H, 9.98; B, 7.91; S, 11.72.

Properties

Bis(1,5-cyclooctanediylboryl)sulfide is a white crystalline solid, which melts at 104–105° without decomposition and can be sublimed *in vacuo* (0.001 torr, bath: 85–110°). The compound must be stored with strict exclusion of moisture and air. The sulfide is very soluble in aliphatic and aromatic hydrocarbons such as toluene or mesitylene and can be crystallized from heptane at very low temperatures.

Analytical Characterization.[1] Mass spectrometry (EI, 70 eV) is suitable for determining the composition of the pure monosulfide [m/z 274 (M^+, 32% relative intensity), 120 (B_1, basepeak)], e.g., the absence of the disulfide [m/z 306 (M^+, 20), 153 (B_1, basepeak)]. ^{1}H NMR (200 MHz, C_6D_6): δ 2.10 (m, 4H), 1.84 (m, 20H), 1.28 (m, 4H) ppm. The ^{11}B NMR spectrum is suitable for determining the absence of BO compounds (64.2 MHz, C_6D_6): δ 83.7 ppm ($h_{1/2} = 200$ Hz). ^{13}C NMR (75.5 MHz, C_6D_6): δ 34.4 (βC), 33.6 (br, αC), 23.5 (γC) ppm.

Characteristic addition compounds are prepared from the components,[1] e.g.: bis(1,5-cyclooctanediylboryl)sulfide(*S–Al*)trichloroaluminium with melting point at 118–119° and ^{11}B, δ 84.ppm ($h_{1/2} = 390$ Hz); bis(4-methylpyridine)(*N-B*)$_2$bis(1,5-cyclooctanediylboryl)sulfide with mp 130–131° and ^{11}B, δ − 7 ppm; trimethylphosphane(*P-B*)bis(1,5-cyclooctanediylboryl)sulfide with mp 205–206°, ^{11}B, δ 78.5 and − 5.3, ^{31}P, δ − 18.9 ppm.

B. 9-MERCAPTO-9-BORABICYCLO[3.3.1]NONANE

Among the known R_2BSH compounds, $C_8H_{14}BSH$ is the reagent of choice as it can be readily obtained in a highly pure form and is a thermally stable liquid having a low vapor pressure. The hydrosulfide can be stored for months at room temperature under an inert atmosphere.

The two methods, A and B, described in this section for the preparation of the 9-BBN-9-thiol (see equations), are efficient and easily reproducible. In both cases one obtains directly a pure product, which can be further purified by distillation under reduced pressure and at bath temperatures not higher than 60°.

According to method A, the pure thiol (9-HS-9-BBN) is obtained from (9-H-9-BBN)$_2$ and H_2S in dried toluene at temperatures up to 100° in about 90% yield. The preparation of 9-HS-9-BBN from the monosulfide and H_2S gas according to method B is extremely simple. No intermediate appears during the procedure. Gas is not evolved. The thiol is obtained with a yield of more than 90%. The intermediate mixtures of the monosulfide and the thiol can be analyzed semiquantitatively by ^{11}B NMR measurements; the signal of the monosulfide ([11]B, δ 84) is shifted to the signal of the thiol ([11]B, δ 78). Side reactions such as BC thiolyses[12] or $C_8H_{14}B$ isomerizations[13] do not take place.

Procedures

■ **Caution.**   *H_2S is an extremely toxic gas and must be handled carefully in a well-ventilated hood. 9-BBN-9-thiol is very reactive to protic solvents (water, alcohols). All manipulations must be carried out under an inert atmosphere (Ar or pure N_2). Before starting the experiment, the glassware apparatus must be evacuated, thoroughly flamed, and filled with dry inert gas (pure N_2 or Ar).*

Method A. The preparation of the thiol from $>BH_2B<$ and H_2S (bp $-60.4°$) is carried out in a three-necked 250-mL flask with magnetic stirrer, thermometer tube, a reflux condenser ($-78°$) with a gas bypass for inert gas, and an inlet tube for gas. To a stirred hot ($>70°$) solution of (9H-9-BBN)$_2$[3,*] (31.11 g, 127.5 mmol) in toluene (140 mL), dried over $NaAl(C_2H_5)_4$, H_2S gas† is introduced slowly (~ 3 h) from the cylinder. H_2 is evolved and is measured in a gasometer. The mixed dimer $C_8H_{14}BS(H)HBC_8H_{14}$ ([11]B, δ 3.4) appears and is enriched in the presence of

---

* Highly pure (9H-9-BBN)$_2$ has mp 157° (DSC: 155°). It is commercially available from Fluka Chemie AG, Buchs, Switzerland or from Aldrich Chemical Co., Milwaukee, WIS 53233.

† Hydrogen sulfide (H_2S) was purchased from Messer Griesheim, 4200 Oberhausen, Germany.

the $>$BH$_2$B$<$ starting material (^{11}B, δ 27) up to about 25%. After about 3 h at 100° H$_2$S is in excess and is refluxed in the condenser ($-$ 78°). The introduction of H$_2$S is terminated. The solution with only one ^{11}B NMR signal (^{11}B, δ 78) is concentrated under vacuum and the toluene is distilled at 12 torr (bath: $\leq$ 40°) to give a pure product (35.4 g, 90%) with bp 40° at 0.1 torr; mp 19°.

Method B. To prepare the thiol from the monosulfide and H$_2$S, a three-necked 500-mL flask is equipped with a magnetic stirrer, a thermometer, a reflux condenser [with gas bypass and (control) bubbler], and an inlet tube for H$_2$S gas.† In the stirred solution of 51.2 g (187 mmole) monosulfide in 250-mL hexane dry H$_2$S gas (bp $-$ 60.4°) was introduced at 20° for about 0.5 h. The temperature rises to about 40°. Then a 1-h introduction of H$_2$S gas at 60° follows. The solvent is completely removed under vacuum (14 torr). The vacuum distillation of the residue in a sublimation tube gives a colorless thiol (51.24 g, 89%) with bp $\sim$ 30° at 0.001 torr. The compound crystallizes in the cooled receiver ($-$ 30°) to give a solid with mp 19°. A viscous, colorless residue (4.73 g) is obtained.

■ **Caution.**   *The thiol decomposes thermally to the monosulfide and H$_2$S, if the compound is distilled off, e.g., at 12 torr. Therefore, it is necessary to distill the borane at < 0.1 torr and at a temperature (bath) lower than 60° (e.g., bp 40° at 0.1 torr).*

Properties

9-borabicyclo[3.3.1]nonane-9-thiol, which is a water-clear, mobile liquid at room temperature, can be vacuum-distilled without decomposition (bp 30° at 0.01 torr). On cooling to about 0° the compound crystallizes, mp 19° (DSC). The extremely moisture-sensitive compound must be stored under an inert gas (Ar, N$_2$).

Analytical Characterization.   IR (hexane): the SH stretching vibration band at 2558 cm^{-1} is characteristic. Mass spectrum (EI, 70 eV): m/z 154 [M$^+$, 53% rel. int. (relative intensity)], 120 (42), 67 (72), 55 (53), 41 (basepeak). ^{1}H NMR (200 MHz, C$_6$D$_6$): δ 2.33 (SH), 1.75 (s, 10H), 1.60 (2H), 1.24 (m, 2H); (200 MHz, CDCl$_3$): 2.50 (SH), 1.83 and 1.69 (12H), 1.34 (2H) ppm. ^{11}B NMR (64.2 MHz): δ 78.4 ppm (C$_6$D$_6$) or 77.9 ppm (CDCl$_3$). ^{13}C NMR (50.3 MHz, C$_6$D$_6$): δ 33.9 (βC), 32.0 (br, αC), 23.3 (γC) ppm; (50.3 MHz, CDCl$_3$): 33.6 (βC), 32.8 (br, αC), 22.9 (γC).

Characteristic addition compounds are prepared from the components, e.g.: pyridine(N-B)9-borabicyclo[3.3.1]nonane-9-thiol,[1] white solid with mp

120–121° and ^{11}B (CD$_2$Cl$_2$), δ + 4.3 ppm ($h_{1/2}$ = 120 Hz); 4-methyl-pyridine(N-B)9-borabicyclo[3.3.1]nonane-9-thiol[1], white solid, mp 110–111°, ^{11}B (CD$_2$Cl$_2$), δ + 3.7 ppm ($h_{1/2}$ = 140 Hz); trimethylphos-phine(P-B)9-borabicyclo[3.3.1]nonane-9-thiol, white solid, mp 85–86°, ^{11}B(C$_6$D$_6$), δ − 6.3 ($h_{1/2}$ = 60 Hz), and ^{31}P(C$_6$D$_6$), δ − 18.3 ppm.

C. BIS(1,5-CYCLOOCTANEDIYLBORYL)DISULFIDE

The white disulfide, only slightly soluble in toluene or mesitylene, is a useful reagent for the sulfurization of phosphines. The range of applicability for the sulfur transfer reagent has not yet been tested. In comparison with the known R$_2$BS$_2$BR$_2$ boranes,[2] the (9-BBN)$_2$S$_2$ compound is more readily available and can be synthesized in a pure form, free of monosulfide and thiol. The disulfide can be prepared in two different ways from (9-H-9-BBN)$_2$ or 9-I-9-BBN, respectively, with elemental sulfur (methods A and B).

The disulfide is obtained from the monosulfide with sulfur* in boiling toluene (method A). The reaction is very slow but gives a high yield, and the disulfide is not contaminated with side products. The pure borane with mp 134–135° is isolated directly as a residue of concentrating the reaction mixture under vacuum or also by filtration the suspension.

The preparation of the disulfide from (9H-9-BBN)$_2$ and S$_8$ is not re-commended because the degradation of the sulfur leads not only to the disulfide but also to the $>$BSH compound. In the absence of further (9H-9-BBN)$_2$, the hydrosulfide then contaminates the$>$BS$_2$B$<$compound and it cannot be separated easily.

---

* Elemental sulfur (S$_8$) was purchased from Reininghaus, Chemische Fabrik, 4300 Essen, Germany.

Starting from 9-iodo-9-borabicyclo[3.3.1]nonane[4,5] and elemental sulfur (method B), one obtains the colorless disulfide in boiling toluene under elimination of iodine after ~ 5 h according to the above equation. Because further 9-iodo-9-BBN reacts only extremely slowly with the disulfide with formation of the monosulfide, the pure disulfide can be isolated nearly quantitatively. In contrast to the reduction of the disulfide with the iodide, the monosulfide is oxidized by bromine in hot toluene, forming the disulfide.[9]

Procedures

■ **Caution.** *See remarks in the preparation of the monosulfide and thiol.*

Method A. Bis(1,5-cyclooctanediylboryl)disulfide is prepared from the monosulfide and sulfur* in a 100-mL flask, equipped with a magnetic stirrer, a thermometer tube, and a reflux condenser with an inert-gas bypass and bubbler. The mixture of the monosulfide (2.16 g, 7.9 mmol) and sulfur (0.26 g, 7.9 mmol) in toluene (30 mL) is heated for 24 h under reflux. Then the solvent is distilled off under vacuum (0.001 torr) at 90–120° (bath temperature) to give a colorless disulfide as residue (2.2 g, 91%), mp 133–135°.

Method B. The disulfide is synthesized from 9-iodo-9-borabicyclo[3.3.1]-nonane[4,5] and sulfur[+] in a 1-L flask with a magnetic stirrer, a thermometer tube, and a reflux condenser, combined with an inert-gas bypass and a bubbler. 9-I-9-BBN (60.36 g, 243 mmol) and sulfur (7.81 g, 243 mmol) in toluene (350 mL) are slowly warmed to reflux temperature. After heating for 5 h under reflux an intensive violet-colored suspension is formed, from which the solvent is removed under vacuum (12 torr; bath: ≤ 40°). After sublimation of about 30 g of iodine at 0.001 torr (bath: ≤ 70°) a colorless, pure disulfide (33 g, 89%) is collected; mp 134–135° (DSC: 133.5°), bp 90–120° at 0.001 torr.

Anal. Calcd. for $C_{16}H_{28}B_2S_2$ (306.2): C, 62.77; H, 9.21; B, 7.06; S, 20.97. Found: C, 63.40; H, 9.54; B, 6.81; S, 20.39.

Properties

Bis(1,5-cyclooctanediylboryl)disulfide is a white crystalline solid, which melts at 134–135° without decomposition (DSC). The compound sublimes under

---

* Elemental sulfur (S_8) was purchased from Reininghaus, Chemische Fabrik, 4300 Essen, Germany.

vacuum (0.001 torr, 110°) and must be stored with strict exclusion of moisture and air. The disulfide is poorly soluble in aromatic hydrocarbons (e.g., C_6D_6, mesitylene) or in halogenated hydrocarbons ($CHCl_3$, CH_2Cl_2).

Characteristic Spectroscopic Data.[1]   Mass spectrum (EI, 70 eV): m/z 306 (M^+, 20% rel. int.), 153 (basepeak). ^1H NMR (200 MHz, C_6D_6): δ 2.0 (m, 4 H), 1.75 (m, 20 H), 1.21 (m, 4 H) ppm. ^{11}B NMR (64.2 MHz, C_6D_6): δ 78.1 ppm ($h_{1/2} = 240$ Hz). ^{13}C NMR (75.5 MHz, C_6D_6): δ 33.9 (βC), 29.4 (br, αC), 23.3 (γC).

Characteristic Addition Compound.[1]   Bis(4-methylpyridine)(N-B)$_2$bis(1,5-cyclooctanediylboryl)disulfide, yellow solid with mp 180–181°, $^{11}B(C_6D_6)$, δ − 6 ppm.

References

1. R. Köster and G. Seidel, *Z. Naturforsch.*, **43b**, 687 (1988).
2. W. Siebert, "Organobor-Schwefel- und Selen-Verbindungen", in *Methoden der Organischen Chemie* (Houben-Weyl-Müller), 4th ed., Vol. XIII/3a, R. Köster (ed.), Thieme, Stuttgart, 1982, p. 880 ff.
3. R. Köster and P. Binger, *Inorg. Synth.*, **15**, 141 (see p. 147) (1974).
4. R. Köster and G. Seidel, *Organometal. Synth.*, **4**, 440 (1988).
5. R. Köster, "Diorgano-halogen-borane," in *Methoden der Organischen Chemie* (Houben-Weyl-Müller), 4th ed., Vol. XIII/3a, R. Köster (ed.), Thieme, Stuttgart, 1982, pp. 388–389.
6. B. M. Mikhailov and Y. N. Bubnov, *Izv. Akad. Nauk SSSR*, 172 (1959); Engl. transl. p. 159; *Chem. Abstr.*, **53**, 15958 (1959).
7. W. Siebert, "Diorgano-organothio-borane," in *Methoden der Organischen Chemie* (Houben-Weyl-Müller), 4th ed., Vol. XIII/3a, R. Köster (ed.), Thieme, Stuttgart, 1982, p. 857.
8. M. Yalpani, R. Boese, and R. Köster, *Chem. Ber.*, **123**, 707 (1990).
9. R. Köster and W. Schüßler, unpublished results, 1986.
10. M. Schmidt, W. Siebert, and F. R. Rittig, *Chem. Ber.*, **101**, 281 (1968).
11. M. Schmidt, F. R. Rittig, and W. Siebert, *Z. Naturforsch.*, **25b**, 1344 (1970).
12. See ref. 7, p. 856.
13. R. Köster and M. Yalpani, *J. Org. Chem.*, **51**, 3054 (1986).

19.   1,5-CYCLOOCTANEDIYLBORYLSELENIDES

Submitted by ROLAND KÖSTER,* GÜNTER SEIDEL,* and MOHAMED YALPANI*
Checked by WALTER SIEBERT† and BERND GANGNUS†

The preparation and characterization of three selenium derivatives of the 9-borabicyclo[3.3.1]nonane (bis(1,5-cyclooctanediylboryl)monoselenide, 9-hydroseleno-9-borabicyclo[3.3.1]nonane, and bis(1,5-cyclooctanediylboryl) diselenide) are described. These representative compounds are exceptional in the class of selenium-containing organoboranes.[1] They are simply and easily prepared in high purity‡ and are well suited for the selenidation or selenation of various organic or inorganic substrates. Other organoboron selenides are thermally not so stable since they easily exchange substituents and/or undergo redox reactions leading to the stabilized $RBSe_3BR$ ring compounds.[2] The reduction of elemental selenium with bis(9-borabicyclo [3.3.1]nonane) is, however, not accompanied by side reactions such as BC-selenidations[3] or $C_8H_{14}B$ isomerizations.[4]

■   **Caution.** *Because of the extreme high reactivity of boron–selenium compounds with moisture, care must be taken in their syntheses to maintain strictly anhydrous conditions.§ Therefore, all parts of the apparatus used must be carefully dried and should be kept under an atmosphere of dry oxygen-free nitrogen or argon. All reactions should be carried out in an efficiently ventilated hood, and provisions must be made for destroying the effluent gas (e.g., traces of H_2Se) by either absorption in a suitable reagent or freezing out. Care must also be taken when cleaning the apparatus. Propanols added under inert gas (Ar, N_2) will efficiently (partially) dissolve and destroy residual compounds before the apparatus is dismantelled. The precipitated red selenium can be dissolved in aqua regia.*

* Max-Planck-Institut für Kohlenforschung, Kaiser-Wilhelm-Platz 1, D-4330 Mülheim an der Ruhr, Germany.
† Anorganisch-Chemisches Institut der Universität Heidelberg, Im Neuenheimer Feld 270, D-6900 Heidelberg, Germany.
‡ Checkers obtained about 2% lower yields of the three organoboron selenides using one-tenth of the procedure scale shown below.
§ For handling and apparatus of air/moisture sensitive compounds, see K. Ziegler, H.-G. Gellert, H. Martin, K. Nagel, and J. Schneider, *Liebigs Ann. Chem.*, **589**, 91, p. 108 ff(1954).

A. BIS(1,5-CYCLOOCTANEDIYLBORYL)MONOSELENIDE

Bis(1,5-cyclooctanediylboryl)monoselenide[1] is a versatile reagent for the selenidation of organic and inorganic compounds. It is very stable when stored for extended periods under a dry argon or nitrogen atmosphere. Because of its ready reactivity it can replace other common Se_1-donating reagents such as H_2Se,[5] $>SiSeSi<$,[6] $\left(>SnSeSn< \right)_3$[7] or Na_2Se.[8] It is readily soluble in nonpolar solvents such as aliphatic and aromatic hydrocarbons.

For its preparation, the commercially available bis(9-borabicyclo-[3.3.1]nonane) $(9H\text{-}9\text{-}BBN)_2$[9] is made to react with elemental selenium at about 150°. In larger-scale preparations [with ≥ 10 g $(9H\text{-}9\text{-}BBN)_2$] it is advantageous to add the elemental selenium in small portions in order to avoid a rapid evolution of hydrogen gas. This is in contrast to the procedure used in the similar preparation of the sulfur analog.[10] The monoselenide, as well as, the by-product H_2 gas are formed quantitatively. The solid selenide is obtained in essentially pure form. If required, it can be further purified without any decomposition by sublimation. This is in contrast to the behavior of the tetraalkyldiboronselenides, which also have a $>BSeB<$ functionality, but readily undergo alkyl substituent exchange and redox reactions during sublimation.[2,11–13] In the case of bis(1,5-cyclo-octanediylboryl)monoselenide, the presence of the two cyclooctanediyl groups imparts an increased stability to this organoboron selenide.

The bis(1,5-cyclooctanediylboryl)monoselenide is also a reagent of choice for the preparation of the corresponding selenol[1] and diselenide,[1] which are described below.

Compounds with R_2BSe functionality having simple alkyl groups are normally available only through the reaction of R_3B or preferably R_2B—X (X = halogen) compounds with a variety of selenium reagents. The reaction of R_2BH compounds in the absence of the 1,5-cyclooctanediyl residue with elemental selenium leads to $(R_2BSe)_2$[7] or $RBSe_3BR$[11,14] boranes.

In the reaction of $(9H\text{-}9\text{-}BBN)_2$[9] with elemental selenium, intermediates of the type $>BSe(H)HB<$ (^{11}B, δ 8) and $>BSe_2B<$ (see below) can be detected. These are, however, quantitatively converted to the monoselenide under the above conditions.

The purity of the monoselenide can be checked by its ^{11}B NMR spectrum in a solvent such as toluene. It shows a relatively narrow ^{11}B NMR signal at δ 87 ($h_{1/2}$ = 200 Hz). The presence of $>$BSe$_2$B$<$ (^{11}B, δ 84) as an impurity is seen by the slight lowering of the frequency of a broadened signal. This is due to averaging of the two chemical shifts. Separate detection and quantification of the monoselenide (^{77}Se, δ 187) in a mixture with the diselenide (^{77}Se, δ 260) is best carried out by ^{77}Se NMR spectroscopy.

Procedure

A three-necked 500-mL flask, equipped with a magnetic-stirrer and a reflux condenser with gas bypass (argon or pure nitrogen) and bubbler, is charged with (9H-9-BBN)$_2$[9,*] (60.96 g, 249.8 mmol) and mesitylene (250 mL), dried and distilled over Na/K alloy. The white suspension is heated to about 150°, and a colorless solution is formed. Then black selenium powder[†] (19.83 g, 251.1 mmol) is added in small portions (2–3 g) during about 2 h. Gas is evolved immediately. The progress of the reaction can be followed by collecting the pure H$_2$ (mass spectrum) in a gasometer [5.53 L (STP)]. A small amount of solid particles is filtered off and after cooling to room temperature, the green-yellow solution is evaporated under vacuum. The solid residue can be sublimed under vacuum (0.001 torr) at 110–120°. The yield is 69.50 g (87%) of a colorless, pure product with mp 79° (DSC: 73°).

Anal. Calcd. for C$_{16}$H$_{28}$B$_2$Se (321.0): C, 59.88; H, 8.79; B, 6.73; Se, 24.60. Found: C, 59.50; H, 8.74; B, 6.93; Se, 24.32.

Properties

Bis(1,5-cyclooctanediylboryl)monoselenide is a white crystalline solid that melts at 79° without decomposition. The compound is extremely moisture-sensitive. The monoselenide is very soluble in aliphatic and aromatic hydrocarbons, such as pentane, toluene, and mesitylene. The compound sublimes *in vacuo* (0.001 torr) at 110–120° without decomposition.

Characteristic Spectroscopic Data.[1] Mass spectrum (EI, 70 eV): *m/z* 322 (M$^+$, 14% rel. int.), 121 (basepeak). ^{1}H NMR (200.1 MHz, C$_6$D$_6$): δ 2.24 (4H), 1.94 (20H), 1.28 (4H) ppm. ^{11}B NMR (64.2 MHz, C$_6$D$_6$): δ 88.3 ppm

---

* Highly pure (9H-9-BBN)$_2$ has mp 157° (DSC: 155°). It is commercially available from Fluka Chemie AG, Buchs, Switzerland or from Aldrich Chemical Co., Milwaukee, WI 53233.

† Selenium powder: 20 mesh, Strem Chemicals, Inc., 7 Mulliken Way, Newburyport, MA 01950.

($h_{1/2}$ = 200 Hz). ^{13}C NMR (75.5 MHz, C_6D_6): δ 34.4 (βC), 33.6 (br, αC), 23.5 (γC) ppm. ^{77}Se NMR: δ 187 ppm (38.2 MHz, C_6D_6) or 184.4 ppm (57.3 MHz, solid).

Characteristic addition compounds are prepared from the components,[1] e.g.: bis(4-methylpyridine)(N-B)$_2$bis(1,5-cyclooctanediylboryl)selenide with mp 169°, ^{11}B, δ + 4.3 ppm (C_6D_6) or + 3.8 ppm (THF-d_8), and ^{77}Se, δ 182.4 (C_6D_6) or 56.2 ppm (solid); trimethylphosphine(P-B)bis (1,5-cyclooctanediylboryl)selenide with mp 190°, ^{11}B (C_6D_6), δ 85.7, and − 4.9 ppm (1 : 1), ^{31}P, δ − 18.6 (br) ppm, ^{77}Se, δ 96 (C_6D_6, 50°) or 105.6 ppm (solid).

B. 9-BORABICYCLO[3.3.1]NONANE-9-SELENOL

It is reasonable that the organoboron selenol can be made in analogy to the procedure used for the preparation of 9-HS-9-BBN[10,15] by employing H_2Se gas. However, the procedure presented here circumvents the use of the highly toxic and disagreeable H_2Se gas.[16] Instead it makes use of the facile protolytic cleavage of one $>$BSe— bond of $>$BSeB$<$ by aniline. The loss of one 9H-9BBN molecule should, however, be accepted.

The product is formed in essentially pure form. If required, further purification can be achieved by vacuum distillation. Since it is extremely moisture-sensitive it should be handled in a dry oxygen-free atmosphere. Its high solubility in nonpolar solvents makes it an attractive HSe donor reagent. Typical applications are the hydroselenoboration of 1-alkynes and carbonyl compounds.[1]

The compound is present as a dimer in solution below − 40°. The ^{77}Se NMR spectrum shows the existence of both *cis* and *trans* dimers. In the crystalline material obtained from toluene solution only the *cis* dimer is found (X-ray analysis).[17] A good and fast probe for the purity of $C_8H_{14}BSeH$ is its melting point (mp + 28°).

Procedure

A three-necked 250-mL flask with magnetic stirrer, reflux condenser (with inert-gas bypass and bubbler), and dropping funnel is charged with a solution of bis(1,5-cyclooctanediylboryl)monoselenide (22.39 g, 69.8 mmol) in pentane

(125 mL). Then 6.50 g (69.8 mmol) of aniline* is added dropwise during approximately 20 min to the stirred solution. The temperature rises slightly and a white solid precipitates (presumably the 1 : 1 addition compound of the starting materials), which redissolves later.

After 3 h of refluxing the solution is cooled and the pentane is removed under vacuum (14 torr). The residue is transferred to an appropriate vacuum distillation apparatus with a 25-mL vessel and an air cooler. The distillation under reduced pressure gives a colorless, pure product (10.96 g, 78%) with bp 30° at 0.001 torr and mp + 28° (DSC).

Anal. Calcd. for $C_8H_{15}BSe$ (201.0): C, 47.81; H, 7.51; B, 5.37; Se, 39.28. Found: C, 47.89; H, 7.77; B, 5.12; Se, 39.06.

The distillation residue [^{11}B, δ 51.6 (92%), 87.8 (8%)] may be dissloved in pentane. On slow cooling in an acetone/CO_2 mixture, crystals of pure 9-phenylamino-9-BBN with mp 37–38° are obtained.

Properties

9-borabicyclo[3.3.1]nonane-9-selenol is a colorless, very oxygen-sensitive solid, which must be handled in an inert-gas atmosphere (Ar, N_2). The compound crystallizes as the *cis*-dimer (X-ray crystal structure),[17] melts at 28° (DSC), and can be vacuum distilled (30°, 0.001 torr) without decomposition. The selenol is completely miscible with aliphatic or aromatic hydrocarbons.[1]

Characteristic Spectroscopic Data.[1] IR (hexane): $v(SeH) = 2300$ cm^{-1}. Mass spectrum (EI, 70 eV): m/z 202 (50% rel. int.), 41 (basepeak). ^1H NMR (200.1 MHz, $CDCl_3$): δ 1.84 (12H), 1.37 (2H), 0.44 (SeH) ppm. ^{11}B NMR (64.2 MHz): δ 84.8 ($CDCl_3$) or 83.8 (C_6D_6). ^{13}C NMR (75.5 MHz, C_6D_6): δ 35.3 (br, αC), 33.8 (βC), 23.1 (γC) ppm. ^{77}Se NMR (38.2 MHz): δ 48.7 ppm (C_7D_8, + 10°) or δ 53 (monomer), − 3.5, and − 38.5 (*cis/trans*-dimer) ppm (C_7D_8, − 45°).

Characteristic addition compounds are prepared from the components:[1] trimethylphosphine(P-B)-9-borabicyclo[3.3.1]nonane-9-selenol, mp 85–86°, ^{11}B (C_6D_6), δ − 6.0 ppm; 4-methylpyridine(N-B)9-hydroseleno-9-borabicyclo[3.3.1]nonane, mp 132° and ^{11}B (C_6D_6), δ + 4.3 ppm; quinuclidine(N-B)9-hydroseleno-9-borabicyclo[3.3.1]nonane, mp 129–130° ^{11}B (C_6D_6), δ + 32 ppm.

---

* Purchased from Bayer AG, 5090 Leverkusen and distilled under vacuum over Na/K alloy.

C.   BIS(1,5-CYCLOOCTANEDIYLBORYL)DISELENIDE

At room temperature, bis(1,5-cyclooctanediylboryl)diselenide C_8H_{14}-BSeSeBC$_8$H$_{14}$ is a solid reagent. It can be used as a selenium donor. However, unlike the monoselenide, it donates in two distinct steps.[1] In the first, it appears to transfer atomic selenium. An example of this type of reaction is the selenidation of phosphanes by the diselenide. A by-product is the monoselenide, which can be isolated as the monophosphane adduct in further reaction with excess phosphane. In the absence of stabilizing donors, the monoselenide product of the first step can act as a selenidation agent, by transferring a selenide ion. The broad scope of applications of the diselenide reagent is, however, not yet fully investigated.

For the preparation of the diselenide, the monoselenide [e.g., made from (9H-9-BBN)$_2$ and selenium powder (see above)] is heated at 120° with an equivalent amount of black selenium powder* until the elemental selenium is dissolved. The one-step synthesis of the diselenide from (9H-9-BBN)$_2$ and selenium is also possible. However, the product is not as pure as in the two-step process, as it is normally contaminated with some selenol and the products thereof (see the preparation of the $C_8H_{14}B$ sulfides).[10,15]

Since the diselenide is less soluble, a product mixture containing both compounds can be readily and quantitatively separated. The purity of the diselenide is determined by its significantly higher melting point (mp 135°). Simple spectroscopic methods such as ^{11}B NMR are uninformative because of the closeness of their ^{11}B chemical shifts. The presence of monoselenide impurity results in only a slight shift of the ^{11}B NMR signal to higher frequencies. When available, ^{77}Se NMR provides the best spectroscopic probe for the detection of either compound as an impurity in the other (^{77}Se, δ 187 for $>$BSeB$<$ and ^{77}Se, δ 260 for $>$BSe$_2$B$<$).

Procedure

In a 250-mL flask (with magnetic stirrer, inside thermometer tube, and reflux condenser with inert-gas bypass and bubbler) a stirred mixture of bis(1,5-cyclooctanediylboryl)monoselenide (BSeB) (10.65 g, 33.2 mmol) and black selenium powder* (2.62 g, 33.2 mmol) in mesitylene (120 mL) is heated 2 h at

---

*Selenium powder: 20 mesh, Strem Chemicals, Inc., 7 Mulliken Way, Newburyport, MA 01950.

120°. On cooling, the product BSeSeB precipitates from the green-yellow dilute solution, which is siphoned off from the solid. The remaining solvent is removed by distillation under vacuum (0.001 torr). The solid residue is washed with toluene and then dried (0.001 torr, 70°). The overall yield of the yellow bis(1,5-cyclooctanediylboryl)diselenide (BSeSeB) with mp 135° (DSC: 134°) is 12.3 g (93%). It can be sublimed at 0.001 torr and 80–120° as a lemon-yellow powder.

Anal. Calcd. for $C_{16}H_{28}B_2Se_2$ (399.9): C, 48.04; H, 7.06; B, 5.41; Se, 39.48. Found: C, 47.94; H, 7.38; B, 5.42; Se, 39.16.

Properties

Bis(1,5-cyclooctanediylboryl)diselenide is a lemon-colored crystalline solid (mp 135° without decomposition). The pure sublimed solid is oxygen and extremely moisture-sensitive, rapidly forming a precipitate of red selenium. Therefore, the inert gas such as pure N_2 or Ar must be thoroughly dried in a tower (~ 40 cm long), filled with an efficient solid desiccant such as $NaAl(C_2H_5)_4$. The diselenide is slightly soluble in hot aromatic hydrocarbons (toluene, mesitylene).

Characteristic Spectroscopic Data. Mass spectrum (EI, 70 eV): m/z 402 (8% rel. int.), 121 (basepeak). ^1H NMR (200.1 MHz, C_7D_8, 80°): δ 2.06 (4H), 1.68 (20H), 1.20 (4H) ppm. ^{11}B NMR (64.2 MHz, C_7D_8, 80°): δ 84.1 ppm. ^{13}C NMR (75.5 MHz, C_7D_8, 80°): δ 34.1 (βC), 33.2 (br, αC), 23.3 (γC) ppm. ^{77}Se NMR: δ 260 ppm (38.2 MHz, C_7D_8) or $\delta \approx$ 262, 242.9 ppm (57.3 MHz, solid).

Characteristic addition compounds are prepared from the components:[1] orange bis(4-methylpyridine)(N-B)$_2$bis(1,5-cyclooctanediylboryl)diselenide, mp 235°, ^{11}B (THF-d_8), δ + 3.1 ppm ($h_{1/2}$ = 200 Hz).

References

1. R. Köster, G. Seidel, and M. Yalpani, *Chem. Ber.*, **122**, 1815 (1989).
2. W. Siebert, "Organobor-Selen-Verbindungen," in *Methoden der Organischen Chemie* (Houben-Weyl-Müller), 4th ed., Vol. XIII/3a, R. Köster (ed.), Thieme, Stuttgart, 1982, p. 890 ff.
3. B. M. Mikhailov and T. A. Shchegoleva, *Izv.Akad. Nauk SSSR, Ser. Khim.*, 357 (1959), Engl. transl. p. 331; *Chem. Abstr.*, **53**, 20041 (1959).
4. R. Köster and M. Yalpani. *J. Org. Chem.*, **51**, 3054 (1986).
5. W. Siebert, E. Gast, F. Riegel, and M. Schmidt, *J. Organometal. Chem.*, **90**, 13 (1975). See ref. 2, p. 892.
6. M. Schmidt and E. Kiewert, *Z. Naturforsch.*, **26b**, 613 (1971).

7. F. Riegel and W. Siebert, *Z. Naturforsch.*, **29b**, 719 (1974). See ref. 2, p. 893.
8. R. Köster, G. Seidel, R. Boese, and B. Wrackmeyer, *Chem. Ber.*, **121**, 1955 (1988).
9. (a) R. Köster and P. Binger, *Inorg. Synth.*, **15**, 141 (see p. 147) (1974); (b) R. Köster, "Diorganohydroborane," in *Methoden der Organischen Chemie* (Houben-Weyl-Müller), 4th ed., Vol. XIII/3a, R. Köster (ed.), Thieme, Stuttgart, 1982, p. 330.
10. R. Köster and G. Seidel, *Inorg. Synth.*, **29**, 60 (1992).
11. W. Siebert and F. Riegel, *Chem. Ber.*, **106**, 1012 (1973).
12. M. Schmidt, W. Siebert, and E. Gast, *Z. Naturforsch.*, **22b**, 557 (1967).
13. M. Schmidt, W. Siebert, and F. R. Rittig, *Chem. Ber.*, **101**, 281 (1968).
14. M. Yalpani, R. Boese, and R. Köster, *Chem. Ber.*, **123**, 707 (1990).
15. R. Köster and G. Seidel, *Z. Naturforsch.*, **43b**, 687 (1988).
16. J. M. Braam, C. D. Carlson, D. A. Stephens, A. E. Rehan, S. J. Compton, and J. M. Williams, *Inorg. Synth.*, **24**, 131ff. (1986).
17. R. Boese and D. Bläser, Universität Essen, Germany.

20. TRIMETHYLAMINE–DIETHYL-1-PROPYNYLBORANE AND DIETHYL-1-PROPYNYLBORANE

$$2B(C_2H_5)_3 + (C_4H_9)_2O—BF_3 \xrightarrow[-(C_4H_9)_2O]{>BH} 3(C_2H_5)_2BF$$

$$(C_2H_5)_2BF + (CH_3)_3N \longrightarrow (CH_3)_3N—BF(C_2H_5)_2$$

$$(CH_3)_3N—BF(C_2H_5)_2 + LiC\equiv CCH_3 \xrightarrow[-LiF]{(C_2H_5)_2O}$$

$$(CH_3)_3N—B(C_2H_5)_2C\equiv CCH_3$$

$$(CH_3)_3N—B(C_2H_5)_2C\equiv CCH_3 \xrightarrow[\substack{-(CH_3)_3N-BF_3 \\ -(C_2H_5)_2O}]{+(C_2H_5)_2O-BF_3} (C_2H_5)_2BC\equiv CCH_3$$

Submitted by ROLAND KÖSTER,* HEINZ JOSEF HORSTSCHÄFER,* and GÜNTER SEIDEL*
Checked by WALTER SIEBERT† and BERND GANGNUS†

The 1-alkynyl-diorganoboranes $R_2BC\equiv CR'$[1] have a special position in the family of the acetylenic triorganoboranes. The π-electrons of the $BC(pp)\pi$ bond interact with the free p_z orbital of the boron atom; therefore, the αC atom will be relatively electron-rich, and this has a direct influence on the dipole additions to the $BC\equiv C$ grouping. Thus, for example, the $BC_{alkynyl}$ bond

* Max-Planck-Institut für Kohlenforschung, Kaiser-Wilhelm-Platz 1, D-4330 Mülheim an der Ruhr, Germany.
† Anorganisch-Chemisches Institut der Universität Heidelberg, Im Neuenheimer Feld 270, D-6900 Heidelberg, Germany.

is extremely sensitive to protolysis.[1] Because of the nucleophilic character of the αC atom of the uncomplexed $R_2BC{\equiv}CR'$ compounds, the boron atom is bonded nearly exclusively to the αC atom in the ${>}BH$ borane hydroboration.[3,4]

This special property of the dialkyl-1-alkynylboranes is applicable to the preparation of certain organocarbaboranes,[5] which will be described in the next section. Because of the high reactivity and the thermal instability of the Lewis base-free $R_2BC{\equiv}CR'$ compounds, care must be taken in their synthesis to maintain the safety conditions mentioned below (see *Properties*).

The described preparative method for diethyl-1-propynylborane $(C_2H_5)_2BC{\equiv}CCH_3$ can also be used for other boranes of this type.[1] The three-step procedure starting from $(C_2H_5)_2BF$ is strictly necessary, because the direct reaction of the uncomplexed compound $(C_2H_5)_2BF$ or $(C_2H_5)_2BCl$ with 1-propynyl alkaline metals does not lead to acceptable results because of the formation of borates,[6] which give further side products.[1,7] If the "direct method" is used, the yield of the diethyl-1-propynylborane decreases drastically.[1]

Procedures

■ **Caution.** *Triethylborane is a spontaneously flammable liquid (bp = 94–95°; mp = − 95°) and must be strictly stored under an inert gas (pure N_2, Ar). All ethyl fluoroboranes and the diethyl-1-propynylborane react vigorously with water and are very sensitive to oxygen. The uncomplexed fluoroboranes are aggressive to ground-glass silicon grease. Thus, the conditions of the following preparation should be strictly maintained.*

The compound $(C_2H_5)_2BF$ is prepared from dibutyl ether–trifluoroborane[8,9] and triethylborane[10,11] in a three-necked 1-L flask equipped with a magnetic stirrer, a thermometer tube, a dropping funnel, and a descending condenser combined with a gas bypass and a bubbler.

Dibutyl ether—BF_3* (220.6 g; 1.11 mol) are added dropwise at 80° in 4 h to stirred 270.3 g (2.76 mol) triethylborane† and about 5 mL of tetraethyldiborane(6) (bp = 110°)[12,13] with $\approx$ 1.4% hydride-H. The temperature rises to about 105° and 250 g of a colorless clear liquid distils off (cooling of the receiver to − 78°). The distillate is then fractionated (30-cm Vigreux column

---

* The compound $(C_4H_9)_2O$—BF_3 is prepared by passing gaseous BF_3 over $(C_4H_9)_2O$ (strongly exothermic); BF_3 is available from Dr. Th. Schuchardt & Co., 8011 Hohenbrunn, Germany or from Eastman Organic Chemicals, Rochester, NY 14650.

† Triethylborane is commercially available from Schering AG, 4619 Bergkamen, Germany or from Callery Chemical Company, Callery, PA 16024.

with 15 mm diameter) under atmospheric pressure giving 177.7 g (61%) pure diethylfluoroborane with bp = 38–39°. The compound should be worked up immediately because of the high aggressivity toward glassware apparatus and silicon fat.

The compound $(CH_3)_3N$—$BF(C_2H_5)_2$[8] is prepared in a 500-mL flask, equipped with a 200-mL cooled pressure-equalizing graduated dropping funnel, thermometer tube, a magnetic stirrer, and a reflux condenser with a gas bypass and a bubbler. Then 112 g (1.89 mol) of trimethylamine is dropped from the dropping funnel (− 35°) to the stirred $(C_2H_5)_2BF$ (146 g, 1.67 mol) at 0°. Intensive mist in the flask shows the formation of the complex compound. After the addition (1.5 h), the product is distilled off under vacuum: 233.7 g (95%) of a colorless, clear liquid with bp = 49° at 14 torr is obtained.

The compound $(CH_3)_3N$—$B(C_2H_5)_2C{\equiv}CCH_3$[1] is prepared in a dried 2-L flask (with magnetic stirrer, thermometer tube, dropping funnel with pressure equalizer, reflux condenser with gas bypass and bubbler) from $(CH_3)_3N$—$BF(C_2H_5)_2$ (216.1 g, 1.47 mol), which is added slowly dropwise (3.5 h) at 0–5° to a suspension of 67.3 g (1.46 mol) $LiC{\equiv}CCH_3$[14] in 800 mL of diethyl ether. The temperature rises to 20°.

After 2 h of additional stirring at room temperature LiF precipitates on cooling to − 78° (without stirring) from the viscous yellow solution. The reaction mixture is then filtered using a cooled coarse glass frit. After drying under reduced pressure (0.001 torr) one obtains 66.5 g of a solid material, which is extracted with pentane for 16 h in a Soxhlet apparatus: 36.7 g (97%) of dry LiF is collected. The pentane solution is then partially (~ $\frac{1}{3}$) concentrated and slowly cooled to − 78° (without stirring): 25.7 g (10.6%) of a white product with mp = 33° is isolated. After removing the ether at reduced pressure (14 torr) the residue is dissolved in pentane and additional crystals are gained on cooling the solution to − 78°. The supernatant clear solution is removed. After the collected crystals are washed with cold pentane, they are dried under vacuum (0.01 torr). Then 211.1 g (86.5%) of $(CH_3)_3N$—$B(C_2H_5)_2C{\equiv}CCH_3$ with mp = 33° is obtained; total yield: 236.9 g (97.1%) of product.

Anal. Calcd. for $C_{10}H_{22}BN$ (167.1): C, 71.81; H, 13.29; B, 6.46; N, 8.38; C_3H_3, 23.4. Found: C. 71.71; H, 13.40; B, 6.42; N, 8.46; C_3H_3, 22.9 (volumetric determination with $5N$ H_2SO_4).[1]

$(C_2H_5)_2BC{\equiv}CCH_3$[1] is prepared from $(CH_3)_3N$—$B(C_2H_5)_2C{\equiv}CCH_3$ and diethyl ether–trifluoroborane in an 1-L flask with a magnetic stirrer, a thermometer inside tube, a pressure-equalized dropping funnel, and a reflux condenser with a gas bypass and a bubbler.

A solution of 71.46 g (503 mmol) $(C_2H_5)_2O$—BF_3* in 100 mL diethyl ether is slowly added dropwise ($\approx$ 2 h) to a solution of 84.09 g (503 mmol) $(CH_3)_3N$—$B(C_2H_5)_2C{\equiv}CCH_3$ in 450 mL of diethyl ether. A white solid precipitates. After 2.5 h stirring at room temperature, 61.7 g (97%) of $(CH_3)_3N$—BF_3 is filtered off, giving an orange-yellow, clear solution. The solvent is quantitatively removed batchwise under vacuum (220–150 torr). Further distillation gives 30.9 g (57%) of a colorless, clear $(C_2H_5)_2BC{\equiv}CCH_3$ product with bp = 27° at 14 torr together with a red, highly viscous residue.

Anal. Calcd. for $C_7H_{13}B$ (108.0); B, 10.00; BC(alkyl), 6.68; C_3H_3, 36.2. Found: B, 9.98; BC(alkyl), 6.71[15]; C_3H_3, 35.3 (volumetric determination by hydrolysis).[1]

Properties

The compound $(C_2H_5)_2BF$ is a colorless clear liquid with bp = 38–39° and is completely miscible with hydrocarbons and chlorinated hydrocarbons. The highly volatile compound must be stored below $-30°$ in closed vessels. The thermally relatively stable borane is very aggressive toward silicon grease used for ground-glas joints.

Characteristic Spectroscopic Data. ^1H NMR spectrum (200.1 MHz, $CDCl_3$): $\delta = 0.88$ ppm. ^{11}B NMR spectrum (64.2 MHz, $CDCl_3$): $\delta = 60.3(d)$ ppm $(J_{FB} = 125.7$ Hz). ^{13}C NMR spectrum (50 MHz, $CDCl_3$): $\delta = \approx 13$ (br, $^2J_{FC} \approx 80$ Hz), 6.1 $(^3J_{FC} = 5.3$ Hz) ppm. ^{13}F NMR spectrum (188.3 MHz, $CDCl_3$): $\delta = -36.2(q)$ ppm.

The compound $(CH_3)_3N$—$BF(C_2H_5)_2$ with mp at $-42°$ is a colorless, slightly viscous liquid at room temperature. It distills at 49°, 14 torr without decomposition. The compound is moisture- and air-sensitive and must be stored under inert gas (Ar, N_2) at room temperature or below. The amine–borane is completely miscible at room temperature with aliphatic or aromatic hydrocarbons and with $CHCl_3$ or CH_2Cl_2.

Characteristic Spectroscopic Data. ^1H NMR spectrum (200.1 MHz, $CDCl_3$): $\delta = 2.3$ (9H), 0.7 (6H), 0.2 (4H) ppm. ^{11}B NMR spectrum (64.2 MHz, $CDCl_3$): $\delta = 10.6(d)$ ppm $(J_{FB} = 79.4$ Hz). ^{13}C NMR spectrum (50.4 MHz, $CDCl_3$): $\delta = 47.3$ (NCH_3), ≈ 10 (br, BCH_2), 9.7 (BCH_2CH_3; $J_{FC} = 3.4$ Hz) ppm. ^{19}F NMR spectrum (188.3 MHz, $CDCl_3$) $\delta = -188.4(q)$ ppm.

---

*Diethyl ether–trifluoroborane is available from BASF AG, 6700 Ludwigshafen/Rhein, Germany or from Eastman Organic Chemicals, Rochester, NY 14650.

$(CH_3)_3N$—$B(C_2H_5)_2C{\equiv}CCH_3$ is a white solid (mp = 33°), which can be crystallized from cold pentane ($-78°$). The air- and moisture-sensitive compound is very soluble in diethyl ether and in pentane. The amine-borane can be stored indefinitely in an inert gas atmosphere (Ar, N_2) in closed vessels at room temperature; 1 mol of $(CH_3)_3N$—$B(C_2H_5)_2C{\equiv}CCH_3$ reacts with an excess of anhydrous trimethylamine N-oxide in boiling toluene liberating 3 mol of $(CH_3)_3N$.[15]

Characteristic Spectroscopic Data. IR ($CHCl_3$): $\nu(C{\equiv}C)$ = 2175 cm^{-1}. Mass spectrum (EI, 70 eV): m/z 167 (M $- 29^+$, 31% rel. int.), 79 (basepeak). ^{1}H NMR spectrum (200 MHz, THF-d_8): δ = 2.44 (9H), 1.71 (3H), 0.82 (6H), 0.25 (4H) ppm. ^{11}B NMR spectrum (64.2 MHz, THF-d_8): δ = -0.9 ppm ($h_{1/2}$ = 60 Hz). ^{13}C NMR spectrum (50 MHz, THF-d_8, $-50°$): δ = 94 ($BC{\equiv}$), 92.4 (CCH_3), 49.0 (NCH_3), 13.3 (BCH_2), 12.9 (BCH_2CH_3), 4.9 ($\equiv CCH_3$) ppm.

$(C_2H_5)_2BC{\equiv}CCH_3$ is a thermally unstable colorless clear liquid with bp = 34–36° at 20 torr. The pure compound with mp = $-66°$ must be stored below $-30°$ in closed vessels. The borane is extremely air-sensitive and can inflame spontaneously. The decomposition of the compound is detectable by a red colorization and by polymerization; 1 mol of the compound reacts with an excess of anhydrous $(CH_3)_3NO$ in boiling toluene liberating 2 mol of $(CH_3)_3N$.[15]

Characteristic Spectroscopic Data. IR (CCl_4): $\nu(C{\equiv}C)$ = 2168 cm^{-1}. Mass spectrum (EI, 70 eV): m/z 108 (M, 15% rel. int.), 79 (basepeak). ^{1}H NMR spectrum (200 MHz, CDCl$_3$): δ = 2.01 (3H), 1.07 (4H), 0.87 ppm (6H). ^{11}B NMR spectrum (64.2 MHz, CDCl$_3$): δ = 73.9 ppm ($h_{1/2}$ = 150 Hz). ^{13}C NMR spectrum (50 MHz, CDCl$_3$, $-50°$): δ = 119.8 (C$\equiv$), 89.0 ($BC{\equiv}$), 21.2 (BCH_2), 9.1 (BCH_2CH_3), 5.4 ($\equiv CCH_3$) ppm.

References

1. R. Köster, H.-J. Horstschäfer, and P. Binger, *Liebigs Ann. Chem.*, **717**, 1 (1968).
2. R. Köster, "Acetylenische Triorganoborane," in *Methoden der Organischen Chemie* (Houben-Weyl-Müller), 4th ed., Vol. XIII/3a, R. Köster (ed.), Thieme, Stuttgart, 1982, p. 235 ff.
3. R. Köster, H.-J. Horstschäfer, P. Binger, and P. K. Mattschei, *Liebigs Ann. Chem.*, 1339 (1975).
4. R. Köster, "Aliphatische Triorganoborane," in *Methoden der Organischen Chemie* (Houben-Weyl-Müller), 4th ed., Vol. XIII/3a, R. Köster (ed.), Thieme, Stuttgart, 1982, p. 75.
5. M. Grassberger and R. Köster, "B-Organocarborane mit drei Bor-Atomen im Gerüst," in

Methoden der Organischen Chemie (Houben-Weyl-Müller), 4th ed., Vol. XIII/3c, R. Köster, (ed.), Thieme, Stuttgart, 1984, p. 162 ff.

6. P. Binger, G. Benedikt, G. W. Rotermund, and R. Köster, *Liebigs Ann. Chem.*, **717**, 21 (1968).
7. P. Binger and R. Köster, *Tetrahedron Lett.*, **1965**, 1901.
8. R. Köster, and M. A. Grassberger, *Liebigs Ann. Chem.*, **719**, 169 (1968).
9. R. Köster, "Diorgano-halogen-borane," in *Methoden der Organischen Chemie* (Houben-Weyl-Müller), 4th ed., Vol. XIII/3a, R. Köster (ed.), Thieme, Stuttgart, 1982, p. 430.
10. R. Köster, P. Binger, and W. V. Dahlhoff, *Synth. React. Inorg. Metal-Org. Chem.*, **3**, 359 (1973).
11. See ref. 4, pp. 147–148.
12. R. Köster and P. Binger, *Inorg. Synth.*, **15**, 141 (1974).
13. R. Köster, "Diorgano-hydro-borane," in *Methoden der Organischen Chemie* (Houben-Weyl-Müller), 4th ed., Vol. XIII/3a, R. Köster (ed.), Thieme, Stuttgart, 1982, p. 333.
14. Preparation of $LiC\equiv CCH_3$ from $LiNH_2$ and propyne or from metallic lithium with propyne in liquid NH_3 in presence of $Fe(NO_3)_3$. See R. A. Raphael, *Acetylenic Compounds in Organic Synthesis*, Butterworth Sci. Publ., London 1955; W. Ziegenbein, "Einführung in die Äthinyl- und Alkinyl-Gruppe in Organischen Verbindungen," *Monographie zu Angew. Chem. und Chemie-Ing.-Techn.* No. 79, Verlag Chemie, Weinheim, 1963.
15. R. Köster and Y. Morita, *Liebigs Ann. Chem.*, **704**, 70 (1967), oxidation with anhydrous $(CH_3)_3NO$ and acidimetric titration of the evolved $(CH_3)_3N$.

21. PENTAETHYL-1,5-DICARBA-*closo*-PENTABORANE(5) AND DECAETHYL-2,6,8,10-TETRACARBA-*nido*-DECABORANE(10)

$$2(C_2H_5)_2BC\equiv CCH_3 + \xrightarrow[-3\ B(C_2H_5)_3]{2[HB(C_2H_5)_2]_2} (C_2H_5)_2C_2B_3(C_2H_5)_3$$

A

$$A \xrightarrow[THF]{+K} [K(THF)_n]^+[(C_2H_5)_2C_2B_3(C_2H_5)_3]^{\cdot-}$$

B*

$$2\ B^{*\cdot} \xrightleftharpoons{} [K_2(THF)_{2n}]^{2+}[(C_2H_5)_2C_2B_3(C_2H_5)_3]^{2-} + A$$

B

$$2B \xrightarrow[-4\ KI]{+2\ I_2,THF} (C_2H_5)_4C_4B_6(C_2H_5)_6$$

C

Submitted by ROLAND KÖSTER,* GÜNTER SEIDEL,* and
HEINZ JOSEF HORSTSCHÄFER*
Checked by WALTER SIEBERT† and BERND GANGNUS†

The preparations of the two perethylated carboranes with C_2B_3 and C_4B_6 skeletons, respectively, starting from B_1 boranes are described in this section.

A. PENTAETHYL-1,5-DICARBA-*closo*-PENTABORANE(5), $Et_2C_2B_3Et_3$

■ **Caution.** *All operations are carried out under inert gas (Ar, pure N_2) with rigorous exclusion of air and moisture. The solvents used are dried and deoxygenated by distillation from sodium/potassium alloy.*

The pentaalkyl-1,5-dicarba-*closo*-pentaboranes(5)[1–5] and the 2,3,4-triethyl-1,5-dicarba-*closo*-pentaborane(5)[3,4] are prepared by elaborate methods in contrast to the syntheses of the unsubstituted $C_2B_3H_5$ carborane.[6–8] The colorless liquid pentaethyl-1,5-dicarba-*closo*-pentaborane(5) **A**, a representative compound of this series, is available from diethyl-1-propynylborane with tetraethyldiborane(6) in triethylborane as solvent [i.e., with ethyldiboranes(6) having a very low hydride content], by hydroboration with subsequent substituent exchanges.[2,4] Compound **A** with its 12 skeletal electrons has the structure of a trigonal bipyramid with the two carbons in the apices and the three boron atoms in the three 2,3,4-basal positions.[5] Compound **A** is the starting material for the synthesis of the dimeric solid compound **C**.[9–11]

Procedure

Pentaethyl-1,5-dicarba-*closo*-pentaborane(5) (**A**) is prepared from diethyl-1-propynylborane and tetraethyldiborane(6)[12,13]/triethylborane[14] mixtures in a 2-L flask, equipped with a magnetic stirrer, a thermometer tube, and a pressure-equalized dropping funnel (1 L).

A mixture of 169.3 g of tetraethyldiborane(6)[12,13] (with 1.4% hydride-H, ~ 2.37 mole >BH) and 711.5 g (7.25 mole) triethylborane‡ is dropped slowly (2.5 h) to stirred undiluted diethyl-1-propynylborane[15,16] (118.7 g, 1.1 mol) at about 0° by cooling with ice water. The temperature of the

* Max-Planck-Institut für Kohlenforschung, Kaiser-Wilhelm-Platz 1, D-4330 Mülheim an der Ruhr, Germany.

† Anorganisch-Chemisches Institut der Universität Heidelberg, Im Neuenheimer Feld 270, D-6900 Heidelberg, Germany.

‡ Triethylborane is available from Schering AG, 4619 Bergkamen, Germany or from Callery Chemical Company, Callery, PA 16024.

reaction mixture only rises a little. After the addition, the mixture is stirred for 3 h at room temperature and the triethylborane (bp = 94–95°) is distilled off quantitatively first under atmospheric pressure (bath: $\leq 125°$) and then together with an excess of tetraethyldiborane(6) under vacuum (7 torr) at $\leq 30°$ (bath). From the residual colorless liquid (~ 108 g) 83.4 g [GLC*: 29% $B(C_2H_5)_3$, 61% **A**] is distilled off at 0.05 torr with bp $\leq 110°$. Then 24.7 g of a solid, dark red, highly viscous residue (found 17.1% B) is obtained. The 83.4 g of colorless liquid is distilled under vacuum through a 50-cm spinning-band column (10 mm diameter) giving 51 g (46%) **A** (GLC* purity: 98%) with bp = 84–86° at 9 torr and mp = $-61.5°$.

Anal. Calcd. for $C_{12}H_{25}B_3$ (201.8): C, 71.43; H, 12.49; B, 16.08. Found: C, 71.25; H, 12.65; B, 15.8.

Properties

Pentaethyl-1,5-dicarba-*closo*-pentaborane(5) (**A**) is a colorless, air- and mois-ture-stable liquid at room temperature. The compound can be distilled *in vacuo* without decomposition (bp = 87° at 14 torr) and also under atmo-spheric pressure (bp = 209° at 744 torr). Compound **A** with $n_D^{20} = 1.4472$ and $d_4^{20} = 0.7978$ (g cm^{-3}) is completely miscible with hydrocarbons and chloro-hydrocarbons.

Characteristic Spectroscopic Data. Mass spectrum (EI, 70 eV): m/z 202 (M$^+$, 68% rel. int.), 187 (43), 173 (82), 41 (basepeak). IR (neat): 2720, 1066, 1005, 958, 800, 762 cm^{-1}. Raman (neat): 1069, 1009, 570, 510 cm^{-1} (polarized lines) and 1036, 996 cm^{-1} (depolarized lines). ^1H NMR spectrum (200 MHz, CDCl$_3$): $\delta = 2.44$ (4H), 1.21 (6H), 0.95 (15H) ppm (J_{CH}; see ref. 5). ^{11}B NMR spectrum (64.2 MHz, CDCl$_3$): $\delta = 13.2$ ppm ($h_{1/2} = 90$ Hz). ^{13}C NMR spectrum (50.3 MHz, CDCl$_3$, $-50°$): $\delta = 105.5$ (C), 18.3 (CCH$_2$), 15.7 (CCH$_2$CH$_3$), 3.9 (br, BCH$_2$), 9.8 ppm (BCH$_2$CH$_3$) (J_{CB}, J_{CC}; see ref. 5).

Compound **A**, stable to oxygen and to alkaline hydroperoxide below 100°, reacts with anhydrous trimethylamine *N*-oxide under liberation of about 9 mol (CH$_3$)$_3$N per mol of **A**.[17] Bromine reacts with neat compound **A** or CCl$_4$ solutions of **A** to give HBr and various brominated C$_2$B$_3$ compounds. Compound **A** does not react with nitric acid up to 85°. Compound **A** is also stable toward pyridine at room temperature, and it shows no reaction with metallic sodium on heating in the absence of any solvent at $\leq 150°$.

---

* GC column: 2.8 m, internal diameter: 5 mm; stationary phase: 7% "apiezon" on "embacel" (silica gur).

B. DECAETHYL-2,6,8,10-TETRACARBA-*nido*-DECABORANE(10), $Et_4C_4B_6Et_6$

Decaethyl-2,6,8,10-tetracarba-*nido*-decaborane(10) **C** (Fig. 2), a valence isomer of the as yet unknown decaethyl derivative in the series of the 2,4,6,8,9, 10-hexaboraadamantanes,[4, 5, 18-20] was the first C_4B_6-carborane synthesized.[9-11] Compound **C** is a representative of the peralkylated carbon-rich carboranes $C_4B_{n-4}R_n$[21] with 24 skeletal electrons ($n = 10$). The skeletal structure of **C** in solution (see NMR spectra) belongs to the nido series. Solid **C** has two reversible phase transformations at $-24.1°$ and $-4.3°$, respectively (DSC measurements*). Because of these phase changes or valence isomerizations it was not possible for a long time to obtain suitable crystals for the determination of the molecular structure(s). Nido structures have been expected for **C** from the determination of the distribution of the skeletal atoms by homo scalar ($^{11}B^{11}B$)-correlated and hetero scalar ($^{13}C^{11}B$)-correlated ^{11}B and ^{13}C NMR spectra, respectively.[9, 11a] The correct structure of the solid **C** at 103 K (LT phase) has now been determined (1990) by X-ray diffraction analysis.[11b, 22]

Compound **C** is synthesized from the pentaethyl-1,5-dicarba-*closo*-pentaborane(5) (**A**) by a dimerization that has not yet been fully clarified mechanistically; 1 mol of **A** ($\delta^{11}B = +13.5$) reacts with ≤ 1.2 mol of metallic potassium in tetrahydrofuran (THF) at room temperature according to the above equation, slowly forming a dark brown solution. The metal does not react faster when the mixture is heated. If ≥ 0.5 mol of potassium per mole of **A** ($\delta^{11}B = 13.5$) are taken up, four new ^{11}B NMR signals of the weakly paramagnetic solution (ESR measurements†) appear at $\delta = 26.8$, -14.2, -16.2, and -37.1 (see Fig. 1). The intensity of the ^{11}B NMR signal of **A** decreases to ca. 50% after reaction of ≥ 1.0–1.2 mol of potassium per mol of **A**. Correspondingly, we presume that a potassium salt of the dianionic species **B** is formed in an equilibrium with **A** and a small amount of the radical anion salt **B*·** (see equation).

The reduced carbaborate **B** reacts with iodine in THF rapidly becoming lighter in color and leading to the quantitative precipitation of potassium iodide. Crude colorless, crystalline **C** is obtained in about 60% yield from the yellow liquid together with about 40% of unreacted **A** according to the above equation.

---

* DSC: DuPont 9900; $10°$ min^{-1}, measured under a pure N_2 atmosphere.
† ESR spectroscopy with V 4500 (Varian), $H_0 = 3330$ G; in X band (9.5 GHz).

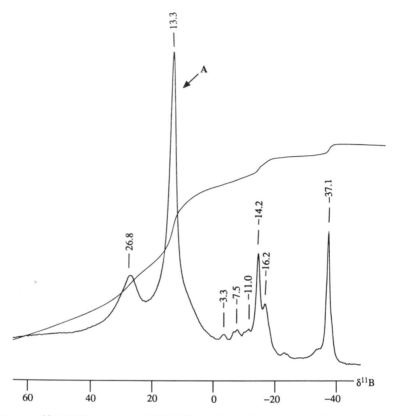

Fig. 1. ^{11}B NMR spectrum (64.2 MHz) of **B** in the presence of **A** (see arrow).

Procedures

A solution of the potassium salt **B** (see equation) is prepared from **A** in a 100-mL flask with a magnetic stirrer, a thermometer tube, and a gas bypass and a bubbler. A piece of 4.76 g (122 mmol) of potassium metal* is stirred for 4 days with a solution of **A** (19.1 g, 95 mmol) in 100 mL of tetrahydrofuran (THF) at 20°. After pipetting off the dark brown liquid from the excess potassium bead (0.51 g, 13 mmol) (consumption: 1.15 mmol K per mmol **A**), the small amount of suspended material therein is filtered off, and a weakly paramagnetic (structured ESR† signal) dark THF solution of **B** is obtained [with 1.18 mmol K per mL solution (acidimetric titration with normal

---

* Purchased from Degussa AG, 6450 Hanau, Germany.
† ESR spectroscopy with V 4500 (Varian), $H_0 = 3330$ G; in X band (9.5 GHz).

H_2SO_4)]. ^{11}B NMR: $\delta = +13.3$ (ca. 50% **A**) and signals of **B** with main peaks at $+26.8$, -14.2, -16.2, and -37.1 ppm (ca. 50%); see Fig. 1.

Compound **C** is prepared from **B** with iodine in a 100-mL flask with a magnetic stirrer, a thermometer tube, and a dropping funnel.

A solution of 5.8 g (22.9 mmol) of iodine* in 80 mL of THF is added dropwise to 38 mL of a stirred solution of **B** (44.8 mmol K) in THF. The color of the solution becomes lighter and the mixture warmer ($T_{max} \approx 30°$) as KI is precipitated. After about 10 hr at $\approx 20°$, 7.5 g (100%) of KI is filtered off and the solution is concentrated to about of a third under vacuum (15 torr). The oily, solid residue is taken up in 9 mL of diethyl ether, suspended particles are filtered off, and **C** is crystallized by cooling the solution to $-78°$. The supernatant solution is pipetted off and the crystals are washed with cold ($\leq -70°$) diethyl ether. After being dried under vacuum, 3.53 g (39%) pure (GC†: 99.6%) **C** are collected. Concentrating the filtrate under vacuum (15 torr) affords about 5 g of a brown, viscous residue, which according to the ^{11}B NMR spectrum, contains $\approx 34\%$ unreacted **A** and $\approx 45\%$ **C**. The total yield of **C** is $\approx 62\%$.

Anal. Calcd. for $C_{24}H_{50}B_6$ (403.5): C, 71.43; H, 12.49; B, 16.08. Found: C, 71.59; H, 12.15; B, 16.32.

Properties of Compounds **B** and **C**

The reduction of **A** leads to a dark brown THF solution with the relatively complex ^{11}B NMR spectrum of **B** (see Fig. 1) in the presence of the remaining **A** with $\delta^{11}B = 13.3$ ppm.

Compound **C** sublimes at $> 270°$ without decomposition and shows endothermic, reversible solid phase transformations at $-24.1†$ and $-4.3°$ (DSC‡: exothermic on cooling). The carborane **C** is very soluble in aliphatic and aromatic hydrocarbons or in many other solvents such as $CHCl_3$, and can be recrystallized from cold diethyl ether and pentane (slowly cooling to $-78°$) or also from hot ethanol. Figure 2 shows the X-ray structure of the C_4B_6 skeleton of one molecule of **C** in the crystal state of the LT phase at 103 K.[11b, 22]

Characteristic Spectroscopic Data of C. Mass spectrum (EI, 70 eV): m/z 404 (M^+, basepeak), 389 ($< 4\%$ rel. int.), 375 (< 3). ^1H NMR spectrum

---

* Purchased from Riedel-de Haën AG, 3016 Seelze, Germany.

† GC column: 2.8 m, internal diameter: 5 mm; stationary phase: 7% "apiezon" on "embacel" (silica gur).

‡ DSC: DuPont 9900; 10° min^{-1}; measured under a pure N_2 atmosphere.

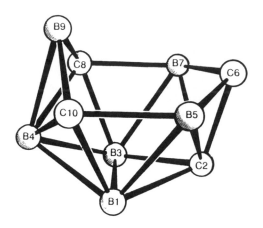

Fig. 2.

(400 MHz, $CDCl_3$): $\delta = 1.94$ (m, 2H), 1.76 (m, 4H), 1.64 (q, 2H), 1.49 (q, 2H), 1.19 (t, 6H), ≈ 1.0 (m, 20H), ≈ 0.8 (m, br, 14H) ppm. ^{11}B NMR spectrum (64.2 MHz, $CDCl_3$): $\delta = + 50.3$ (1B), $+ 6.7$ (2B), $- 6.3$ (1B), $- 20.7$ (2B) ppm. $^{13}C\{^{11}B\}$NMR spectrum (50.3 MHz, CD_2Cl_2): skeletal atoms: $\delta = 37.9$ (br, 1C), 29.2 (s, 2C), $- 6.7$ (s, 1C); ethyl residues: $\delta = 20.64$ (t, 1C), 20.45 (t, 2C), 20.30 (t, 1C), 16.50 (q, 1C), 14.40 (q, 2C), 12.60 (q, 1C), 11.97 (q, 2C), 11.15 (q, 1C), 10.85 (q, 2C), 10.36 (q, 1C), 9.7 (t, 1C), 6.5 (t, 2C), 4.2 (t, 3C) ppm.

The solid white decaethyl-2,6,8,10-tetracarbadecaborane(10) (**C**) can be exposed to air and moisture for only a short time without changing its appearance and composition at room temperature. Compound **C** is stable to iodine at room temperature. The solution of **C** in THF is attacked by metallic potassium to form a pale-yellow-colored mixture.[22] Compound **C** reacts analogously to **A** with anhydrous $(CH_3)_3NO$[17] above 75°. In refluxing toluene 1 mol of **C** liberates more than 16 mol of trimethylamine from the N-oxide reagent.

References

1. R. Köster and G. Rotermund, *Tetrahedron Lett.*, **1964**, 1667.
2. R. Köster, H.-J. Horstschäfer, and P. Binger, *Angew. Chem.*, **78**, 777 (1966); *Angew. Chem. Int. Ed. Engl.*, **5**, 730 (1966).
3. R. Köster, H.-J. Horstschäfer, P. Binger, and P. K. Mattschei, *Liebigs Ann. Chem.*, **1975**, 1339; H. J. Horstschäfer, Dissertation Technische Hochschule Aachen 1967.
4. M. Grassberger and R. Köster, "B-Organocarborane mit drei Bor-Atomen im Gerüst," in *Methoden der Organischen Chemie* (Houben-Weyl-Müller), 4th ed., Vol. XIII/3c, R. Köster (ed.), Thieme, Stuttgart 1984, pp. 160–163.
5. R. Köster and B. Wrackmeyer, *Z. Naturforsch.*, **36b**, 704 (1981).

6. I. Shapiro, C. D. Good, and R. E. Williams, *J. Am. Chem. Soc.*, **84**, 3837 (1962).
7. R. N. Grimes, *J. Am. Chem. Soc.*, **88**, 1070 (1967).
8. J. F. Ditter, E. B. Klusmann, J. D. Oakes, and R. E. Williams, *Inorg. Chem.*, **9**, 889 (1970).
9. R. Köster, G. Seidel, and B. Wrackmeyer, *Angew. Chem.*, **96**, 520 (1984); *Angew. Chem. Int. Ed. Engl.*, **23**, 512 (1984).
10. M. Grassberger and R. Köster, "B-Organocarborane mit sechs bis acht Bor-Atomen im Gerüst," in *Methoden der Organischen Chemie* (Houben-Weyl-Müller), 4th ed., Vol. XIII/3c, R. Köster (ed.), Thieme, Stuttgart 1984, p. 178.
11a. B. Wrackmeyer and R. Köster, "Analytik der B-Organocarborane," in *Methoden der Organischen Chemie* (Houben-Weyl-Müller), 4th ed., Vol. XIII/3c, R. Köster (ed.), Thieme, Stuttgart 1984, p. 605.
11b. X-ray structure analysis of the LT phase of solid C by R. Boese and D. Bläser, University of Essen, Germany, November 1990; see ref. 22.
12. R. Köster and P. Binger, *Inorg. Synth.*, **15**, 141 (1974).
13. R. Köster, "Diorgano-hydro-borane," in *Methoden der Organischen Chemie* (Houben-Weyl-Müller), 4th ed., Vol. XIII/3a, R. Köster (ed.), Thieme, Stuttgart 1982, p. 333.
14. R. Köster, P. Binger, and W. V. Dahlhoff, *Synth. React. Inorg. Metal-Org. Chem.*, **3**, 359 (1973).
15. R. Köster, H. J. Horstschäfer, and G. Seidel, *Inorg. Synth.*, Vol. 29, 77 (1992).
16. R. Köster, H. J. Horstschäfer, and P. Binger, *Liebigs Ann. Chem.*, **717**, 1 (1968).
17. R. Köster and Y. Morita, *Liebigs Ann. Chem.*, **714**, 70 (1967).
18. R. Köster, G. Seidel, and B. Wrackmeyer, *Angew. Chem.*, **97**, 600 (1985); *Angew. Chem. Int. Ed. Engl.*, **24**, 572 (1985).
19. M. P. Brown, A. K. Holliday, and G. M. Way, *J. Chem. Soc. Dalton Trans.*, **1975**, 148.
20. I. Rayment and H. M. M. Shearer, *J. Chem. Soc. Dalton Trans.*, **1977**, 136.
21. R. N. Grimes, *Adv. Inorg. Chem. Radiochem.*, **26**, 57, 114 (1983).
22. R. Köster, G. Seidel, B. Wrackmeyer, D. Bläser, R. Boese, M. Bühl, and P. v. R. Schleyer, *Chem. Ber.*, **124**, in press (1991).

22. TRIMETHYLSILYL-SUBSTITUTED DERIVATIVES OF 2,3-DICARBA-*nido*-HEXABORANE(8) AND LITHIUM, SODIUM, AND DILITHIUM SALTS OF THE CARBORANE DIANION 2,3-BIS(TRIMETHYLSILYL)-2,3-DICARBA-*nido*-HEXABORATE(2 −)

$$B_5H_9 + 4(CH_3)_3SiC{\equiv}CSi(CH_3)_3 \rightarrow [(CH_3)_3Si]_2C_2B_4H_6$$
$$+ [(CH_3)_3Si(H)C{=}CSi(CH_3)_3]_3B$$

$$[(CH_3)_3Si]_2C_2B_4H_6 + HCl \rightarrow [(CH_3)_3Si]C_2B_4H_7 + (CH_3)_3SiCl$$

$$[(CH_3)_3Si]_2C_2B_4H_6 + NaH \rightarrow Na^+[((CH_3)_3Si)_2C_2B_4H_5]^{-*} + H_2$$

$$Na^+[((CH_3)_3Si)_2C_2B_4H_5]^- + Li^+[(CH_3)_3C]^- \rightarrow$$
$$Na^+Li^+[((CH_3)_3Si)_2C_2B_4H_4]^{2-} + (CH_3)_3CH$$

$$[(CH_3)_3Si]_2C_2B_4H_6 + 2Li^+[(CH_3)_3C]^- \rightarrow Li_2^+[((CH_3)_3Si)_2C_2B_4H_4]^{2-}$$
$$+ 2(CH_3)_3CH$$

Submitted by REYNALDO D. BARRETO† and NARAYAN S. HOSMANE†
Checked by MARK A. BENVENUTO‡ and RUSSELL N. GRIMES‡

Small carboranes of the type $nido$-2,3-$R_2C_2B_4H_6$ (R = H, alkyl, or aryl) and their mono- and dianions are of considerable interest in inorganic chemistry as these species are important building blocks for a variety of extended metallacarboranes.[1-5] However, the synthesis of these $nido$-carboranes has presented a challenge to chemists owing to the difficulties involved in producing the desired compounds in sufficient quantity and purity. Added to this, the safety considerations involved in the use of B_5H_9 have made the synthesis of these compounds a significant undertaking. The reports by Grimes and coworkers of convenient routes to these carboranes using a strong Lewis base have made their production in good yields and large quantities possible.[6]

These procedures, however, are not convenient for the synthesis of C-trimethylsilyl and C-phenyl-substituted derivatives because of the difficulties arising in the separation of the desired carboranes from side products.[7] A new route, one that is both safe and that does not require the use of a Lewis base, has been developed to produce the trimethylsilyl-substituted derivatives of the carboranes in large quantities and with minimal separation problems. However, the yield and quantity isolated for the mono-C-trimethylsilyl derivative are much lower.[8] Since it is expected that the chemistry of the mono-C-trimethylsilyl-substituted derivative will be different from that of the disubstituted cases, owing to the acidity of the $C_{(cage)}$-hydrogen,[9] a high yield synthetic route to this compound is also desired. Previous work has shown that $C_{(cage)}$–Si bonds are susceptible to attack by acids.[10] When such an approach, using HCl, is employed on the bis-C-trimethylsilyl carborane derivative, the desired mono-trimethylsilyl carborane product is obtained in high yield.[11] It should also be mentioned that variations of this process have successfully been employed to produce the parent $nido$-2,3-$C_2B_4H_8$ carborane[12] and may also be useful in the production of other derivatives. The procedures necessary to make the bis-C-trimethylsilyl- and mono-C-trimethylsilyl-substituted C_2B_4-carborane derivatives are described here.

Recently, the structure of the monosodium salt of C-SiMe$_3$-substituted $nido$-C_2B_4-carborane has been determined by X-ray analysis.[13] The structure shows that the monoanion* is an ion cluster consisting of two C_2B_4 carborane cages and two tetrahydrofuran (THF)-solvated sodium ions with a

---

*The monoanion actually exists in the solid state as $(C_4H_8O\cdot Na^+)_2[2,3\text{-}((CH_3)_3Si)_2\text{-}2,3\text{-}C_2B_4H_5^-]_2$ (see Scheme 1).

† Department of Chemistry, Southern Methodist University, Dallas, TX 75275.

‡ Department of Chemistry, University of Virginia, Charlottesville, VA 22901.

crystallographic inversion center half way between the sodium ions. Although the preparation of the disodium salt of the carbons adjacent *nido*-C_2B_4-carborane is not yet known, the synthesis of its first stable dianion, $Na^+Li^+[2\text{-}(Si(CH_3)_3)\text{-}3\text{-}(R)\text{-}2,3\text{-}C_2B_4H_4]^{2-}$ (R = $Si(CH_3)_3$, CH_3, H), was reported in 1986.[14] The discovery of this stable double salt has led to the production of a number of sila and germacarboranes that could not be produced from the corresponding monosodium salts (see Scheme 1).[14–16] Recently, Grimes and coworkers have found that the preparation of the lithium sodium and dilithium double salts is generally applicable to C-alkyl and C-aryl substituted *nido*-C_2B_4-carboranes as well.[17] Also presented here

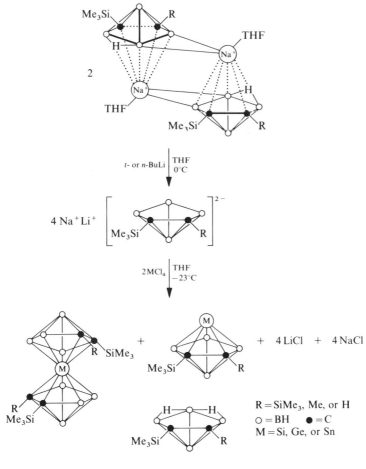

Scheme 1

are the synthetic procedures necessary to produce the Na^+Li^+ and Li_2^+ salts of the bis(C-trimethylsilyl)-substituted-C_2B_4-carborane dianions starting from the neutral *nido*-carboranes. These procedures may also be applied to produce the C-methyl-C-trimethylsilyl, mono-C-trimethylsilyl, and C-H-substituted carborane derivatives.[18]

A. 2,3-BIS(TRIMETHYLSILYL)-2,3-DICARBA-*nido*-HEXABORANE (8)

Procedure

■ **Caution.** B_5H_9 *is highly toxic and explodes spontaneously on exposure to air. This procedure should be carried out only on a high-vacuum line by a chemist experienced in its use.*

A 100-g (588-mmol) sample of bis(trimethylsilyl)acetylene (1,2-ethynediylbis(trimethylsilane)) $[(CH_3)_3SiC\equiv CSi(CH_3)_3]$* is placed into a 500-mL flask fitted with a Teflon stopcock and degassed on a vacuum line. No further purification is necessary. The pentaborane(9) B_5H_9† is used as received.

A 500-mL single-ended stainless steel cylinder‡ fitted with a high-vacuum Swagelok shutoff valve (Fig. 1*A*), is attached to a vacuum line at 10^{-4} torr. Since $(CH_3)_3SiC\equiv CSi(CH_3)_3$ is not very volatile at room temperature *in vacuo*, the condensation of this material into the stainless steel cylinder is facilitated by using the shortest possible path on the vacuum line that is well-wrapped with a heating tape and heated to 60–70°C. It is important that the flask containing the degassed $(CH_3)_3SiC\equiv CSi(CH_3)_3$ is also heated to 70°C during the condensation. The $(CH_3)_3SiC\equiv CSi(CH_3)_3$ and 10.6 g (168 mmol) of B_5H_9 are condensed sequentially into the cylinder at -196°C. Care must be taken to keep the upper portion of the cylinder warm during condensation of the reactants (gentle heating with a heat gun) in order to prevent clogging of the valve. After the condensation of the reactants the shutoff valve is closed, the cylinder is detached from the line and allowed to warm to room temperature. The *lower half* of the cylinder is then immersed in an oil bath at 140–150°C for 48–72 h. At the end of this time, the cylinder is reattached to the vacuum line through a 100-mL glass bulb (Fig. 1*B*), cooled to -78°C (Dry-ice bath), and pumped out over a period of 24 h through a series of traps at -196°C to remove H_2 and to collect $(CH_3)_3SiH$. Since the bis(trimethylsilyl)-substituted carborane and other byproducts are handled in multigram quantities in high-vacuum lines, the traps used for this procedure

---

* Hüls America, Inc., P.O. Box 456, Piscataway, NJ 08855,
† Callery Chemical Co., Mars-Evans City Road, Callery, PA 16024.

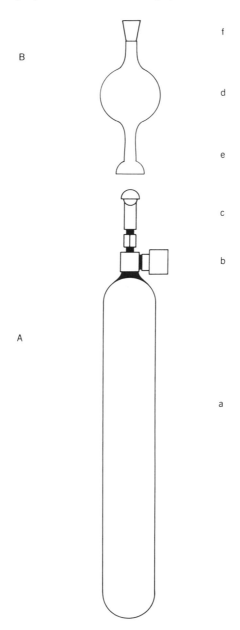

Fig. 1. (*A*) Carborane reactor consisting of a 500-mL stainless steel cylinder (a), a Swagelok high-vacuum shutoff valve (b), and a ball joint to the vacuum line (c), (*B*) Glass bulb trap consisting of a 100-mL glass bulb (d), and glass joints to connect it to the reactor cylinder (e) and the vacuum line (f).

are of $\sim$ 100–110-mL capacity. The cylinder is then slowly warmed to 120°C using an oil bath and the volatile products of the reaction are fractionated through two traps at 0°, and one each at $-$ 78 and $-$ 196°C. The initial warming of the cylinder must be carefully monitored and controlled to prevent the migration of a viscous brown polymeric by-product into the traps. The task is facilitated by the glass bulb that acts as a room temperature trap. The fractionation is continued at this temperature for 24–48 h, at which point the temperature of the oil bath is increased to 140–150°C and the fractionation continued for an additional 24–48 h. During this time, the carborane 2,3-[(CH$_3$)$_3$Si]$_2$C$_2$B$_4$H$_6$ (12.49 g, 56.9 mmol) is collected in the 0°C traps. The contents of the $-$ 78° and $-$ 196°C traps are then combined and refractionated through traps at $-$ 45° (C$_6$H$_5$Cl slush), $-$ 93° (toluene slush), and $-$ 196°C. These traps ultimately yielded unreacted (CH$_3$)$_3$SiC$\equiv$CSi(CH$_3$)$_3$, unreacted B$_5$H$_9$ (1.75 g, 27.7 mmol), and the product (CH$_3$)$_3$SiH, respectively. The brown polymeric material, mentioned previously, is retained in the cylinder. The unreacted starting materials may be collected into a separate container for reuse later. The yield for this process, based on the B$_5$H$_9$ consumed is 40.7%. When the preparation of this carborane is scaled down to one-tenth for all reagents, yields as high as 30.4% are obtained.

Properties

2,3-Bis(trimethylsilyl)-2,3-dicarba-*nido*-hexaborane(8) (Fig. 2) is a clear, colorless liquid at room temperature that is slightly air-sensitive (**Caution:** *This carborane ignites in air when in contact with cloth.*) The boiling point of the liquid at 23 torr is 143.5 $\pm$ 0.5°C. The IR (CDCl$_3$ vs CDCl$_3$) exhibits major absorption bands at 2955 (vs), 2900 (s), 2590 (vs), 1948 (m), 1922 (w, br), 1530 (w), 1495 (w), 1465 (w), 1410 (s), 1375 (ms), 1330 (ms), 1250 (vs), 1155 (ms), 1100 (m), 1070 (m), 1030 (m), 1000 (s), 960 (m), 930 (vs), 845 (vs), 680 (vs), 655

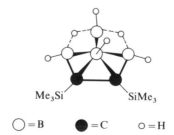

$\bigcirc$ = B $\bullet$ = C   o = H

Fig. 2.   Structure of *nido*-2,3-[(CH$_3$)$_3$Si]$_2$-2,3-C$_2$B$_4$H$_6$.

(ms), 620 (vs), 520 (s), 480 (m), 400 (vs), 368 (vs), and 275 (w) cm^{-1}. The proton-coupled ^{11}B NMR spectrum of the product gives doublets at + 1.99 ppm ($J = 156$ Hz) and − 49.9 ppm ($J = 173$ Hz). The ^{1}H NMR spectrum gives a broad quartet at 3.99 ppm (basal B—H), a singlet at 0.36 ppm [(CH$_3$)$_3$Si], a broad quartet at − 0.72 ppm (apical B—H), and a broad peak at − 1.66 ppm (B—H—B bridge).

B.   2-(TRIMETHYLSILYL)-2,3-DICARBA-*nido*-HEXABORANE(8)

Procedure

■ **Caution.** *Anhydrous hydrogen chloride stored under high pressure is corrosive, and causes severe burns if it is inhaled or comes in contact with the skin or eyes. All appropriate safeguards for the use of such a reagent should be applied.*

The anhydrous hydrogen chloride (obtained from Matheson)* is fractionated through a series of traps held at − 78°C (Dry Ice–isopropanol) − 120°C (ethanol slush), and − 196°C (liquid N$_2$) that collected a trace quantity of H$_2$O, a trace quantity of Cl$_2$ impurity, and pure HCl gas, respectively.

A 500-mL single-ended stainless steel cylinder† fitted with a high-vacuum Swagelok shutoff valve (Fig. 1*A*) is attached to a high-vacuum line at 10^{-4} torr. Since *nido*-2,3-[(CH$_3$)$_3$Si]$_2$-2,3-C$_2$B$_4$H$_6$ is not very volatile at room temperature *in vacuo*, the condensation of this material into the stainless steel cylinder is facilitated by using the shortest possible path on the vacuum line that is well wrapped with a heating tape and heated to 60–70°C. It is necessary that the flask containing *nido*-2,3-[(CH$_3$)$_3$Si]$_2$-2,3-C$_2$B$_4$H$_6$ is also heated to 70°C during the condensation. A 11.8-g (53.7-mmol) sample of *nido*-2,3-[(CH$_3$)$_3$Si]$_2$-2,3-C$_2$B$_4$H$_6$[8], prepared in the previous synthesis, is transferred under high vacuum into the cylinder. The anhydrous HCl gas (Matheson*, 75 mmol) is then condensed into the cylinder at − 196°C and the shutoff valve is closed. The cylinder is detached from the vacuum line and allowed to warm to room temperature. The *lower half* of the cylinder is then immersed in an oil bath at 140°C for 40–50 h. At the end of this time, the cylinder is cooled to room temperature and reattached to the vacuum line. Since the mono(trimethylsilyl)-substituted carborane and the byproduct, Me$_3$SiCl, are handled in multigram quantities in high-vacuum lines, the traps used for this procedure are of ∼ 100–110-mL capacity. This cylinder is again

---

* Matheson, P.O. Box 908, La Porte, Texas 77571.
† Tech Controls, Inc., 4238 Spring Valley Road, Dallas, TX 75234.

slowly warmed to 100°C using an oil bath and the volatile products of the reaction are collected and fractionated through traps held at $-23°$(CCl$_4$ slush), $-45°$ (C$_6$H$_5$Cl slush), $-78°$ (Dry Ice–isopropanol), and $-196°$C for 24 h. During this time, the carborane 2-[(CH$_3$)$_3$Si]C$_2$B$_4$H$_7$ (3.77 g, 25.6 mmol) is collected in the $-45°$C trap. The traps at -23, -78, and $-196°$C ultimately yielded unreacted 2,3-[(CH$_3$)$_3$Si]$_2$C$_2$B$_4$H$_6$ (1.21 g, 5.5 mmol), (CH$_3$)$_3$SiCl, and a trace quantity of unreacted HCl, respectively. The yield for this process, based on the 2,3-[(CH$_3$)$_3$Si]$_2$C$_2$B$_4$H$_6$ consumed, is 53.1%. When the preparation of this carborane is scaled down to one-tenth for all reagents, yields as high as 37.6% are obtained.

Properties

2-(Trimethylsilyl)-2,3-dicarba-*nido*-hexaborane(8) (Fig. 3) is a clear, colorless liquid at room temperature that is slightly air-sensitive. The boiling point of the liquid at 39 torr is 83–84°C. The IR (gas phase; 5 torr) exhibits major absorption bands at 2965 (vs), 2908 (ms), 2860 (sh), 2600 (vs), 1975 (sh), 1940 (m), 1518 (m), 1490 (sh), 1415 (w), 1340 (m), 1260 (vs), 1103 (s), 1085 (sh), 978 (w), 916 (ms), 844 (vvs), 748 (m), 707 (w), 649 (w), 618 (w), 602 (vw), 481 (vw), 400 (w), 359 (s) cm^{-1}. The proton-coupled ^{11}B NMR spectrum of the product gives doublets at $+0.49$ ppm ($^1J = 160$ Hz), -0.11 ppm ($^1J = 160$ Hz, $^2J = 43$ Hz), and at -51.82 ppm ($J = 176$ Hz). The ^{1}H NMR spectrum gives a broad singlet at 6.1 ppm [C$_{(cage)}$-H], a broad quartet at 3.2 ppm (basal B—H), a singlet at 0.40 ppm [(CH$_3$)$_3$Si], a broad quartet at -0.72 ppm (apical B—H), and a broad peak at -1.75 ppm (B—H—B bridge).

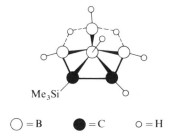

Me$_3$Si

○ = B   ● = C   ○ = H

Fig. 3.   Structure of *nido*-2-[(CH$_3$)$_3$Si]-2,3-C$_2$B$_4$H$_7$.

C. LITHIUM SODIUM 2,3-BIS(TRIMETHYLSILYL)-2,3-DICARBA-*nido*-HEXABORATE(2 −)

Procedure

■ **Caution.** *Sodium hydride (NaH) and t-butyllithium (t-BuLi) react explosively with water, violently with lower alcohols, and ignite spontaneously in moist air.*

A 0.40-g (17.4-mmol) sample of sodium hydride (NaH)* is placed in a two-necked flask of 250-mL capacity with ground-glass or Teflon stopcocks. This is washed three (3) times with dry hexane in an evacuable glove-bag and stored under dry nitrogen. Hexane and tetrahydrofuran (THF) were distilled over sodium benzophenone and stored over Li[AlH$_4$] under vacuum. The 1.7 M t-BuLi in pentane (Aldrich Chemical Co.) is used as received. The *nido*-2,3-((CH$_3$)$_3$Si)$_2$-2,3-C$_2$B$_4$H$_6$ is prepared by the method described above.

A 2.25-g (10.2-mmol) sample of *nido*-2,3-((CH$_3$)$_3$Si)$_2$-2,3-C$_2$B$_4$H$_6$ is transferred under vacuum into a 25-mL ampule that is fitted with a Teflon stopcock. This ampule is then attached in an inert atmosphere to the two-necked flask that contained NaH (flask A in Fig. 4a). This flask is now attached to a sintered-glass filter unit with a sidearm that is also attached to another flask (flask B) equipped with a rubber septum and a magnetic stirring bar. The entire apparatus (Fig. 4a) is attached to a vacuum line and evacuated. Then 50 mL of THF is condensed under vacuum into flask A at − 196°C. The carborane in the ampule is then poured onto the heterogeneous mixture of NaH and THF in flask A at − 78°C (Dry Ice–isopropanol) and the reaction mixture is slowly warmed to room temperature. Care must be taken at this point to control the evolution and buildup of H$_2$ by occasional cooling of flask A to − 196°C, measuring evolved H$_2$, and pumping of the hydrogen gas, sequentially. The reaction is allowed to continue until H$_2$ evolution ceases. At this point, the reaction flask is cooled to − 196°C and the evolved H$_2$ is measured and pumped out of the apparatus. The reaction mixture in flask A is allowed to settle and the THF solution of the salt is decanted through the glass frit of the filter unit into flask B. The THF in flask B is removed under vacuum and 50 mL of hexane is then condensed into this flask at − 196°C. At this point, 6.5–7.0 mL (11–12 mmol) of 1.7 M t-Buli in pentane is injected through the septum onto the solution in flask B (Fig. 4a) at − 78°C. Flask B is then allowed to warm to room temperature and the mixture is stirred for 1 h. The solvents in flask B are then removed under vacuum to give 3.20 g (10.2 mmol) of [Na$^+$Li$^+$]

---

* Aldrich Chemical Co., P.O. Box 2060, Milwaukee, WIS 53210.

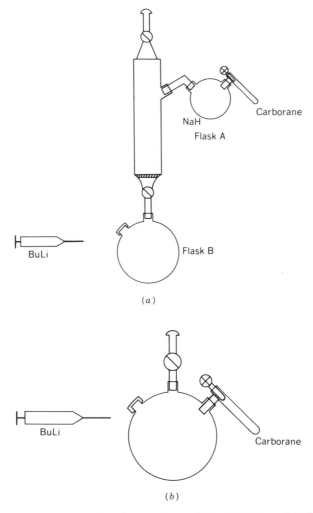

Fig. 4. Reaction apparatus for the synthesis of Na^+Li^+ (a) and Li_2^+ (b) salts of 2,3-bis(trimethylsilyl)-2,3-dicarba-*nido*-hexaborate(2 −).

$(C_4H_8O)[2,3-((CH_3)_3Si)_2-2,3-C_2B_4H_4]^{2-}$. The yield, based on *nido*-2,3-$((CH_3)_3Si)_2-2,3-C_2B_4H_6$ consumed, is essentially quantitative.[19]

Properties

The compound $[Na^+Li^+](C_4H_8O)[2,3-((CH_3)_3Si)_2-2,3-C_2B_4H_4]^{2-}$ is an off-white solid at room temperature that is highly air-sensitive. The IR (in

CD_3CN) exhibits major absorption bands at 2982 (vs), 2912 (s), 2878 (ms), 2509 (s), 2450 (s), 2350 (m), 2285 (m), 2150 (m), 1620 (w), 1508 (ms), 1491 (sh), 1421 (ms), 1389 (ms), 1365 (m), 1270 (sh), 1258 (vvs), 1215 (sh), 1204 (ms), 1168 (m), 1078 (w), 868 (vs), 852 (vs), 838 (vs), 770 (s), 688 (ms), 643 (ms), 635 (sh) cm^{-1}. The proton coupled ^{11}B NMR spectrum (in CD_3CN) of the product gives doublets at $+20.85$ ppm (2B, $J = 112$ Hz), 2.39 ppm (1B, $J = 89$ Hz) and at -44.53 ppm (1B, $J = 161$ Hz). The ^1H NMR spectrum gives a singlet at 3.65 ppm (4H, THF), a quartet at 3.45 ppm (2 basal B—H), a quartet at 2.49 ppm (1 basal B—H), a singlet at 1.80 ppm (4H, THF), a singlet at 0.14 ppm [$(CH_3)_3Si$], and a quartet at -1.97 ppm (apical B—H).

D.   DILITHIUM 2,3-BIS(TRIMETHYLSILYL)-2,3-DICARBA-*nido*-HEXABORATE(2 −)

Procedure

■ **Caution.** *t-butyllithium (t-BuLi) reacts explosively with water, violently with lower alcohols, and ignites spontaneously in moist air.*

A 1.54-g (7.0-mmol) sample of *nido*-2,3-((CH_3)$_3$Si)$_2$-2,3-$C_2B_4H_6$ is transferred under vacuum into a 25-mL ampule that is fitted with a Teflon stopcock. The ampule is then attached to the three-necked flask that is also equipped with a rubber septum, a connecter to a vacuum line, and a magnetic stirring bar (Fig. 4b). This apparatus is attached to a vacuum line and evacuated. Hexane (50 mL) is then condensed under vacuum into this flask at $-196°C$. The carborane in the ampule is then poured onto hexane in the flask at room temperature. The flask is then cooled to $-78°C$ (Dry Ice–isopropanol) and 8.5–9.0 mL (14.5–15.3 mmol) of 1.7 M t-BuLi in pentane is now injected through the septum onto the solution in the flask (Fig. 4b). The homogeneous mixture in the flask is allowed to warm to room temperature and stirred for 1 h. The solvents in the flask are then removed under vacuum to give 1.62 g (7.0 mmol) of Li_2^+ [2,3-((CH_3)$_3$Si)$_2$-2,3-$C_2B_4H_4$]$^{2-}$. The yield, based on *nido*-2,3-((CH_3)$_3$Si)$_2$-2,3-$C_2B_4H_6$ consumed, is essentially quantitative.[19]

Properties

The compound Li_2^+ [2,3-((CH_3)$_3$Si)$_2$-2,3-$C_2B_4H_4$]$^{2-}$ is an off-white solid at room temperature that is extremely air-sensitive. The IR (CD_3CN) exhibits major absorption bands at 2981 (s), 2930 (m), 2875 (m), 2548 (ms), 2509 (ms), 2469 (sh), 2312 (s), 2282 (ms), 2165 (m), 1681 (w), 1520 (s), 1422 (br, ms), 1369 (m), 1272 (ms), 1259 (vs), 1215 (m), 1116 (mw), 990 (m), 870 (sh), 850 (vvs),

770 (m), 688 (mw), 643 (w), 631 (mw) cm^{-1}. The proton coupled ^{11}B NMR spectrum (in CD_3CN) of the product gives doublets at $+20.94$ ppm (2B, $J = 119$ Hz), 2.37 ppm (1B, $J = 89$ Hz) and at -44.50 ppm (1B, $J = 160$ Hz). The ^{1}H NMR spectrum gives a quartet at 3.70 ppm (2 basal B—H), a quartet at 2.48 ppm (1 basal B—H), a singlet at 0.14 ppm [$(CH_3)_3Si$], and a quartet at -2.00 ppm (apical B—H).

Acknowledgment

This work was supported by the National Science Foundation, the Robert A. Welch Foundation, and the donors of the Petroleum Research Fund Administered by the American Chemical Society.

References and Note

1. (a) R. N. Grimes, in *Carboranes*, Academic Press: New York, 1970; (b) R. N. Grimes, in *Comprehensive Organometallic Chemistry*, G. Wilkinson, F. G. A. Stone, and E. W. Abel (eds.), Pergamon Press, Oxford, 1982; Vol. 1. (c), *Metal Interactions with Boron Clusters*, R. N. Grimes (ed.), Plenum Press, New York, 1982; (d) R. N. Grimes, in *Advances in Boron and the Boranes*, Vol. 5 in the series *Molecular Structure and Energetics*, J. F. Liebman, A. Greenberg, and R. E. Williams, (eds.), VCH, New York, 1988, p. 235; (e) N. S. Hosmane and J. A. Maguire, in *Advances in Boron and the Boranes*, Vol. 5 in the series *Molecular Structure and Energetics*, J. F. Liebman, A. Greenberg, and R. E. Williams (eds.), VCH, New York, 1988, p. 297; (f) N. S. Hosmane and J. A. Maguire, *Adv. Organomet. Chem.*, **30**, 99 (1990).
2. R. G. Swisher, E. Sinn, and R. N. Grimes, *Organometallics*, **2**, 506 (1983).
3. L. Barton and P. K. Rush, *Inorg. Chem.*, **24**, 3413 (1985).
4. R. N. Grimes, W. J. Rademaker, M. L. Denniston, R. F. Bryan, and P. T. Greene, *J. Am. Chem. Soc.*, **94**, 1865 (1972).
5. L. G. Sneddon, *Pure & Appl. Chem.*, **59**, 837 (1987).
6. (a) N. S. Hosmane and R. N. Grimes, *Inorg. Chem.*, **18**, 3294 (1979); (b) R. B. Maynard, L. Borodinsky, and R. N. Grimes, *Inorg. Synth.*, **22**, 211 (1983).
7. N. S. Hosmane and R. N. Grimes, unpublished results; N. S. Hosmane, N. N. Sirmokadam, and J. D. Buynak, *Abstracts*, Imeboron-V; *Chem. Abstr.*, **22**, 39 (1983).
8. (a) N. S. Hosmane, N. N. Sirmokadam, and M. N. Mollenhauer, *J. Organomet. Chem.*, **279**, 359 (1985); (b) W. A. Ledoux and R. N. Grimes, *J. Organomet. Chem.*, **28**, 37 (1971).
9. T. Onak, in *Comprehensive Organometallic Chemistry*, G. Wilkinson, F. G. A. Stone, and E. W. Abel (eds.), Pergamon Press, Oxford, 1982; Vol. 1, Chapter 5.4.
10. N. N. Schwartz, E. O'Brien, S. Karlan, and M. M. Fein, *Inorg. Chem.*, **4**, 661 (1965).
11. N. S. Hosmane, N. N. Maldar, S. B. Potts, D. W. H. Rankin, and H. E. Robertson, *Inorg. Chem.*, **25**, 1561 (1986).
12. N. S. Hosmane, M. S. Islam, and E. G. Burns, *Inorg. Chem.*, **26**, 3237 (1987).
13. N. S. Hosmane, U. Siriwardane, G. Zhang, H. Zhu, and J. A. Maguire, *J. Chem. Soc. Chem. Commun.*, 1128 (1989).
14. N. S. Hosmane, P. de Meester, U. Siriwardane, M S. Islam, and S. S. C. Chu, *J. Chem. Soc., Chem. Commun.*, 1421 (1986).
15. U. Siriwardane, M. S. Islam, T. A. West, N. S. Hosmane, J. A. Maguire, and A. H. Cowley, *J. Am. Chem. Soc.*, **109**, 4600 (1987).

16. N. S. Hosmane, M. S. Islam, B. S. Pinkston, U. Siriwardane, J. J. Banewicz, and J. A. Maguire, *Organometallics*, **7**, 2340 (1988).
17. J. H. Davis Jr., E. Sinn, and R. N. Grimes, *J. Am. Chem. Soc.*, **111**, 4776 (1989).
18. Owing to the acidity of the C–H hydrogen, the use of excess BuLi should be avoided in the preparation of the dianions of the mono-C-trimethylsilyl and C–H-substituted derivatives.
19. The preparations presented here are for the isolation of the dianions. In actual practice, where the dianions are used as intermediates in the syntheses of other carborane species, these reactions may be run strictly in THF solution and the dianion products left unisolated in the solution.

23.   1,2-(1,2-ETHANYLDIYL)-1,2-DICARBADODECABORANE(12) (1,2-ETHANO-*o*-CARBORANE)

Submitted by MAITLAND JONES, Jr.,* ZHEN-HONG LI,* and ROBERT P. L'ESPERANCE*
Checked by KENNETH WADE†

1,2-(1,2-Ethanyldiyl)-1,2-dicarbadodecaborane(12) [1,2-ethano-*o*-carborane] (**1**) has been made through the intramolecular insertion reaction of (2-methyl-*o*-carboran-1-yl)carbene.[1] The reported synthesis of this cage compound is impractical, as it involves several steps with low yields, the last of which produces a number of products in comparable amounts. The new route to 1,2-ethano-*o*-carborane reported here uses an intramolecular displacement of a leaving group by a cage carbanion to produce **1** in 40% yield. The synthesis can also be accomplished through a one-step reaction in the gas phase in 64% yield.[2] In both cases, 1,2-ethano-*o*-carborane is the predominant product. This relatively easy synthesis provides practical access to **1** for the first time, and should facilitate further study of this molecule which contains the smallest carbocyclic ring fused to a carborane frame.

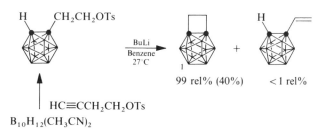

Ts = *p*-toluenesulfonate (4-methylbenzenesulfonate)

* Department of Chemistry, Princeton University, Princeton NJ 08544.
† Department of Chemistry, University of Durham, Durham, DH1 3LE, United Kingdom.

A. 2-(1,2-DICARBADODECABORAN(12)-1-YL)ETHYL-4-METHYLBENZENESULFONATE [2-(*o*-CARBORAN-1-YL)-ETHANOL TOSYLATE]

Procedure

In a 250-ml, three-necked, round-bottomed flask is placed 4 g (32.7 mmol) of decaborane(14) (Dexsil) in 50 mL of acetonitrile. The solution is refluxed under an argon atmosphere for 2 h. A yellow solid is precipitated. The excess acetonitrile is removed by distillation to yield the $B_{10}H_{12}(CH_3CN)_2$ complex, which is used directly in the next step. The complex is dissolved in 50 mL of toluene, and 10.5 g (46.8 mmol) of 3-butyn-1-ol tosylate (3-butynyl 4-methylbenzenesulfonate)[3] is added. The reaction mixture is refluxed under argon for 12 h. The toluene is removed by distillation to leave a gummy yellow solid. This material is placed in a Soxhlet apparatus and extracted with hexane for 12 h followed by dichloromethane for 6 h. The organic solutions are combined and the solvent removed on a rotatory evaporator to leave a whitish solid. This solid is purified by column chromatography on EM Science grade 62 silica gel (CHCl₃) to give 4.5 g (40%) of a white solid, mp 112–114°C.

Precise Mass. Calcd. for $C_{11}H_{22}{}^{11}B_8{}^{10}B_2SO_3$: 342.2292, Found: 342.2246.

Properties

2-(*o*-Carboran-1-yl)ethanol tosylate is a white solid readily soluble in chlorinated solvents, benzene, and diethyl ether, but sparingly soluble in hexane. Treatment with conventional bases serves to eliminate *p*-toluenesulfonic acid.

^1H NMR (CDCl₃, 300 MHz) δ 7.78, 7.40 (AA'BB', 4H), 4.11 (t, 2H, $J = 6.1$ Hz), 3.69 (br s, 1H), 2.63 (t, 2H, $J = 6.1$ Hz), 2.49 (s, 3H), 1.4–3.0 (br m, 10H, B–H). IR (KBr): 3058, 2577, 1597, 1426, 1355, 1190, 1173, 815 cm^{-1}.

B. 1,2-ETHANO-*o*-CARBORANE

Procedure

In a 100-mL, three-necked, round-bottomed flask is placed 1.5 g (4.4 mmol) of 2-(*o*-carboran-1-yl)ethanol tosylate in 50 mL of benzene. The flask is placed under an argon atmosphere, cooled with an ice bath, and 2.1 mL of 2.5 *M* butyllithium (Aldrich) in hexane is added at once. The solution is allowed to stir for 14.5 h at room temperature. The reaction mixture is poured into 25 mL of 5% HCl and the organic layer separated. The aqueous layer is extracted with 2×10 mL ether. The combined organic layers are washed with saturated $NaHCO_3$ solution and dried over $MgSO_4$. The solvent is removed on a rotatory evaporator to leave a white solid. Analytic gas chromatography on a 6-ft $\times \frac{1}{8}$-in 15% OV 17 on Chromosorb W-HP (80–100 mesh) column operated at 160°C with an injector temperature of 200°C and detector temperature of 220°C showed **1** at 3.3 min. On occasion, approximately 1% of 1-vinyl-*o*-carborane also appeared at ~ 2.5 min. The crude material is purified by column chromatography on EM Science grade 62 silica gel using hexane as solvent to give 0.3 g (40%) of a white waxy solid, mp 274–275° (sealed).[5]

Properties

1,2-Ethano-*o*-carborane is a white solid, which survives unchanged after heating at 350°C for 24 h. This four-membered carbon ring compound is remarkably stable and chemically inert. It remains untouched when treated with *N*-bromosuccinimide, $LiN(SiMe_3)$, Br_2, or DDQ.[4] The boron cage can be chlorinated with $AlCl_3/CCl_4$.[2]

  Spectral data are as follows: ^1H NMR ($CDCl_3$, 300 MHz), δ 2.85 (s, 4H), 1.4–3.6 (br m, 10H, B–H). IR (KBr): 2975, 2580, 1442, 1250, 1230, 1150, 1060, 720 cm^{-1}; MS (70 ev): m/e 170.5, 169.5 (basepeak).

References and Note

1. S. L. Chari, S. H. Chiang, and M. Jones, Jr., *J. Am. Chem. Soc.*, **104**, 3138 (1982).
2. R. P. L'Esperance, Z.-H. Li, D. Van Engen, and M. Jones, Jr., *Inorg. Chem.*, **28**, 1823 (1989).
3. L. Brandsma, *Preparative Acetylene Chemistry*, Elsevier, New York, 1971, p. 159.
4. R. P. L'Esperance, unpublished work.
5. The checkers suggest purification by sublimation at approximately 0.05 mm and 30°C.

24.   7,9-DISELENA-*NIDO*-UNDECABORANE(9)

$$B_{10}H_{14} + KOH + 2H_2O \rightarrow K[B_9H_{14}] + B(OH)_3 + H_2$$

$$K[B_9H_{14}] + K_2(Se)_n + H_2O \rightarrow B_9H_9Se_2 + K_2(Se)_{n-2} + 3H_2 + KOH$$

Submitted by G. D. FRIESEN,* T. P. HANUSA,* and L. J. TODD*
Checked by L. G. SNEDDON†

The diselenaborane was obtained as an unexpected minor product during the reaction of decaborane(14) with polyselenide ion in aqueous base. The major product was the $[B_{10}H_{11}Se]^{-1}$ ion.[1] It was reported previously that aqueous base degraded $B_{10}H_{14}$ to form the $[B_9H_{14}]^-$ ion.[2] Subsequent study proved that polyselenide ion reacted rapidly with $[B_9H_{14}]^-$ to form $B_9H_9Se_2$ in good yield.[3] The procedure for the preparation of this diselenaborane in two steps from decaborane(14) is described here.

Procedure

■ **Caution.** *This preparation should be conducted in a well-ventilated hood. Decaborane is a very toxic material and should be handled with gloves.*

The dipotassium polyselenide reagent is prepared in a three-necked 250-mL flask equipped with a magnetic stirrer, reflux condenser topped with a nitrogen inlet, and a heating mantle. A nitrogen atmosphere is maintained during the course of the entire preparation. Powdered gray selenium (8.31 g, 0.105 g-atom), potassium hydroxide pellets (5.92 g, 0.105 mol), and 45 mL of deoxygenated water† are added to the flask, the magnetic stirrer is started, and the mixture is brought to reflux temperature. After refluxing for 20 minutes, the deep red liquid is allowed to cool to room temperature.

The potassium tetradecahydrononaborate(1 −) is prepared in a separate two-necked 100-mL flask equipped with a magnetic stirrer and a nitrogen inlet. Decaborane(14)† (2.57 g, 21.0 mmol), potassium hydroxide pellets (2.36 g, 42.1 mmol), and 35 mL of deoxygenated water§ are added to the flask, and stirring begun. The stirring is continued for 1 h, during which time the initially deep yellow solution fades to a very pale yellow. The solution is then neutralized by the cautious addition of concentrated aqueous hydrochloric acid (approx. 37% HCl). This requires about 0.5 mL, but neutrality should be checked with pH-indicating paper, as the addition of excessive acid will decrease the product yield. When the gas evolution has ceased, trace amounts of regenerated decaborane(14) are removed by adding 30 mL of hexanes to the $[B_9H_{14}]^-$ solution. The mixture is stirred for a few minutes

* Department of Chemistry, Indiana University, Bloomington, IN 47405.
† Department of Chemistry, University of Pennsylvania, 231 South 34th Street, Philadelphia, PA 19104.
‡ The water was deoxygenated by boiling for a few minutes, and then bubbling a stream of nitrogen gas through it as it cooled.
§ The decaborane is purified by sublimation at 60°C and 10^{-2} mm.

and the hexane layer is removed with a syringe. The extraction is repeated with a second 30-mL portion of hexanes.

The solution containing the $[B_9H_{14}]^-$ ion is added as rapidly as possible to the stirred polyselenide solution. A white precipitate is formed immediately. Stirring is continued at room temperature for 1 h. The reaction mixture is then extracted in a separatory funnel with 600 mL of hexanes divided into several portions. The combined hexane extracts are rotary evaporated to dryness, and the crude $B_9H_9Se_2$ is purified by sublimation at 60°C $(10^{-2}$ torr) onto a water-cooled probe. The yield of white $B_9H_9Se_2$ is 4.32 g (78% based on decaborane).

Properties

7,9-Diselena-*nido*-undecaborane(9) is a white crystalline solid, mp 340–342°C. It is soluble in dichloromethane and hexanes and insoluble in water. In the presence of air, the compound is stable for several months in the solid state, but in solution, red decomposition products are observed within a day or two. The infrared spectrum (KBr pellet) contains major absorption bands at 2590 (vs), 2555 (vs), 1400 (w), 997 (s), 974 (s), 906 (m), 880 (w), 857 (w), 819 (w), 783 (m), 766 (m), 739 (w), 692 (w), 563 (w), and 500 (w) cm^{-1}. This compound reacts with strong base to give an anionic intermediate that can then be reacted with cyclopentadiene monomer and anhydrous cobalt chloride in the presence of triethylamine to give complexes such as $Co(\eta^5\text{-}C_5H_5)(B_9H_9Se_2)$ and $Co_2(\eta^5\text{-}C_5H_5)_2(B_9H_9Se).$[4]

References

1. J. L. Little, G. D. Friesen, and L. J. Todd, *Inorg. Chem.*, **16**, 869 (1977).
2. L. E. Benjamin, S. F. Stafiej, and E. A. Takacs, *J. Am. Chem. Soc.*, **85**, 2674 (1963).
3. G. D. Friesen, A. Barriola, and L. J. Todd, *Chem. Ind.* (*London*), 631 (1978).
4. G. D. Friesen, A. Barriola, P. Daluga, P. Ragatz, J. C. Huffman, and L. J. Todd, *Inorg. Chem.*, **19**, 458 (1980).

25.   6,9-DISELENA-*ARACHNO*-DECABORANE(10)

$$B_9H_{13} \cdot S(CH_3)_2 + Na_2(Se)_2 + 4H_2O \longrightarrow Na[B_8H_9Se_2] + NaB(OH)_4$$
$$+ S(CH_3)_2 + 4H_2$$

$$Na[B_8H_9Se_2] + H^+ \xrightarrow{\text{ion exchange}} B_8H_{10}Se_2 + Na^+$$

Submitted by G. D. FRIESEN,* R. L. KUMP,* and L. J. TODD*
Checked by L. G. SNEDDON†

Reaction of $[B_9H_{14}]^-$ with the polyselenide ion formed $B_9H_9Se_2$ in good yield.[1] Treatment of the closely related borane $B_9H_{13} \cdot S(CH_3)_2$ with the polyselenide reagent gave quite different results. In this case, both 6,9-$B_8H_{10}Se_2$ and the $[B_9H_{12}Se]^-$ ion were formed.[2] The synthesis of the novel diselenaborane is described here.

Procedure

■ **Caution.** *This synthesis must be conducted in a well-ventilated hood. Some of the possible by-products (e.g., H_2Se) are toxic and malodorous.*

Under a nitrogen atmosphere that is maintained during the course of the preparation, 0.87 g (0.038 mol) of sodium is dissolved in 50 mL of liquid ammonia in a three-necked 100-mL flask equipped with a Dry Ice condenser topped with a nitrogen inlet and a magnetic stirring bar. Powdered gray elemental selenium, 6.0 g (0.076 mol) is added slowly (if addition is too rapid, vigorous splattering results) by means of a solid addition tube over a 45-min period. The mixture is stirred for an additional 30 min at Dry Ice temperature, at which point the solution is dark olive in color. The ammonia is evaporated to a volume of approximately 25 mL. Ice-cold water, 25 mL, previously deoxygenated and saturated with nitrogen, is slowly syringed into the reaction flask. The color of the solution changes to red.

The reagent $B_9H_{13} \cdot S(CH_3)_2$ is prepared using the literature method used for the synthesis of $B_9H_{13} \cdot S(C_2H_5)_2$.[3] First, 2 g (0.011 mol) of $B_9H_{13} \cdot S(CH_3)_2$ is rapidly added to the reaction mixture, which is stirred at room temperature for 12 h. Hexane (10 mL) that has been deoxygenated and saturated with N_2 is added to the dark green solution and stirred for 15 min. The hexane layer is drawn off by syringe to remove any hexane-soluble impurities. Excess tetramethylammonium chloride in deoxygenated water is added to the reaction flask, which gives a white precipitate. The solid is filtered, washed with water, and dried under nitrogen. The dried precipitate is dissolved in acetonitrile and rapidly placed on an H^+ ion exchange column (50 g of Rexyn 101 resin) and eluted with 150–200 mL of acetonitrile. Deoxygenated water (50 mL) is added to the eluant and the acetonitrile is removed by rotary evaporation at 30°C. The resulting solid is removed by filtration in air and dried under vacuum. Crystallization from

* Department of Chemistry, Indiana University, Bloomington, IN 47405.
† Department of Chemistry, University of Pennsylvania, 231 South 34th Street, Philadelphia, PA 19104.

CH_2Cl_2/CH_3CN yields 0.68 g (24% based on $B_9H_{13} \cdot S(CH_3)_2$) of white $6,9\text{-}B_8H_{10}Se_2$, mp 197–199°C (dec.).

Properties

6,9-Diselena-*arachno*-decaborane(10) is air-sensitive in solution but, as a solid, is stable under anhydrous conditions. The compound is very soluble in CH_2Cl_2 and tetrahydrofuran, slightly soluble in acetonitrile, and insoluble in water. The infrared spectrum (KBr pellet) contains major absorption bands at 2570 (vs), 2550 (s), 1036 (w), 1010 (w), 990 (s), 976 (sh), 958 (w), 950 (w), 912 (m), 900 (m), 863 (m), 800 (s), 728 (w), and 712 (m) cm^{-1}. The ^{11}B NMR spectrum (CHCl$_3$ solvent) contains resonances at 14.9(176); $-$ 16.9(160); $-$ 30.7 ppm ($J_{^{11}B-^{1}H}$ 155 Hz) with peak areas of 2 : 4 : 2, respectively [$BF_3 \cdot O(C_2H_5)_2 = 0$ ppm]. Bridge hydrogen coupling is observed on the $-$ 16.9-ppm signal. The ^{11}B NMR spectrum indicates that the two selenium atoms are mostly likely in the 6 and 9 positions of an arachno 10-atom framework. Treatment of $6,9\text{-}B_8H_{10}Se_2$ with $CoCl_2$ and cyclopentadiene monomer in the presence of triethylamine forms $Co(\eta^5\text{-}C_5H_5)(B_8H_8Se_2)$.

References

1. G. D. Friesen, A. Barriola, and L. J. Todd, *Chem. Ind. (London)*, 631 (1978).
2. G. D. Friesen, R. L. Kump, and L. J. Todd, *Inorg. Chem.*, **19**, 1485 (1980).
3. B. M. Graybill, J. K. Ruff, and M. F. Hawthorne, *J. Am. Chem. Soc.*, **83**, 2669 (1961).

Chapter Three

TRANSITION METAL COORDINATION COMPOUNDS

26. SOLVATED AND UNSOLVATED ANHYDROUS METAL CHLORIDES FROM METAL CHLORIDE HYDRATES

$$MCl_n \cdot xH_2O + 2x(CH_3)_3SiCl \rightarrow MCl_n + x\ [(CH_3Si]_2O + 2x\,HCl$$

Submitted by PHILIP BOUDJOUK* and JEUNG-HO SO*
Checked by MARLIN N. ACKERMANN,† SUSAN E. HAWLEY, and
BENJAMIN E. TURK†

Although there are several methods for preparing anhydrous metal halides,[1] thermal and chemical methods of removing water from hydrated metal halides are the most frequently employed. The pyrolysis of metal halide hydrates has been studied extensively and can lead to anhydrous salts, although temperature control is important for many hydrates because water is released stepwise and mixtures of hydrates can be obtained.[2] Dehydrating agents such as 2,2-dimethoxypropane and thionyl chloride are efficient dehydrating agents and have been widely used. The former has the disadvantage of producing methanol and acetone, which often associate with metal halides;[3] thus thionyl chloride has been used as the standard dehydrating agent for metal chlorides.[4] On refluxing, it reacts with water to evolve hydrogen chloride and sulfur dioxide. Even though these by products are removed from the reaction mixture, there are drawbacks involved with thionyl chloride; it is a severe lachrymator that must be freshly distilled before use, and, because the reaction is slow, it must be used in excess to achieve

* Department of Chemistry, North Dakota State University, Fargo, ND 58105.
† Department of Chemistry, Oberlin College, Oberlin, OH 44074.

convenient rates of dehydration. Removing the last traces of thionyl chloride is sometimes difficult.

A new dehydration method for metal chlorides is described here by which anhydrous transition metal chlorides can be prepared conveniently and in high yields under mild conditions. The large solubilities of some metal chloride hydrates in tetrahydrofuran (THF) allow homogeneous reactions and short reaction times ($\sim$ 1 h). The formation of HCl and of the very strong silicon–oxygen bonds in hexamethyldisiloxane drive the reaction. The reactions are easily monitored by characteristic color changes and work-up is a relatively simple procedure because low-boiling compounds [i.e., HCl ($-$ 84°C), trimethylchlorosilane (57°C), and hexamethyldisiloxane (101°C)] are efficiently removed from the reaction under reduced pressure. Since trimethylchlorosilane and hexamethyldisiloxane are very soluble in hydrocarbons, the last traces of both can be removed by washing with hexane. The halides are isolated as the THF complexes.

The THF adducts were characterized by IR spectroscopy by comparing the C—O—C symmetric and asymmetric stretches with those reported elsewhere.[6a] Excellent agreement was obtained in each case. The reported yields are an average of two trials.

Some salts could not be dehydrated in THF. Iron(III) chloride for example, polymerizes THF[6a] and must be prepared using neat trimethylchlorosilane. Cobalt(II) chloride dihydrate gave a similar result requiring neat trimethylchlorosilane. In both cases, nearly quantitative yields of the anhydrous salt were obtained. Hydrated zinc chloride, which we prepared by adding 10 wt % water to the anhydrous chloride because well-defined hydrates are not commercially available, was very efficiently dehydrated in neat trimethylchlorosilane to give a 96% yield of zinc chloride. In contrast, the THF/trimethylchlorosilane mixture afforded a comparatively modest 71% yield of the tetrahydrofuranate. Chromium(III) chloride hexahydrate, on the other hand, could not be completely dehydrated in neat trimethylchlorosilane and required THF for an efficient reaction giving 89% yield of chromium(III) chloride tris(tetrahydrofuran). Our results are summarized in Table I.

Materials

Commercially available metal chloride hydrates and trimethylchlorosilane can be used without further purification. Tetrahydrofuran should be freshly distilled from sodium benzophenone. Wet THF gives similar results but necessarily requires more trimethylchlorosilane. IR spectra were obtained as nujol mulls on a Beckman Model 4240 spectrometer.

TABLE I. Dehydration of Metal Chlorides with Trimethylchlorosilane

$MX_n \cdot H_2O$	Color	Product	Color	Yields (%)
$CrCl_3 \cdot H_2O^{a,b}$	Green	$Cr(THF)_3Cl_3$	Purple	89
$CuCl_2 \cdot 2H_2O^b$	Blue	$Cu(THF)_{0.8}Cl_2$	Yellow	95
$BaCl_2 \cdot 2H_2O^b$	White	$BaCl_2$	White	95
$ZnCl_2 \, n(H_2O)^c$	White	$Zn(THF)_2Cl_2$	White	71
$ZnCl_2 \, n(H_2O)^{c,d}$	White	$ZnCl_2$	White	96
$BaCl_2 \cdot 2H_2O^d$	White	$BaCl_2$	White	95
$CuCl_2 \cdot 2H_2O^d$	Blue	$CuCl_2$	Yellow	90
$CoCl_2 \cdot 6H_2O^{d,e}$	Red	$CoCl_2$	Blue	95
$FeCl_3 \cdot 6H_2O^{d,e}$	Orange	$FeCl_3$	Dark. green	95

[a] Requires THF for complete dehydration.
[b] Approximate composition of reaction mixture: 30 mL trimethylchlorosilane + 20 mL THF + 10 mmol hydrate.
[c] Samples of hydrated $ZnCl_2$ were made by adding water (10% by weight) to anhydrous $ZnCl_2$.
[d] Dehydrations were performed in 30 mL trimethylchlorosilane + 10 mmol hydrate.
[e] $FeCl_3$ and $CoCl_2$ react with THF, reactions must be run in neat trimethylchlorosilane.

Procedure

■ **Caution.**   *HCl is corrosive. All reactions should be carried out in a well-ventilated hood.*

A.   Chromium(III) Chloride: A Typical Procedure

(Also applicable to Cu, Ba, Zn, and Co dichlorides). Chromium(III) chloride hexahydrate (2.66 g, 10 mmol), 20 mL of THF, and a magnetic stir bar were placed in a 100-mL, three-necked, round-bottomed flask equipped with a condenser fitted with a drying tube. At room temperature, 32 mL of trimethylchlorosilane (253 mmol) was added dropwise with stirring to the slurry, causing the evolution of heat. The color of the reaction mixture changed from dark green to deep purple. The purple solid that precipitated was washed with hexane followed by evaporation of residual solvent at reduced pressure to give 3.34 g (8.9 mmol) 89% yield of chromium(III) chloride tris(tetrahydrofuran). The C—O—C stretches were observed at 850 cm^{-1} (symmetric) and 1010 cm^{-1} (asymmetric).[6a] No O—H absorptions (3500–3300 cm^{-1}) were detected.

B.   Iron(III) Chloride

Treatment of iron(III) chloride hexahydrate by the above procedure leads to polymerization of THF. However, the reaction can be conveniently carried out without solvent. Thus, 30 mL of trimethylchlorosilane was added to powdered iron(III) chloride hexahydrate (2.70 g, 10 mmol) at room temperature. Stirring for 30 min followed by refluxing for an additional 3 h led to 95% yield of the green anhydrous salt after removal of the liquid under reduced pressure. The IR spectrum showed no O—H absorptions nor Si—C—H peaks (1225–1275 cm^{-1}), indicating that all water and organosilicon compounds were removed.

Acknowledgement

Financial support from the Air Force Office of Scientific Research through Grant No. 88-0060 is gratefully acknowledged.

References

1. (a) S. Y. Tyree, *Inorg. Synth.*, **4**, 105 (1953); (b) F. A. Cotton and G. Wilkinson, *Advanced Inorganic Chemistry*, 4th ed., Wiley, New York, 1980, pp. 549–550.
2. (a) M. Hassanein, *Thermochim. Acta*, **61**, 121 (1983); (b) J. R. Williams, W. W. Wendlandt, *Thermochim. Acta*, **7**, 275 (1973); (c) W. K. Grindstaff and N. Fogel, *J. Chem. Soc. Dalton, Trans.*, 1972, 1476.
3. K. Starke, *J. Inorg. Nucl. Chem.*, **11**, 77 (1959).
4. (a) J. Shamir, *Inorg. Chim. Acta*, **156**, 163 (1989); (b) A. R. Pray, *Inorg. Synth.*, **5**, 153 (1957); (c) H. Z. Hecht, *Anorg. Chem.*, **254**, 37 (1947).
5. (a) R. J. Kern, *J. Inorg. Nucl. Chem.*, **24**, 1105 (1962); (b) P. J. Jones, A. L. Hale, W. Levason, and F. P. McCullough, Jr., *Inorg. Chem.*, **22**, 2642 (1983).

27. SOLID SOLVATES: THE USE OF WEAK LIGANDS IN COORDINATION CHEMISTRY

Submitted by WILLEM L. DRIESSEN* and JAN REEDIJK*
Checked by KIM R. DUNBAR† and LAURA E. PENCE†

Hydrated transition metal salts are commonly used as precursors for other coordination compounds with organic ligands. Water is a common solvent for the preparation and the characterization of many coordination compounds. However, the presence of water precludes the formation of coordination compounds with weak ligands, which are not able or are barely able to compete with water as a ligand. For instance, when anhydrous MgI_2 is allowed to react with dry acetone, the solid solvate[1] $[Mg(acetone)_6]I_2$ is

* Department of Chemistry, Gorlaeus Laboratories, Leiden University, P.O. Box 9502, 2300 RA Leiden, The Netherlands.
† Department of Chemistry, College of Natural Science Michigan State University, East Lansing, MI 48823.

formed,[2] but the slightest contact with water, even exposure to the open air, rapidly transforms this compound to the corresponding hydrate $[Mg(H_2O)_6]I_2$.

To synthesize coordination compounds with weak ligands, methods have been developed whereby water is either absent from the start[3] or is removed through a chemical reaction.[4,5] In this contribution the preparation of coordination compounds of some divalent metals with nitromethane,[3] ethanol,[4] acetone,[6] diphenyl sulfoxide,[7] and acetonitrile[8] are described. These descriptions are merely examples of simple general methods for the preparation of coordination compounds of Mg^{2+}, Mn^{2+}, Fe^{2+}, Co^{2+}, Ni^{2+}, Cu^{2+}, Zn^{2+}, and Cd^{2+} with weak ligands, such as those mentioned above and acetic acid,[5] nitrobenzene,[9] hydrogen cyanide,[10] tetrahydrofurane,[11] dioxane,[12] diglyme (1,1-oxybis[2-methoxyethane]),[13] 1,4,7,10,13,16-hexa-oxacyclooctadecane (18-*crown*-6),[14] ethyl acetate,[15] and 2,4-pentanedione (acetylacetone in the neutral ketonic form).[16]

To obtain coordination compounds of nitromethane, the weakest ligand known so far,[3] two conditions must be fulfilled. First, no other potential ligand must be present. This means, in practice, that all manipulations must be done under rigorously anhydrous circumstances. Second, a large counter-ion is required[17] to induce the cation to become completely solvated by only the nitromethane molecules. This demand is met by an elegant process called "the chloride-ion transfer method." In this process a very large anion is formed through the incorporation of the relatively small chloride ion of a divalent metal chloride into the coordination sphere of a strong Lewis acid, like antimony pentachloride or iron trichloride.

Starting Materials

Extra care must be given to exclude any moisture in equipment, reagents, air (or N_2), and solvents that are used for the preparation and isolation of the coordination compounds with complex chloroanions (procedures A, B, and E), and for the isolation of all other compounds:

- Air (or N_2 gas) can best be dried by passing it through a gas jar charged with P_2O_5 on silica.
- For filtration, glass filters must be used throughout, as paper filters contain some water. Also, during the filtering process, totally dry air (or dry N_2 gas) must be passed through the glass filter.
- All glassware, including the glass filters, must be thoroughly dried by heating at *ca.* 200°C for at least half an hour and then cooled to ambient temperatures in a desiccator charged with P_2O_5.

- The rotatory evaporator must be flushed with warm dry air, and in the connection to the water pump a gas jar charged with P_2O_5 (or P_2O_5 on silica) must be inserted. When releasing the vacuum, dry air or dry N_2 gas must be admitted.

- Furthermore, all contacts of solutions or solids with the open air should be kept to a minimum.

All solvents, ligands, metal hydrates, Lewis acids, and drying agents are available from the common commercial sources.

The white solid $AlCl_3$ can also be prepared[18] freshly and quite easily by standard methods.[19] Diphenyl sulfoxide $(C_6H_5)_2SO$ can be prepared in very good yields through a Friedel–Crafts reaction with $SOCl_2$ and $AlCl_3$ in sodium-dried benzene, analogous to the synthesis of diphenyl sulfide.[20] Equally good results can be obtained with ditolyl sulfoxide, $(C_7H_7)_2SO$, which can be prepared in the same way using toluene instead of benzene.

Anhydrous divalent metal chlorides MCl_2, in which M is Mn, Fe, Co, Ni, Cu, and Zn can be obtained by heating the corresponding, commercially available, hydrated divalent metal chlorides $MCl_2(H_2O)_n$, $n = 4,6$ at 200°C for about 1 h in a gentle stream of dry HCl gas.[19] Other, more elaborate, methods have been reported[21,22] for the preparation of anhydrous divalent transition metal chlorides.

Acetone, acetonitrile, 1,2-dichloroethane, and nitromethane are best dried by storage over anhydrous calcium sulfate, and *n*-hexane over CaH_2 or Na wire.[23]

Antimony pentachloride, a yellow liquid that hydrolyzes immediately when in contact with moist air, is preferentially used as the solid adduct with nitromethane.

■ **Caution.** *The corrosive and toxic liquid $SbCl_5$ must be handled in a well-ventilated hood.*

Preparation of the solid $SbCl_5(CH_3NO_2)$, pentachloro(nitromethane)antimony(V), is best done in relatively large quantities.

Procedure

A thoroughly dry 250-mL round-bottomed flask is charged with 40 mL of dry 1,2-dichloroethane and 30.0 g (0.1 mol) of antimony pentachloride.* To this solution, a solution of 6.1 g (0.1 mol) of dry nitromethane in 15 mL of dry

---

*Weigh the stoppered flask with 1,2-dichloroethane. Add rapidly an estimated quantity of $SbCl_5$ directly from its bottle. Weigh again, and calculate the amount of dry nitromethane (a small excess is all right) to be added.

1,2-dichloroethane is rapidly added. Immediately, some heat will evolve, which will raise the temperature of the flask by $\sim 10°C$. After cooling, crystals of $SbCl_5(CH_3NO_2)$ will have formed, which can be used without further purification. A second crop can be obtained by evaporating most of the solvent under reduced pressure at elevated temperatures. Eventually, the product can be washed with a little dry *n*-hexane. $CH_3Cl_5NO_2Sb$: Pale yellow crystalline solid, very hygroscopic, decomposes with water. Yield: $\sim 70\%$.

A. HEXAKIS(NITROMETHANE)COBALT(II) BIS[HEXA-CHLOROANTIMONATE(V)], $[Co(CH_3NO_2)_6][SbCl_6]_2$

$$CoCl_2 + 2\,SbCl_5 + 6\,CH_3NO_2 \xrightarrow{CH_3NO_2} [Co(CH_3NO_2)_6]\,[SbCl_6]_2$$

Procedure

A thoroughly dry 100-mL conical flask is *rapidly* charged with 1.3 g (0.01 mol) of $CoCl_2$, 7.2 g (0.02 mol) of $SbCl_5(CH_3NO_2)$, 40 mL of dry nitromethane, and a magnetic stirring bar (1 cm) and then immediately tightly stoppered. This mixture is stirred at room temperature for several hours until almost all solid, blue $CoCl_2$ has disappeared. If not clear, the pink-red solution must be filtered without suction through a thoroughly dry glass filter. The clear solution is then concentrated under reduced pressure with the aid of a thoroughly dry rotatory evaporator to a volume of about 10 mL. The temperature of the solution should not be raised above 50°C. Crystallization of the desired product will occur when this solution is stored at low temperatures (< 5°C). The crystals must be collected and stored under rigorously anhydrous circumstances. In general, recrystallization is not necessary. Eventually, the product can be washed with a little dry *n*-hexane. $C_6H_{18}Cl_{12}CoN_6O_{12}Sb_2$; mp $= 102°C$ dec.; pink-red solid. Yield $\sim 60\%$.

B. HEXAKIS(ACETONE)COBALT(II) BIS[HEXACHLORO-ANTIMONATE(V)], $[Co(CH_3COCH_3)_6][SbCl_6]_2$

$$[Co(CH_3NO_2)_6][SbCl_6]_2 + 6CH_3COCH_3 \xrightarrow{CH_3NO_2}$$
$$[Co(CH_3COCH_3)_6][SbCl_6]_2$$

Nitromethane,[3] the weakest known ligand for transition metal ions, is readily replaced by other weak ligands such as esters, aldehydes, ethers, and ketones.[6,11-16] Using this ligand substitution reaction, transition metal solvates of acetone may be prepared.

Procedure

To a solution of 5.5 g (0.005 mol) of $[Co(CH_3NO_2)_6][SbCl_6]_2$ in 40 mL of thoroughly dry nitromethane is added 1.8 g (0.03 mol) of dry acetone. (Instead of starting with solid $[Co(CH_3NO_2)_6][SbCl_6]_2$, compound **B** can also be prepared through the addition of the appropriate amount of dry acetone to the pink-red solution of procedure **A**). The colour of the solution becomes slightly more red. The solution is then concentrated by evaporation of the solvent nitromethane under reduced pressure at moderate temperatures ($\leqslant 50°C$) to about one-third its volume. Storage of the resulting solution below 5°C will produce pink crystals, which can be isolated by filtration and stored under rigorously anhydrous circumstances. Although recrystallization is usually not necessary, it can nevertheless be effected using a mixture of dry nitromethane and dry acetone in a 9 : 1 ratio. Eventually, the product can be washed with a little dry *n*-hexane. $C_{18}H_{36}Cl_{12}CoO_6Sb_2$; pink solid. Yield: $\sim 60\%$.

C. HEXAKIS(ETHANOL)NICKEL(II) BIS[TETRAFLUORO-BORATE(1 −)], $[Ni(C_2H_5OH)_6][BF_4]_2$

$$[Ni(H_2O)_6][BF_4]_2 + 6 HC(OC_2H_5)_3 \rightarrow$$

$$[Ni(C_2H_5OH)_6][BF_4]_2 + 6 C_2H_5OH + 6 HCOOC_2H_5$$

When hydrated transition metal salts are allowed to react with certain dehydrating agents, such as triethyl orthoformate[4] or acetic anhydride,[5] then coordination compounds of ethanol and acetic acid, respectively, are formed. These complexes may then react with other weak ligands, yielding coordination compounds that cannot be formed through a direct reaction of these ligands with the metal hydrates.

Procedure

A 100-mL Erlenmeyer flask is charged with 3.4 g (0.01 mol) of $[Ni(H_2O)_6][BF_4]_2$ and 10 mL of absolute ethanol. Then, 9.0 g (0.06 mol) of triethyl orthoformate is added, and the mixture is stirred. Dissolution of the nickel salt is accompanied by a very slight rise in temperature. If the solution is not clear, it should be filtered through a fine-pore filter paper. The resulting yellow-green solution is concentrated to about one-third of its volume (reduced pressure, rotatory evaporator) and then stored below 5°C to allow crystallization. The solid is collected on a glass filter under anhydrous conditions and may be washed with a little dry *n*-hexane. The solid may be recrystallized from absolute ethanol or dry nitromethane, although this is seldom necessary. $C_{12}H_{36}B_2F_8NiO_6$; green solid. Yield: $\sim 60\%$.

D. HEXAKIS(DIPHENYL SULFOXIDE)NICKEL(II) BIS[TETRAFLUOROBORATE(1−)], [Ni(dpso)₆][BF₄]₂

$$[Ni(H_2O)_6][BF_4]_2 + 6HC(OC_2H_5)_3 + 6DPSO \xrightarrow{acetone}$$

$$[Ni(dpso)_6][BF_4]_2 + 12C_2H_5OH + 6HCOOC_2H_5$$

Diphenyl sulfoxide (DPSO) is a stronger ligand than ethanol and a weaker ligand than water. It can thus form coordination compounds from hydrates through the intermediate action of a dehydrating agent.[4, 5]

Procedure

A clear yellow-green solution of $[Ni(C_2H_5OH)_6][BF_4]_2$ is obtained by dissolving, while stirring, 0.65 g (0.002 mol) of $[Ni(H_2O)_6][BF_4]_2$ in a mixture of 10 mL of acetone and 5 mL of triethyl orthoformate, followed by filtration through analytic-grade filter paper. At room temperature a solution of 2.8 g (0.014 mol, which is a slight excess) of diphenyl sulfoxide in 15 mL of acetone is added. Yellow-green crystals start to precipitate within an hour. If not, the solution may be reduced in volume and then chilled. The next day these crystals are collected and washed once with 5 mL of acetone and twice with 10-mL portions of diethyl ether. The product is $C_{72}H_{60}B_2F_8NiO_6S_6$; yellow-green solid. Yield: $\sim 80\%$.

E. HEXAKIS(ACETONITRILE)IRON(III) BIS[TETRACHLOROALUMINATE(1−)], [Fe(CH₃CN)₆][AlCl₄]₂

$$FeCl_2 + 2AlCl_3 + 6CH_3CN \xrightarrow{CH_3CN} [Fe(CH_3CN)_6][AlCl_4]_2$$

Solvates of acetonitrile can be obtained through the method of chloride-ion transfer. This can be performed through the nitromethane precursors, but also directly in acetonitrile itself. Acetonitrile is inert, unlike acetone, toward strong Lewis acids and is also a fairly good solvent for this type of coordination compound. In addition, Lewis acids weaker than $SbCl_5$ can be used as chloride-ion acceptors to form the chloroanions in acetonitrile, because acetonitrile is a stronger solvate-forming ligand than nitromethane.

Procedure

A dry 100-mL Erlenmeyer flask is *rapidly* charged with a Teflon-coated 1-cm bar magnet, 1.3 g (0.01 mol) of anhydrous $FeCl_2$, 2.7 g (0.02 mol) of $AlCl_3$, and 40 mL of dry acetonitrile, after which it is immediately tightly stoppered. The mixture is stirred until (almost) all of the solid has disappeared. If the solution is not clear, it should be filtered through a dry glass filter without suction in a moisture-free environment. The clear solution is concentrated to about one-half its volume (reduced pressure, $T \leqslant 50°C$). The product crystallizes at low temperature ($\leqslant 5°C$) and may be washed with a little dry *n*-hexane. The product is $C_{12}H_{18}Al_2Cl_8FeN_6$; brown-yellow solid. Yield: $\sim 60\%$.

Some General Properties

All the compounds described in this contribution are sensitive to moisture. When in contact with water, they decompose—as has already been stressed above—yielding hydrated metal salts and/or hydroxides or oxides.

These compounds also contain large, noncoordinating, complex anions. As a consequence of their composite nature, these anions show characteristic infrared absorptions.[24] The tetrahedral anion $[BF_4]^-$ has a very strong absorption band at 1070 cm^{-1} (with in some cases a high-energy shoulder originating from ^{10}B) and an adsorption band at 530 cm^{-1}. The octahedral $[SbCl_6]^-$ anion has a very strong absorption band in the far-infrared region (at 345 cm^{-1}). The tetrahedral $[AlCl_4]^-$ anion has a very strong absorption band at 495 cm^{-1}.

The ligands acetonitrile and acetone have characteristic groups directly involved in the bonding to the metal ions. The bond orders of the nitrile bond and the carbonyl bond are influenced by the metal ions. The bond order of the nitrile group is increased on coordination. Consequently the CN-stretching absorption ($v_{C\equiv N} = 2257$ cm^{-1} in liquid acetonitrile) shifts to higher frequencies.[25] The bond order of the carbonyl group is lowered on coordination, which is reflected in a shift to lower frequencies of the C=O stretching vibration ($v_{C=O} = 1718$ cm^{-1} in liquid acetone).[6] The magnitude of the shift depends on the "inductive effect" of the particular metal ion. Therefore, an Irving–Williams sequence[26] in the magnitudes of frequency shifts is observed[6,25] for many complexes of the transition metal ions Mn^{2+}, Fe^{2+}, Co^{2+}, Ni^{2+}, Cu^{2+} and Zn^{2+}.

The divalent metal ions in the compounds obtained with the methods described above are solvated by six molecules of one kind[1], and, consequently, the reflectance spectra in the visible–near-infrared (VIS–IR) region of the solid compounds containing metal ions with partly filled d orbitals are typical for a regular octahedral environment.[27,28]

For further details, one is referred to the original literature.[24,27,28]

References and Notes

1. A *solvate* is defined as a solid coordination compound in which the central (transition) metal ion is surrounded by ligands of one kind only.
2. B. N. Menshutkin, *Chem. Zentr., II,* **1906,** 1838.
3. W. L. Driessen and W. L. Groeneveld, *Rec. trav. chim. Pays-Bas,* **88,** 491 (1969).
4. P. W. N. M. van Leeuwen, *Rec. trav. chim. Pays-Bas,* **86,** 247 (1967).
5. P. W. N. M. van Leeuwen, *Rec. trav. chim. Pays-Bas,* **87,** 86 (1968).
6. W. L. Driessen and W. L. Groeneveld, *Rec. trav. chim. Pays-Bas,* **88,** 977 (1969).
7. P. W. N. M. van Leeuwen and W. L. Groeneveld, *Rec. trav. chim. Pays-Bas,* **85,** 1173 (1966).
8. J. Reedijk and W. L. Groeneveld, *Rec. trav. chim. Pays-Bas,* **87,** 513 (1968).
9. W. L. Driessen, L. M. van Geldrop, and W. L. Groeneveld, *Rec. trav. chim. Pays-Bas,* **89,** 1271 (1970).
10. P. L. A. Everstijn, A. P. Zuur, and W. L. Driessen, *Inorg. Nucl. Chem. Lett.,* **12,** 227 (1976).
11. W. L. Driessen and M. den Heyer, *Inorg. Chim. Acta,* **33,** 261 (1979).
12. M. den Heyer and W. L. Driessen, *Inorg. Chim. Acta,* **39,** 43 (1980).
13. W. L. Driessen, M. den Heyer, N. van Gils, and W. L. Groeneveld, *Trans. Met. Chem.,* **6,** 296 (1981).
14. W. L. Driessen and M. den Heyer, *Trans. Met. Chem.,* **6,** 338 (1981).
15. W. L. Driessen, W. L. Groeneveld, and F. W. van der Weij, *Rec. trav. chim. Pays Bas,* **89,** 353 (1970).
16. W. L. Driessen and P. L. A. Everstijn, *Rec. trav. chim. Pays-Bas,* **99,** 238 (1980).
17. W. E. Hatfield, R. Whyman, R. C. Fay, K. N. Raymond, and F. Basolo, *Inorg. Synth.,* **11,** 47 (1968).
18. Sometimes, the $AlCl_3$, which is used as a Friedel–Crafts catalyst, contains water and is, therefore, unsuitable.
19. G. Brauer, *Handbook of Preparative Inorganic Chemistry,* 2nd ed., Academic Press, New York, 1965.
20. W. W. Hartmann, L. A. Smith, and J. B. Dickey, *Org. Synth., Coll. Vol.,* **2,** 242 (1943).
21. A. R. Pray, *Inorg. Synth.,* **5,** 153 (1957).
22. P. Kovacic and N. O. Brace, *Inorg. Synth.,* **6,** 172 (1960).
23. W. L. Jolly, *The Synthesis and Characterization of Inorganic Compounds,* Prentice Hall, Englewood Cliffs, NJ, 1970.
24. K. Nakamoto, *Infrared and Raman Spectra of Inorganic and Coordination Compounds,* 3rd ed., Wiley, New York, 1978.
25. J. Reedijk, A. P. Zuur, and W. L. Groeneveld, *Rec. trav. chim. Pays-Bas,* **86,** 1126 (1967).
26. H. Irving and R. J. P. Williams, *J. Chem. Soc.,* **1953,** 3192.
27. J. Reedijk, W. L. Driessen, and W. L. Groeneveld, *Rec. trav. chim., Pays-Bas,* **88,** 1095 (1969).
28. J. Reedijk, P. W. M. N. van Leeuwen, and W. L. Groeneveld, *Rect. trav. chim. Pays-Bas,* **87,** 129 (1968).

28. NIOBIUM(III) AND (IV) HALIDE COMPLEXES

Submitted by STEVEN F. PEDERSEN,* JACK B. HARTUNG, Jr.,*
ERIC J. ROSKAMP* and PETER S. DRAGOVICH*
Checked by CHARLES J. RUFFING† and BRUCE A. KLEIN†

The structural inorganic chemistry of ligated niobium(III) and niobium(IV) halides have been intensively studied over the years.[1] More recently some d^2 and d^1 niobium reagents have been used in organic synthesis.[2] In general, low-valent niobium complexes have been synthesized via reduction of a d^0 precursor employing reducing agents such as sodium–mercury amalgam or aluminium powder.[1] In all of these cases, removal of some or all of the reaction solvent is required in order to isolate the niobium complex. This requirement makes large-scale syntheses of these compounds cumbersome. Tributyltin hydride is an efficient reagent for reducing niobium pentachloride or pentabromide to lower oxidation states in high yields and on large scales.[2a,b,3] Furthermore, syntheses employing this reducing agent have been designed such that the niobium containing product precipitates from the reaction mixture, circumventing the aforementioned isolation problems.

In addition to their applications in organic synthesis, all of the niobium halide complexes described here serve as convenient precursors to a variety of known coordination complexes of niobium(III) and (IV)[4] as well as organometallic compounds.[3]

■ **Caution.** *All reactions and manipulations should be performed under an atmosphere of nitrogen, either in a dry box or using standard Schlenk techniques.*

All solvents used in these reactions were purified by standard methods and purged with nitrogen immediately before use. The tributyltin hydride was either purchased from Aldrich Chemical Company and distilled prior to use (the checkers noted that they did not distill this material) or prepared from $(Bu_3Sn)_2O$/polymethylhydrosiloxane.[5] In the latter case we recommend that the tributyltin hydride be immediately redistilled (through a 2-cm × 30-cm Vigreux column) after the initial distillation from the reaction mixture. The niobium pentachloride (Aldrich or Cerac) and the niobium pentabromide (Cerac) were used as received.

■ **Caution.** *Tributyltin hydride is toxic and readily absorbed through the skin. Niobium pentachloride and niobium pentabromide are corrosive solids.*

---

* Department of Chemistry, University of California, Berkeley, CA 94720.
† Aldrich Chemical Company, P.O. Box 355, Milwaukee, WI 53201.

A. NbCl$_4$(thf)$_2$

$$NbCl_5 + Bu_3SnH \xrightarrow[\text{toluene, 0°C}]{\text{THF}} NbCl_4(thf)_2 + Bu_3SnCl + 0.5\,H_2$$

A three-necked 5-L Morton flask* is fitted with an overhead mechanical stirrer, a nitrogen inlet adapter, and a rubber septum and charged with toluene ($\sim$ 3 L) and niobium pentachloride (187.3 g, 0.693 mol). The mixture is cooled in an ice bath. A 1-L Schlenk flask is charged with tetrahydrofuran (315 mL, 3.88 mol) and tributyltin hydride (202 g, 0.694 mol) and fitted with a rubber septum. This solution is added, via cannula,† over 20–30 min to the vigorously stirred toluene suspension. Shortly after the addition is completed ($\leqslant$ 10–20 min), the solution is filtered and the solid is washed with toluene (200 mL) and pentane (2 × 300 mL) and dried *in vacuo* for 12 h (< 1 mm). Yield, 249 g (95%)‡ of a free-flowing yellow powder is obtained. Melting point (sealed capillary) = 127–137°C dec. (lit. 145°C); IR (cm^{-1}, Nujol mull) v 990, 820.

Anal. The checkers analyzed for niobium. Calcd: Nb, 24.52. Found: Nb, 24.7.

Properties

An alternative synthesis of this material employing aluminum powder as the reducing agent has been described.[1f,6] NbCl$_4$(thf)$_2$ appears to be indefinitely stable at room temperature if stored under an inert atmosphere. Only on a couple of occasions have we observed that the solid turns gray on standing. The yellow complex is easily retrieved by adding the discolored solid to tetrahydrofuran and stirring the suspension for 20 min followed by filtration and drying *in vacuo*.

B. NbCl$_3$(dme)

$$NbCl_5 + 2\,Bu_3SnH \xrightarrow[-78°C]{\text{DME}} NbCl_3(dme) + 2\,Bu_3SnCl + H_2$$

A three-necked 5-L Morton flask§ is fitted with an overhead mechanical stirrer and a nitrogen inlet adapter and charged with 1,2-dimethoxyethane

---

* The checkers used a regular 5-L flask.
† The checkers added the tributyltin hydride via an addition funnel.
‡ The checkers obtained 244 g (93%). They have also run this reaction on two times this scale and obtained the same yield.
§ The checkers used a regular 5-L flask.

(3 L) and tributyltin hydride (555 g, 1.91 mol). A 500-mL round-bottomed flask containing niobium pentachloride (250 g, 0.93 mol) is attached to the setup via a piece of Latex tubing (30 cm long, 1.8 cm in diameter) that is adapted with two male 24/40 joints on each end. The reaction mixture is then cooled to $-78°C$ and the solution is vigorously stirred while the niobium pentachloride is added over a 1-h period.*

■ **Caution.** *Vigorous stirring is essential and should be maintained in such a manner that any dark residue formed near the joint where the $NbCl_5$ is introduced is constantly being washed into the reaction mixture.*

After the addition is complete the bath temperature is maintained between $-78°$ and $-60°C$ for 4 h. At this point, the cold bath is removed and the mixture is stirred an additional 2 h. The solution is filtered† and the solid is washed with 1,2-dimethoxyethane (3 × 300 mL) and pentane (3 × 300 mL) followed by drying *in vacuo* (12 h at < 1 mm). Yield, 252 g (94%)‡ of a brick-red solid is obtained. Melting point (sealed capillary) = 116–130°C dec.; IR (cm^{-1}, Nujol mull) v 1283, 1238, 1178, 1069, 1028, 1007, 846.

Anal. Calcd. for $C_4H_{10}Cl_3O_2Nb$: C, 16.60; H, 3.48; Cl, 36.76. Found: C, 16.94; H, 3.71; Cl. 36.65; N, 0.00. The checkers analyzed for niobium. Calcd.: Nb, 32.11. Found: Nb, 31.1.

Properties

The molecular structure of $NbCl_3(dme)$ is currently unknown. The solid is indefinitely stable at room temperature when stored under an inert atmosphere.

C. $NbBr_4(thf)_2$

$$NbBr_5 + Bu_3SnH \xrightarrow[\text{toluene, 0°C}]{\text{THF}} NbBr_4(thf)_2 + Bu_3SnBr + 0.5\,H_2$$

A dry three-necked 1-L flask is fitted with an overhead mechanical stirrer, an addition funnel, and a nitrogen inlet adapter and charged with toluene (600 mL) and $NbBr_5$ (40.0 g, 81.2 mmol). The addition funnel is charged with tetrahydrofuran (120 mL, 1.48 mol) and tributyltin hydride (23.6 g, 81.1 mmol). The reaction is cooled using an ice bath and the solution is

---

* The checkers added the $NbCl_5$ via a powder addition funnel.

† If the reaction is to be filtered in a dry box, we have found it convenient to let the reaction stand for 30 min, allowing the product to settle, and then removing a large portion of the solvent (~ 2.5 L) via cannula, directly into a waste bottle.

‡ The checkers obtained 240 g (88%).

stirred vigorously while the tributyltin hydride solution is added over 0.5 h. After the addition is completed, the mixture is stirred an additional 45 min and then filtered. The solid is washed with toluene (2 × 150 mL) and pentane (2 × 150 mL) and dried *in vacuo* (6 h, < 1 mm) to give 36.3 g (80%)* of a red-orange powder. Melting point (sealed capillary) = 97°C dec.; IR (cm^{-1}, Nujol mull) ν 990, 816.

Anal. The checkers analyzed for niobium. Calcd.: Nb, 16.69. Found: Nb, 16.9.

Properties

The product has been previously prepared by reaction of $NbBr_4$ with tetrahydrofuran.[1i] This complex must be stored in a low temperature freezer ($\leqslant -30°C$).

D. $NbBr_3(dme)$

$$NbBr_5 + 2\,Bu_3SnH \xrightarrow[-78°C]{DME} NbBr_3(dme) + 2\,Bu_3SnBr + H_2$$

A three-necked 1-L flask is fitted with an overhead mechanical stirrer and a nitrogen inlet adapter and charged with dry 1,2-dimethoxyethane (600 mL) and tributyltin hydride (59.0 g, 0.203 mol). A 250-mL flask containing niobium pentabromide (50.0 g, 0.102 mol) is attached to the setup via a piece of Latex tubing (30 cm long, 1.8 cm in diameter) that is adapted with two male 24/40 joints on each end. The reaction mixture is cooled to ∼ − 78°C (Dry Ice/2-propanol) and the solution is stirred vigorously while the $NbBr_5$ is added in portions over a 30-min period.†

■ **Caution.** *Vigorous stirring is essential to ensure that any NbBr$_5$ clinging near the top of the ground-glass joint is washed into the reaction mixture.*

After the addition is complete, the mixture is left to stir with no further addition of Dry Ice to the cold bath. After 3.5 h, the bath is removed and the material is allowed to warm to ambient temperature (∼ 1 h). During this time the heterogeneous reaction mixture turns from a brownish hue to a light purple. The solution is filtered and the resulting solid is washed with 1,2-dimethoxyethane (2 × 150 mL) and pentane (2 × 150 mL) and dried *in vacuo* (12 h at < 1 mm). Yield, 37.8 g (88%)‡ of a bright purple, free-flowing solid is

---

* The checkers obtained 39 g (86%).

† The checkers added the $NbBr_5$ via a powder addition funnel.

‡ The checkers ran this reaction on four times this scale and obtained 172 g (86%) of product.

obtained. Melting point (sealed capillary) = 130–160°C dec.; IR (cm^{-1}, Nujol mull) v 1070, 1021, 1006, 978, 845.

Anal. Calcd. for $C_4H_{10}Br_3O_2Nb$: C, 11.36; H, 2.38; Br, 56.70. Found: C, 11.34; H, 2.33; Br, 56.70; N, 0.00. The checkers analyzed for niobium. Calcd.: Nb, 21.98. Found: Nb, 21.9.

Properties

The molecular structure of $NbBr_3$(dme) is currently unknown. The solid appears to be stable at room temperature when stored under an inert atmosphere. However, storage in a low-temperature freezer ($\leq -30°C$) is recommended.

References and Notes

1. For examples, see the following works and references cited therein: (a) F. A. Cotton, *Polyhedron*, **6**, 667 (1987); (b) J. A. M. Canich and F. A. Cotton, *Inorg. Chem.*, **26**, 3473 (1987); (c) J. A. M. Canich and F. A. Cotton, *ibid.*, *idem.*, p. 4236; (d) F. A. Cotton, M. P. Diebold, and W. J. Roth, *Inorg. Chim. Acta*, **105**, 41 (1985); (e) J. L. Templeton, W. C. Dorman, J. C. Clardy, and R. E. McCarley, *Inorg. Chem.*, **17**, 1263 (1978); (f) L. E. Manzer, *ibid.*, **16**, 525 (1977); (g) D. A. Miller and R. D. Bereman, *Coord. Chem. Rev.*, **9**, 107 (1972/73); (h) E. T. Maas, Jr. and R. E. McCarley, *Inorg. Chem.*, **12**, 1096 (1973); (i) G. W. A. Fowles, D. J. Tidmarsh, and R. A. Walton, *ibid.*, **8**, 631 (1969).
2. (a) E. J. Roskamp, P. S. Dragovich, J. B. Hartung, Jr. and S. F. Pedersen, *J. Org. Chem.*, **54**, 4736 (1989); (b) J. B. Hartung, Jr. and S. F. Pedersen, *J. Am. Chem. Soc.*, **111**, 5468 (1989); (c) E. J. Roskamp and S. F. Pedersen, *ibid.*, **109**, 6551 (1987); (d) *ibid.*, *idem.*, p. 3152; (e) A. C. Williams, P. Sheffels, D. Sheehan, and T. Livinghouse, *Organometallics*, **8**, 1566 (1989).
3. J. B. Hartung, Jr. and S. F. Pedersen, *Organometallics*, **9**, 1414 (1990).
4. S. F. Pedersen, unpublished results.
5. K. Hayashi, J. Iyoda, and I. Shiihara, *J. Organomet. Chem.*, **10**, 81 (1967).
6. L. E. Manzer, *Inorg. Synth.*, **21**, 135 (1982).

29. CHROMIUM(V) FLUORIDE AND CHROMIUM(VI) TETRAFLUORIDE OXIDE

Submitted by S. P. MALLELA* and JEAN'NE M. SHREEVE*
Checked by DARRYL D. DesMARTEAU†

A. CHROMIUM(V) FLUORIDE

$$2CrO_2F_2 + 3F_2 \xrightarrow{190°} 2CrF_5 + O_2$$

Chromium(V) fluoride has been prepared by the static fluorination of chromium powder by elemental fluorine at 400° and 200 atm[1] and by the high temperature, high-pressure static fluorination of chromium(III) fluoride.[2,3] The procedure described below involves fluorination of chromium(VI) difluoride dioxide, CrO_2F_2, by elemental fluorine under rather mild conditions, which is the most convenient method. Chromium(V) fluoride is a useful precursor to a large number of complex salts that contain CrF_6^- and CrF_7^{2-}.

Procedure

■ **Caution.** *Fluorine is a highly oxidizing and very reactive gas. It must not be allowed to come in contact with easily oxidized materials or any organic compounds. All metal components, especially valves, must be free of hydrocarbon grease before use with fluorine. Eye protection must be worn, and skin and clothing should be protected with gloves and a lab coat. Information is available on the safe handling of fluorine and on the treatment in the event of personal exposure.*[4]

A 100-mL Hoke‡ spun Monel vessel (rated for 5000 psi) prepassivated with either F_2 or ClF and equipped with a Whitey stainless steel valve (SSD IRM4) (Seattle Valve§) is loaded with 15 mmol (1.83 g) of CrO_2F_2[5] (by PVT techniques) using a well-constructed Monel or nickel metal vacuum line[6] (prepassivated with ClF) and equipped with Whitey valves and Swagelok connections at − 196° (liquid N_2). An excess of fluorine (Matheson¶) (120 mmol) is introduced into the Monel vessel by a pressure difference

* Department of Chemistry, University of Idaho, Moscow, ID 83843.
† H. L. Hunter Chemistry Laboratory, Clemson University Clemson, SC 29634.
‡ Hoke, Inc., One Tenakill Park, Cresskill, NJ 07626.
§ Seattle Valve and Fitting Co., 13417 N.E. 20th St., Bellevue, WA 98005.
¶ Matheson Gas Products, P.O. Box 85, 932 Paterson plank Road, East Rutherford, NJ 07073.

technique[6] and the valve closed. The line is evacuated slowly through a soda lime trap to destroy the residual fluorine.

■ **Caution.** *The reaction between fluorine and soda lime is highly exothermic. Therefore, the fluorine must be removed very slowly.*

The Monel vessel is allowed to warm to room temperature and then heated at 190° for 46 h. The vessel is then allowed to cool to ambient temperature and is cooled finally to $-196°$. The unreacted F_2 and other volatile compounds are removed under dynamic vacuum first at $-196°$ (the pump must be protected by a fluorine scrubber that contains soda lime or alumina) and later at $-60°$ (Dry Ice–ethanol slush).

At low temperatures ($\sim 170°$) and low fluorine pressure (~ 77 mmol) small amounts of unreacted CrO_2F_2 are recovered. Under the conditions described, CrO_2F_2 is converted quantitatively to CrF_5 (2.12 g). The purity is checked by electron diffraction measurements[7] and by infrared spectra (solid as well as gas).

Properties

Chromium(V) fluoride is a red sticky solid that melts at 35°. The infrared spectrum of liquid CrF_5 has strong absorption bands at 799, 730, 684, and 533 cm^{-1}.[8,9] The liquid-phase Raman spectrum[8,9] has bands at 780, 755, 668, 390, 320, 305, and 204 cm^{-1}. The gas-phase (2 torr) IR has a very strong absorption band at 771 cm^{-1} with a shoulder at 800 cm^{-1}. The mass-spectrometric cracking pattern of CrF_5 shows several peaks; these are assigned to single-charged positive ions: CrF_5^+, CrF_4^+, CrF_3^+, CrF_2^+, and Cr^+.[10]

B. CHROMIUM(VI) TETRAFLUORIDE OXIDE

$$2CrO_2F_2 + 2F_2 \xrightarrow[200°]{CsF} 2CrF_4O + O_2$$

Chromium(VI) tetrafluoride oxide CrF_4O was first synthesized via fluorination of either metallic chromium or CrO_3,[10,11] and the route has been modified to use CrO_3 at lower temperature.[12] Very recently, fluorination of CrO_2F_2 with KrF_2 in anhydrous HF was reported.[13] However, this route requires the use of a rather exotic fluorinating agent. The procedure described below involves the catalytic fluorination of CrO_2F_2 in the presence of CsF.

Procedure

■ **Caution.** *Fluorine is a highly oxidizing and extremely reactive gas. Easily oxidized materials and all organic compounds must be absent. Eye* *protection must be worn. Skin and clothing must be protected with gloves and lab coat.*

In a dry box, about 1 mg of dry CsF is introduced into a 150-mL Monel vessel (test pressure 5000 psi) (Hoke) (prepassivated with either F_2 or ClF) equipped with a stainless steel Whitey valve (SSD, IRM4) (Seattle Valve). The vessel is connected to a well-constructed Monel or nickel vacuum line prepassivated with ClF,[6] and is evacuated and cooled to $-196°$. Then 10 mmol (1.22 g) of CrO_2F_2[5] is transferred (using *PVT* techniques) into the vessel. Then ~ 54 mmol (2.05 g) of fluorine (Matheson) is introduced using a pressure-difference technique.[6] After closing the valve, the vessel is allowed to warm to ambient temperature and the vessel is placed in a tube furnace. The valve is cooled by circulating cold tap water through a coiled copper tube placed around it. The vessel is heated at $\sim 200°$ for 62 h. In the unlikely event of a fluorine leak, the reaction is carried out in an oven in a good fume hood. After the reaction, the vessel is allowed to cool to ambient temperature and attached to the vacuum line and cooled to $-196°$. The unreacted fluorine is slowly removed under dynamic vacuum via a soda-lime trap. All other volatile compounds are removed at $-60°$. The majority of the product is found in the vicinity of the valve. The dark red solid CrF_4O remaining in the vessel is vacuum-transferred at $25°$ to a Kel-F tube (Zeus*) equipped with a Whitey valve. The product may also be removed by scraping it into a Kel-F container in a dry box. The yield is 1.3 g (9.0 mmol, 90%).

Anal.[12] Calcd. for CrF_4O: Cr, 36.1; F, 52.8. Found: Cr, 35.9; F, 53.2.

Properties

Chromium(VI) tetrafluoride oxide (mp 55°) is a dark red solid that exists in equilibrium with its purple vapor at $25°$.[10, 13, 14] It is highly soluble in BrF_5, SO_2ClF, and anhydrous HF.[13] The gas-phase IR spectrum[13] has a very strong characteristic absorption band at 755 cm^{-1} assigned to $\gamma_7(E)$. Medium absorption bands are at 1028 $\gamma_1(A_1)$ and at 696 $\gamma_2(A_1)$ cm^{-1}. The mass spectrometric cracking pattern is $CrOF_4^+$, $CrOF_3^+$, $CrOF_2^+$, CrF_2^+, $CrOF^+$, CrO^+, and Cr^+.[10]

References

1. O. Glemser, H. W. Roesky, and K. H. Hellberg, *Angew. Chem. Int. Ed. Engl.*, **2**, 266 (1963).
2. J. Slivnik and B. Zemva, *Z. Anorg. Allgem. Chem.*, **385**, 137 (1971).

*Zeus Industrial Products, Inc., Thompson St., Raritan, NJ 08869.

3. E. G. Hope, P. J. Jones, W. Levason, J. S. Ogden, M. Tajik, and J. W. Turff, *J. Chem. Soc. Dalton Trans.*, **1985**, 1443.
4. T. A. O'Donnell, in *Comprehensive Inorganic Chemistry*, Vol. II, A. F. Trotman-Dickson, (ed.), Pergamon Press, London, 1973, pp. 1014–1019, 1028–1030; *Matheson Gas Data Book*, 6th ed., Matheson Co., East Rutherford, NJ, 1980.
5. G. L. Gard, *Inorg. Synth.*, **24**, 67 (1986).
6. D. F. Shriver, *The Manipulation of Air-Sensitive Compounds*, McGraw-Hill, New York, 1986; K. O. Christie, R. D. Wilson, and C. J. Schack, *Inorg. Synth.*, **24**, 5 (1986).
7. E. J. Jacob, L. Hedberg, K. Hedberg, H. Davis, and G. L. Gard, *J. Phys. Chem.*, **88**, 1935 (1984).
8. R. Bougon, W. W. Wilson, and K. O. Christe, *Inorg. Chem.*, **24**, 2286 (1985).
9. S. D. Brown, T. M. Loehr, and G. L. Gard, *J. Chem. Phys.*, **64**, 260 (1976).
10. A. J. Edwards, W. E. Falconer, and W. A. Sunder, *J. Chem. Soc. Dalton Trans.*, **1974**, 541.
11. A. J. Edwards, *Proc. Chem. Soc. (London)*, **1963**, 205.
12. E. G. Hope, P. J. Jones, W. Levason, J. S. Ogden, M. Tojik, and J. W. Turff, *J. Chem. Soc. Dalton Trans.*, **1985**, 529.
13. K. O. Christe, W. W. Wilson, and R. A. Bougon, *Inorg. Chem.*, **25**, 2163 (1986).
14. J. Huang, K. Hedberg, J. M. Shreeve, and S. P. Mallela, *Inorg. Chem.*, **27**, 4633 (1988).

30. AMMONIUM HEXACHLOROMOLYBDATE(III)

Submitted by TAKASHI SHIBAHARA* and MIKIO YAMASAKI*
Checked by ABDUL K. MOHAMMED† and ANDREW W. MAVERICK†

Two methods have been used to obtain hexachloromolybdate(III), $[MoCl_6]^{3-}$, and/or aquapentachloromolybdate(III), $[MoCl_5(H_2O)]^{2-}$: electrolytic reduction of MoO_3 in concentrated hydrochloric acid[1] (method 1) and oxidation of tetrakis(acetato)dimolybdenum, $Mo_2(O_2CCH_3)_4$, in concentrated hydrochloric acid[2] (method 2). Method 1 requires an electrolytic apparatus and complicated procedures, and the yield is not high. Method 2 is simpler than 1 and gives good yield; however, the starting material $Mo_2(O_2CCH_3)_4$ is expensive and must be prepared from $Mo(CO)_6$.

We report here a facile preparation of ammonium hexachloromolybdate(III). This method employs reduction of ammonium molybdate by tin metal in concentrated hydrochloric acid without air-free techniques.

Material

Ammonium molybdate $(NH_4)_6Mo_7O_{24} \cdot 4H_2O$ was purchased from Nacalai Tesque. The use of sodium molybdate instead of ammonium molybdate gives

* Department of Chemistry, Okayama University of Science, 1-1 Ridai-cho, Okayama 700, Japan.
† Department of Chemistry, Louisiana State University, Baton Rouge, LA 70803.

a somewhat impure product, due, probably, to contamination with NaCl. The use of MoO_3 and H_2MoO_4 is inadequate, since these scarcely dissolve in concentrated HCl.

Procedure

$$2[Mo_7O_{24}]^{6-} + 96H^+ + 84Cl^- + 21Sn \rightarrow$$
$$14[MoCl_6]^{3-} + 21Sn^{2+} + 48H_2O$$

■ **Caution.** *Hydrogen chloride gas is toxic. Operations should be carried out in a well-ventilated hood.*

First, 5 g (4.05 mmol) of $(NH_4)_6Mo_7O_{24} \cdot 4H_2O$ is dissolved in concentrated HCl (100 mL) in an Erlenmeyer flask (200 mL). Tin metal (shot; 20 g, 0.17 mol) is added, and the loosely stoppered flask is heated in a boiling-water bath for 8 min with vigorous stirring. The color of the solution turns deep red brown via yellow green, deep green, and red brown. While heating, a yellow precipitate appears within a minute, which disappears, however, within 8 min on continuous heating. After further heating for 3 min, ammonium chloride (5.0 g, 0.094 mol) is added and dissolved by additional heating (2 min). The solution is cooled in an ice bath for 1 min (temperature of the solution is $\sim 55°C$) and filtered immediately using a sintered-glass filter (No. 3) by means of suction. Longer cooling will give rise to coprecipitation of the desired product. The amount of unchanged tin metal is ~ 13 g, and the yellow precipitate (~ 1 g) is discarded. The filtrate which is transferred into an Erlenmeyer flask (100 mL), is cooled with ice again and a fairly vigorous stream of gaseous hydrogen chloride is bubbled through the solution until it is saturated [~ 5 min; filtration at this stage gives ~ 4.3 g (42%) of the desired product], and then the flask is tightly stoppered. The solution is kept in a refrigerator for ~ 24 h and the red crystals that deposit are collected by filtration and washed with ethanol, then dried in a vacuum desiccator. Yield: ~ 6.5 g (64%).

Anal. Calcd. for $H_{12}N_3MoCl_6$: Mo, 26.44; N, 11.58; H, 3.33%. Found: Mo, 26.4; N, 11.45; H, 3.38%.

The filtrate contains an appreciable amount of Mo(III) species: addition of ethanol (~ 25 mL) followed by introduction of HCl gas (~ 5 min) with cooling in an ice bath gives an additional product after a week's storage in a refrigerator; it is somewhat impure, though. Yield: ~ 1.2 g (12% based on the starting material $(NH_4)_6Mo_7O_{24} \cdot 4H_2O$).

Properties

The solid is stable in air; the solution, however, is air-oxidized, as described previously.[1, 2] Somewhat different electronic spectra have been reported on $[MoCl_6]^{3-}$,[3] and our finding $[\lambda_{max}, nm\ (\varepsilon, M^{-1}\ cm^{-1})\ 419\ (40.5), 523\ (30.5)$, and 677 (1.5)] obtained in concentrated HCl solution under nitrogen atmosphere*, is very similar to the reported values in refs. 3a and 3b. Even a very small amount of oxygen will cause the spectrum to change around 310 nm. However, the shapes of the three peaks change only a little, and the spectrum can be measured in the air, if it is measured promptly. In $6\ M$ HCl, $[MoCl_6]^{3-}$ undergoes aquation immediately to give $[MoCl_5(H_2O)]^{2-}$.[3b]

References

1. K. H. Lohmann and R. C. Young, *Inorg. Synth.*, **4**, 97 (1953).
2. J. B. Brencic and F. A. Cotton, *Inorg. Synth.*, **13**, 170 (1972).
3. (a) Q. Yao and A. W. Maverick, *Inorg. Chem.*, **27**, 1669 (1988); (b) W. Andruchow, Jr. and J. DiLiddo, *Inorg. Nucl. Chem. Lett.*, **8**, 689 (1972); (c) C. Furlani and O. Piovesana, *Mol. Phys.*, **9**, 341 (1965); (d) J. Lewis, R. S. Nyholm, and P. W. Smith, *J. Chem. Soc. A*, **1969**, 57.

31. DIOXOBIS(2,4-PENTANEDIONATO)MOLYBDENUM(VI), MoO₂(acac)₂ AND μ-OXODIOXOTETRAKIS(2,4-PENTANEDIONATO)DIMOLYBDENUM)(V), Mo₂O₃(acac)₄

**Submitted by MUKUL C. CHAKRAVORTI† and
DEBASIS BANDYOPADHYAY†**
Checked by MALCOLM H. CHISHOLM‡ and CHARLES E. HAMMOND‡

Extensive studies on the preparative and structural aspects of 2,4-pentanedionato and other β-diketonato complexes of transition metals have been made over the last 100 years. However, such complexes of molybdenum have not received much attention. The study of the coordination chemistry of molybdenum in general has had an impetus in recent years because of its key role in many biological systems. Two 2,4-pentanedionato complexes of oxomolybdenum(VI, V) are well characterized. $MoO_2(acac)_2$ has been prepared starting from either MoO_3 or $(NH_4)_6Mo_7O_{24}\cdot 4H_2O$.[1] A method for

---

* In order to get rid of trace amounts of oxygen in nitrogen gas to be used to purge air, nitrogen gas is passed through a bottle containing concentrated HCl and $SnCl_2\cdot 2H_2O$ ($\sim 0.5\ M$) prior to use.

† Department of Chemistry, Indian Institute of Technology, Kharagpur 721 302, India.

‡ Department of Chemistry, Indiana University, Bloomington, IN 47405.

the preparation of this complex by refluxing MoO_3 with 2,4-pentanedione for 18 h has appeared[2] in *Inorganic Syntheses*. X-ray crystal structure[3] and mass spectrometric studies[4] of the complex have been made. The binuclear complex $Mo_2O_3(acac)_4$ has been prepared by the reduction of $MoO_2(acac)_2$ with zinc in 2,4-pentanedione,[5] heating $MoO_2(acac)_2$ with 2,4-pentanedione in a sealed tube,[6] or refluxing $MoO(OH)_3$ with 2,4-pentanedione in an inert atmosphere.[7] However, no detailed method of preparation is available. The binuclear moiety $Mo_2O_3^{4+}$ offers for study a variety of structural arrangements.

A rapid method for the preparation of $MoO_2(acac)_2$ at room temperature starting from ammonium *para*-molybdate, $(NH_4)_6Mo_7O_{24} \cdot 4H_2O$, is described below. Two methods for the preparation of $Mo_2O_3(acac)_4$ are also given. The first method is based on the refluxing of $MoO(OH)_3$ with 2,4-pentanedione, while the other uses the electrochemical reduction of $MoO_2(acac)_2$.

A. DIOXOBIS(2,4-PENTANEDIONATO)MOLYBDENUM(VI), $MoO_2(acac)_2$

$$(NH_4)_6Mo_7O_{24} + 14Hacac + 6H^+ \rightarrow 7MoO_2(acac)_2 + 6NH_4^+ + 10H_2O$$

Procedure

A 3.0-g (2.4-mmol) quantity of powdered ammonium *para*-molybdate, $(NH_4)_6Mo_7O_{24} \cdot 4H_2O$, is dissolved in 6.0 mL of aqueous ammonia (15%) by stirring.* To this clear solution 7.0 mL (6.8 mmol) of 2,4-pentanedione is added with stirring. A light yellow solution results. A 5.0-mL quantity of concentrated nitric acid (density 1.42) is added slowly to the mixture with vigorous stirring. The mixture warms and turns greenish yellow in color. The solution is allowed to cool to room temperature when a yellow precipitate appears. The solid is separated after half an hour by filtration under suction. It is washed with water and then with ethanol. The yield is about 4.0 g (72%).† The compound is of moderate purity and can be used for ordinary purposes. It is recrystallised by dissolving 2.0 g of the crude product in 6.0 mL of hot (about 90°) 2,4-pentanedione followed by cooling to − 10° in a freezer for 18 h. The yellow crystalline substance is separated by filtration under suction, washed with ethanol, and dried *in vacuo*. The yield of the recrystallized

---

* The checkers used 6.0 mL of 58% aqueous ammonia. All the solid did not dissolve. After the addition of 2,4-pentanedione with agitation, some precipitate is formed which dissolved on the addition of 5.0 mL of concentrated nitric acid.

† The checkers report a yield of 93% on cooling the hot solution in a freezer at − 10° for 18 h.

product is 1.8 g. (The checkers note that all the 2,4-pentanedione is not removed and the product slowly becomes tinged with red.)

Anal. Calcd. for $MoO_2(acac)_2$: Mo, 29.4; C, 36.8; H, 4.3. Found: Mo, 29.2; C, 36.5; H, 4.1.

Properties

The compound is a light yellow crystalline substance, which decomposes over time when stored in air under fluorescent light, affording a light blue solid. But no decomposition occurs when stored *in vacuo* or under nitrogen. It melts at 179°. It is insoluble in water, but very slightly soluble in ethanol and chloroform. In the infrared spectrum, it gives strong bands at 930 and 900 cm^{-1}, showing the symmetric and antisymmetric Mo=O stretches, respectively. Other important bands occur at 1560, 1500, 1350, 1260, 1015, 795, 665, 570, and 445 cm^{-1}. The visible, ultraviolet, and NMR spectra have been reported.[8] ^{1}H NMR spectrum in CD_2Cl_2 (yellow solution) at 25° has signals at δ 2.12, 2.14 and 5.83 ppm. The ^{13}C NMR spectrum in CD_2Cl_2 at 25° has signals at δ 196.7, 185.2, 104.9, 28.0, and 25.7 ppm.

B. μ-OXODIOXOTETRAKIS (2,4-PENTANEDIONATO) DIMOLYBDENUM(V), $Mo_2O_3(acac)_4$

$$2MoO(OH)_3 + 4Hacac \rightarrow Mo_2O_3(acac)_4 + 5H_2O$$

Procedure

Chemical Method. The compound $MoO(OH)_3$ has been prepared by the method given in *Inorganic Syntheses.*[9] After 1.0 g of finely ground $MoO(OH)_3$ (6.1 mmol) and 6.0 mL (58.5 mmol) of 2,4-pentanedione are placed in a 25-mL round-bottomed flask, the mixture is heated under reflux for 6 h, during which time an intense brown solution forms. This is filtered warm by suction to remove a small amount of undissolved $MoO(OH)_3$.* The filtrate is kept in a freezer at $-10°$ overnight and then the brown micro-crystalline precipitate is separated by filtration under suction. This is washed with water and then with ethanol. The product is dried *in vacuo*. The yield is

---

* The checkers noted that the residue contains both the product and unreacted $MoO(OH)_3$ after hot filtration. The residue was digested with 10 mL of 2,4-pentanedione at 120° and the filtrate cooled in a freezer. After washing and drying as described above, 0.9 g of the compound was obtained.

about 1.2 g (61%). The compound is of moderate purity. It can be re-crystallized by dissolving in 10 mL of hot (~ 120°) 2,4-pentanedione fol-lowed by cooling in a freezer at − 10° overnight. The product is separated by filtration under suction, washed with ethanol, and dried *in vacuo*. The yield of the recrystallized product is 0.6 g.

Electrochemical Method. The electrolytic cell used for this purpose is a 100-mL tall-form beaker fitted with a rubber stopper through which four glass tubes are inserted. Two platinum foils each connected with platinum wire electric leads are inserted through two of the tubes. (The checkers used a platinum gauze as cathode and a nichrome coiled wire as anode.) The other two tubes allow input and exit of nitrogen gas (Fig. 1).

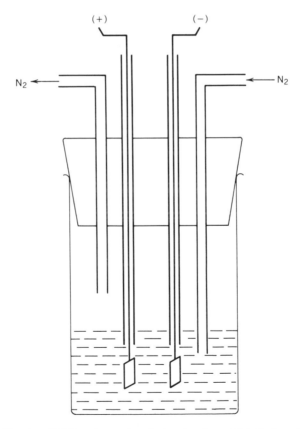

Fig. 1.   Cell for electrochemical reduction of $MoO_2(acac)_2$.

A 2.0-g (6.1-mmol) quantity of $MoO_2(acac)_2$ is placed in the electrolytic cell. To it, 5.0 mL of 2,4-pentanedione, 40 mL of acetonitrile and 0.05 g of tetraethylammonium perchlorate are added. A clear yellow solution is obtained on stirring. The electrodes are dipped in the solution, and nitrogen gas is passed through the inlet tube. The platinum foils are then connected to the two terminals of a variable-voltage power supply[10] and a DC voltage of 25 V is applied, giving a current of 60 mA. The solution gradually turns brown in the vicinity of the cathode. Finally, the whole solution becomes dark brown. The electrolysis is continued for 6 h. After the circuit is switched off, the solution (with a little brown precipitate deposited inside the beaker) is slowly evaporated on a water bath to one-third of the original volume. The nitrogen supply is continued during this evaporation. The rubber stopper is removed, and the solution is allowed to cool to room temperature. A deep brown microcrystalline precipitate separates from the solution. This is removed by filtration, washed, dried, and recrystallized as described in the chemical method. Yield of the recrystallized sample is about 0.7 g (36%).

Anal. Calcd. for $Mo_2O_3(acac)_4$: Mo, 30.1; C, 37.7; H, 4.4. Found: Mo, 29.8; C, 37.4; H, 4.5.

For the analysis of molybdenum the samples are decomposed by repeated fuming with a few drops of nitric acid each time in a platinum crucible and finally heated to about 500° and weighed as MoO_3. Carbon and hydrogen are determined by standard microchemical methods.

Properties

The compound is a dark brown crystalline substance. It melts at 169°. It is insoluble in water and common organic solvents except for a slight solubility in chloroform. It is very weakly paramagnetic (μ_{eff} is 0.33 BM at 24°). An Mo—O—Mo bridge was proposed[11] to account for the observed low magnetic moment. The infrared spectrum gives a band at 950 cm^{-1} due to Mo—O$_t$ stretching. Other major IR bands occur at 1550, 1350, 1270, 1020, 930, 800, 675, and 445 cm^{-1}. The visible, UV, ESR, and mass spectra were reported earlier.[4,7]

References

1. M. M. Jones, *J. Am. Chem. Soc.*, **81**, 3188 (1959).
2. W. C. Fernelius, K. Terada, and B. E. Bryant, *Inorg. Synth.*, **6**, 147 (1960).
3. B. Kamenar, M. Penavic, and C. K. Prout, *Cryst. Struct. Commun.*, **2**, 41 (1973).
4. M. C. Chakravorti, D. Bandyopadhyay, and M. K. Chaudhuri, *Int. J. Mass. Spectr. Ion Processes*, **68**, 1 (1986).

5. D. Grdenic and B. Kopar-Colig, *Proc. Chem. Soc.*, **1963**, 308.
6. H. Gehrke, Jr. and J. Veal, *Inorg. Chim. Acta*, **4**, 623 (1969).
7. J. Sobczak and J. J. Zio'lkowski, *Transition Met. Chem.*, **8**, 333 (1983).
8. F. W. Moore and R. E. Rice, *Inorg. Chem.*, **7**, 2510 (1968).
9. M. C. Chakravorti and S. C. Pandit, *Inorg. Synth.*, **21**, 170 (1982).
10. D. C. Power Supply, Aplab, India, Model 7612.
11. M. L. Larson and F. W. Moore, *Inorg. Chem.*, **2**, 881 (1963).

32. DECAKIS(ACETONITRILE)DIMOLYBDENUM(II) TETRAFLUOROBORATE(1−)

F. ALBERT COTTON* and KENNETH J. WIESINGER*

■ **Caution.** *The acetonitrile must be freshly and rigorously purified. Extreme caution should be used in handling the tetrafluoroboric acid as it is corrosive.*

$$Mo_2(O_2CCH_3)_4 + H[BF_4] \cdot Et_2O \xrightarrow[\text{heat}]{CH_3CN/CH_2Cl_2}$$

$$Mo_2(NCCH_3)_8 + [ax\text{-}CH_3CN]_2[BF_4]_4 \cdot 2CH_3CN$$

drying, 8 h at 35°C

$$\downarrow$$

$$[Mo_2(NCCH_3)_8(ax\text{-}CH_3CN)_{0.5}][BF_4]_4$$

(1)

Complexes with multiple, especially quadruple, bonds between two metal atoms have been the subject of much research for at least 25 years.[1] There have been numerous reports of the synthesis of $[M_2]^{n+}$ species that contain weakly coordinating neutral and/or anionic ligands. Several examples of this class of species have been reported for M = Mo: $[Mo_2(H_2O)_4(CF_3SO_3)_2]$ $[CF_3SO_3]_2$,[2] $[Mo_2(NCCH_3)_8][CF_3SO_3]_4$,[2] $cis\text{-}[Mo_2(O_2CCH_3)_2(NCCH_3)_6]$ $[X]_2$ (X = BF_4^-, $CF_3SO_3^-$),[3,4] $[Mo_2(EtO_2CCH_3)_4][CF_3SO_3]_4$,[5] $[Mo_2(CF_3SO_3)_4$,[5] $[Mo_2(en)_4][Cl]_4$,[6,7] Mo_2^{4+}(aq),[7] $trans\text{-}[Mo_2(O_2CCH_3)_2(dmpe)_2]$ $[BF_4]_2$,[8] and $[Mo_2(O_2CCH_3)_2(NCCH_3)_5][BF_3OH]_2$.[9] It is unfortunate that

* Laboratory for Molecular Structure and Bonding, Department of Chemistry, Texas A&M University, College Station, TX 77843.

many of these reported procedures lead to impure products or uncertain formulations, and only in the cases of $[Mo_2(O_2CCH_3)_2(NCCH_3)_6][X]_2$ (X = BF_4^-, $CF_3SO_3^-$),[3,4] and $[Mo_2(O_2CCH_3)_2(dmpe)_2][BF_4]_2$[8] were the actual structures determined crystallographically. Some analogous studies have been reported for the $[Ru_2]^{n+}$ and $[Rh_2]^{n+}$ cations.[10-13] The most recent examples are $[Rh_2(NCCH_3)_{10}][BF_4]_4$,[14] and $[Rh_2(H_2O)_2(NCCH_3)_8]$ $[PF_6]_4 \cdot 2H_2O$.[15]

We report here the synthesis of $[Mo_2(NCCH_3)_8(ax\text{-}NCCH_3)_{0.5}]$-$[BF_4]_4$ (**1**). [Abbreviations used here are: ax = axial and dmpe = 1,2-bis-(dimethylphosphino)ethane.]

Procedures

Tetrakis(acetato)dimolybdenum(II) is prepared as reported.[16] Acetonitrile (2L) is passed through a column (3 cm × 40 cm) of alumina gel (dried at 120°C for 2 days) and then distilled from calcium hydride under an atmosphere of dry nitrogen. Dichloromethane is distilled from phosphorus pentoxide under nitrogen. Tetrafluoroboric acid is used as purchased from Aldrich as an 85% $H[BF_4]$ diethyl ether solution. It is handled with a pretreated syringe in a similar manner to alkyl-lithium reagents. All manipulations arc performed using standard vacuum-line and Schlenk techniques under a dry and oxygen free atmosphere of argon. All glassware is oven dried at 120°C for > 18 h, assembled while hot and then evacuated. The reaction is carried out in a vented fume hood.

The reaction is performed in a three-necked 250-mL flask equipped with a 5-in. spiral-coil condenser.

Tetrakis(acetato)dimolybdenum(II) (1.20 g, 2.80 mmol) is dissolved in acetonitrile (20 mL) and CH_2Cl_2 (100 mL). To this vigorously stirring yellow solution is added $H[BF_4] \cdot Et_2O$ (6.0 mL, 85% $H[BF_4]$ solution) with a subsequent color change to a red solution. The solution is continuously stirred at room temperature and progresses through color changes from red to purple to blue-purple within 30 min of the acid addition. During this period, a large crop of bright blue-purple microcrystals precipitates from the reaction solution. The reaction mixture is then heated to reflux temperature and gently refluxed for 40 min. Since the blue-purple material forms so quickly, the solution is refluxed for a short period of time in order to coagulate the precipitate and remove any impurities. As the reaction mixture is cooled to room temperature, the solution becomes less intensely blue as a result of product precipitation from the solution. The supernatant liquid is decanted off.

The solid is rinsed with dichloromethane (4 × 10 mL) until the wash solution is clear. The solid is further rinsed with diethyl ether (3 × 10 mL) to remove any residual $H[BF_4] \cdot Et_2O$. On drying under vacuum for 8 h at 35°C, 2.24 g (93% yield) of a bright blue solid, $[Mo_2(NCCH_3)_8$ $(ax-NCCH_3)_{0.5}][BF_4]_4$ (1), is isolated. Prolonged drying at elevated temperatures causes decomposition to a black intractable material.

The purest form of the material can be obtained by dissolving 200 mg of the product in acetonitrile (15 mL), filtering, and precipitating out large dark blue crystals by slow diffusion of diethyl ether or dichloromethane (20 mL) into the acetonitrile layer. These crystals have been shown to be $[Mo_2(NCCH_3)_8(ax-NCCH_3)_2][BF_4]_4 \cdot CH_3CN$ by X-ray crystallography.[17] We believe that the species in solution is $[Mo_2(NCCH_3)_8(ax-NCCH_3)_2][BF_4]_4$.

Properties

The compound $[Mo_2(NCCH_3)_8(ax-NCCH_3)_2][BF_4]_4$ is not stable in acetonitrile in air, as the solution turns purple and then decomposes to a brown color. Furthermore, the solid is hygroscopic. The solid dissolves in acetonitrile and ethanol:acetonitrile solutions. It is virtually insoluble in other common solvents and slowly decomposes in acetone. The pure compound is easily identified by UV–VIS, ^1H NMR, and IR spectra. The electronic spectrum has one absorption at 597 nm. The ^1H NMR (CD_3CN, 22°C) spectrum contains only one singlet at 1.95 ppm due to free CH_3CN. This indicates that the CH_3CN ligands in this complex are labile and exchange with the CD_3CN solvent. The infrared spectrum (CH_3CN solution, cm^{-1}) contains three stretches at 2360 (s), 2338 (m), and 2306 (w), and a strong BF_4^- band at 1071 cm^{-1}. Similar bands are seen at 2325 (m), 2293, (s) 2247 (w), and 1059 (vs) cm^{-1} for the Nujol mull.

References

1. F. A. Cotton and R. A. Walton, *Multiple Bonds between Metal Atoms*, Wiley-Interscience: New York, (1982); and references cited therein.
2. J. M. Mayer and E. H. Abbott, *Inorg. Chem.*, **22**, 2774 (1983).
3. G. Pimblett, D. Garner, and W. Clegg, *J. Chem. Soc. Dalton Trans.*, **1986**, 1257.
4. F. A. Cotton, A. H. Reid, Jr., and W. Schwotzer, *Inorg. Chem.*, **24**, 3965 (1985).
5. E. H. Abbott, F. Schoenewolf, Jr., and T. J. Backstrom, *J. Coord. Chem.*, **3**, 255 (1974).
6. A. R. Bowen and H. J. Taube, *J. Am. Chem. Soc.*, **93**, 3287 (1971).
7. A. R. Bowen and H. J. Taube, *Inorg. Chem.*, **13**, 2245 (1974).
8. L. J. Farrugia, A. McVitie, and R. D. Peacock, *Inorg. Chem.*, **27**, 1257 (1988).
9. J. Telser and R. S. Drago, *Inorg. Chem.*, **23**, 1798 (1984).
10. F. Maspero and H. Taube, *J. Am. Chem. Soc.*, **90**, 7361 (1968).
11. P. Legzdins, G. L. Rempel, and G. Wilkinson, *Chem. Comm.*, **1969**, 825.
12. P. Legzdins, R. W. Mitchell, G. L. Rempel, J. D. Ruddick, and G. Wilkinson, *J. Chem. Soc. (A)*, **1970**, 3322.

13. C. R. Wilson and H. Taube, *Inorg. Chem.*, **14**, 2276 (1975).
14. K. R Dunbar, *J. Am. Chem. Soc.*, **110**, 8247 (1988).
15. L. M. Dikareva, V. I. Andrianov, A. N. Zhilyaev, and I. B. Baranovskii, *Russ. J. Inorg. Chem.*, **34**, 240 (1989).
16. T. A. Stephenson, E. Bannister, and G. Wilkinson, *J. Chem. Soc.*, **1964**, 2538.
17. F. A. Cotton and K. J. Wiesinger, *Inorg. Chem.*, **30**, 871 (1991).

33. HEXAKIS(DIMETHYLAMIDO)DITUNGSTEN AND TUNGSTEN(IV) CHLORIDE

Submitted by MALCOLM H. CHISHOLM* and JAMES D. MARTIN*
Checked by JOHN E. HILL† and IAN P. ROTHWELL†

Crucial to the development of the chemistry of the triple bonds between molybdenum and tungsten have been high yield syntheses of the dinuclear complexes, $M_2(NMe_2)_6$, that provide excellent starting materials (see Scheme 1). These compounds sublime *in vacuo*, allowing a complete separation from the halide salts generated during their preparation. $Mo_2(NMe_2)_6$[1] may be prepared in high yields from the metathetic reaction involving $MoCl_3$ and $LiNMe_2$ according to the following equation:

$$MoCl_3 + 6LiNMe_2 \rightarrow Mo_2(NMe_2)_6 \qquad (1)$$

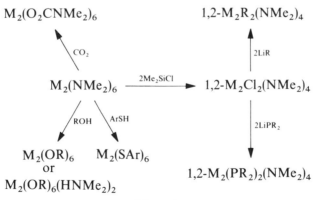

Scheme 1

* Department of Chemistry, Indiana University, Bloomington, IN 47405.
† Department of Chemistry, Purdue University, West Lafagette IN 47907.

For tungsten no simple halide, WCl_3, is known. Nevertheless, $W_2(NMe_2)_6$ can be prepared in high yields from the metathetic reactions involving WCl_4[2] or $NaW_2Cl_7(thf)_5$[3] (the product of a Na/Hg reduction of WCl_4) and $LiNMe_2$. Both reactions yield $W_2(NMe_2)_6$ in $> 70\%$ yield based on tungsten. However, we have found the former reaction, which we report here, to be more useful and to give less of the undesired $W(NMe_2)_6$ by-product. The yield and purity of $W_2(NMe_2)_6$ appear to be strongly correlated to the quality of the WCl_4 starting material. Our preparation of WCl_4 employs the reduction of WCl_6 by red phosphorus (Eq. 2), similar to that described previously.[4]

$$WCl_6 + \tfrac{2}{3}P \xrightarrow{\Delta} WCl_4 + \tfrac{2}{3}PCl_3 \qquad (2)$$

Our general procedures and techniques have been described in detail.[5] The starting materials and products are air- and moisture-sensitive; therefore, an inert atmosphere of dry and oxygen-free nitrogen is used throughout all experimental procedures. WCl_6 is used as purchased from Pressure Chemical Co. All hydrocarbon solvents are dried and distilled from sodium-benzophenone and stored over 4-Å molecular sieves under a nitrogen atmosphere. Lithium dimethylamide is prepared from $HNMe_2$ and *n*-butyl-lithium.

A. TUNGSTEN(IV) CHLORIDE, WCl_4

Tungsten(VI) chloride (50.0 g, 0.126 mol) and two-thirds of an equivalent of red phosphorous (2.60 g, 0.084 mol) are placed into chamber C of the reaction vessel shown in Fig. 1. The reaction vessel is then evacuated ($\sim 10^{-2}$ torr)

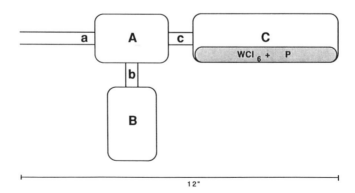

Fig. 1.   Three-chamber reaction vessel for preparation of WCl_4. Chambers A, B, and C are $1\tfrac{1}{2}$ in. diameter and constrictions a, b, and c are $\tfrac{1}{4}$ in. diameter.

and flame-sealed at constriction a. The two solids must be thoroughly mixed together and dispersed laterally in chamber C (see Fig. 1). Chamber C of the reaction vessel is placed into a tube furnace and heated to 225°C. Chambers A and B should remain at room temperature. As the reaction proceeds, PCl_3 begins to distill into chamber B. After 1 h the temperature is raised to 275°C. At this temperature impurities such as $WOCl_4$ (red), WO_2Cl_2 (yellow), WCl_5 (green), and any excess WCl_6 (purple) sublime into chambers A and B. It is helpful to periodically heat the constrictions between the three chambers using a gas torch to clear any buildup of sublimate. After 2 h at 275°C chamber B is cooled with liquid nitrogen and flame-sealed at constriction b. The reaction vessel (now only chambers A and C) is inserted farther into the tube furnace such that only about one-third of chamber A remains exposed and the furnace temperature is raised to 300°C. After 12 h, chamber A is flame-sealed at constriction c, leaving the black, nonvolatile powder, WCl_4 in chamber C (39.0 g, 0.120 mol, 95%). This material is pure enough for most purposes but has been found to contain up to 0.15% phosphorous.[4, 6]

Reviewers noted some reactions on mixing WCl_6 and red phosphorus and advise keeping the time between mixing the reactants and beginning to heat the reaction mixture to a minimum.

B.   HEXAKIS(DIMETHYLAMIDO)DITUNGSTEN, $W_2(NMe_2)_6$

A two necked 500-mL flask, equipped with a nitrogen inlet and a large Teflon-coated magnetic stirring bar, is charged with WCl_4 (32.0 g, 0.098 mol). The WCl_4 is taken up in Et_2O (80 ml), resulting in a gray ether slurry that is cooled to 0°C. A second two-necked 500 mL flask, equipped with a nitrogen inlet and a Teflon-coated magnetic stirring bar, is charged with $LiNMe_2$ (20.0 g, 0.392 mol). The $LiNMe_2$ is taken up in tetrahydrofuran (THF) (200 ml), giving an off-white slurry. The $LiNMe_2$ slurry is slowly transferred to the magnetically stirred WCl_4 slurry via cannula, causing the gray slurry to become a yellow brown solution with precipitated LiCl. When the addition of lithium amide is complete, the reaction mixture is stirred at 0° for an additional 3 h and then allowed to warm to room temperature. After about 20 h at room temperature a reflux condenser is attached to the reaction flask and the reaction is refluxed for 2 h. The yellow solution of $W_2(NMe_2)_6$ is then filtered through Celite to separate the product from LiCl. To ensure complete extraction, the original flask and Celite pad are washed with THF (2 × 50 mL). The solvent is then removed *in vacuo*, leaving a yellow-brown residue.[7] The dried residue, on sublimation onto a water-cooled cold finger (100°C, 10^{-4} torr), yields bright yellow crystals of $W_2(NMe_2)_6$ (23.3 g, 0.037 mol, 75%).[8] The first isolated fraction may occasionally be discolored slightly by impurities including $W(NMe_2)_6$.

Anal. Calcd. for $W_2N_6C_{12}H_{36}$: C, 22.78; H, 5.74; N, 13.34. Found: C, 22.53; H, 5.51; N, 13.02.

Properties

$W_2(NMe_2)_6$, ($W≡W$), is an air- and moisture-sensitive yellow crystalline solid that may be stored for long periods under an inert atmosphere. It is appreciably soluble in hydrocarbon solvents. The ^1H NMR spectrum shows one broad resonance at 3.4 ppm (C_6D_6, 25°C). Other physical and chemical properties are described in the literature.[2]

Acknowledgment

We thank the National Science Foundation for support.

References and Notes

1. M. H. Chisholm, F. A. Cotton, B. A. Frenz, L. W. Shire, and B. R. Stults, *J. Am. Chem. Soc.*, **98**, 4469 (1976).
2. M. H. Chisholm, F. A. Cotton, M. Extine, and B. R. Stults, *J. Am. Chem. Soc.*, **98**, 4477 (1976).
3. M. H. Chisholm, B. W. Eichhorn, K. Folting, J. C. Huffman, C. D. Ontiveros, and W. E. Streib, *Inorg. Chem.*, **26**, 3183 (1987).
4. G. I. Novikov, N. V. Andreeva, and O. G. Polyachenok, *Russ. J. Inorg. Chem.*, **6**, 1019 (1961).
5. M. H. Chisholm and D. A. Haitko, *Inorg. Synth.*, **21**, 51 (1982).
6. R. E. McCarley and T. M. Brown, *Inorg. Chem.*, **3**, 1232 (1964).
7. Reviewers have commented that the dark brown color resulting from some minor impurity may be removed by adding activated charcoal to a toluene solution of crude $W_2(NMe_2)_6$, which is then filtered, yielding a clear yellow solution. After removal of solvent, the product was obtained in a 38% yield.
8. Yields of this reaction may vary between 40 and 80% depending on reaction conditions. We find the yields to be strongly dependent on the quality of the WCl_4. We also have found that best yields are obtained when (1) the THF slurry of $LiNMe_2$ is added slowly, 45 min to 1 h, and (2) the reactions are performed under conditions no more concentrated than those described here.

34.   TETRACARBONYL[1,2-ETHANEDIYLBIS-(DIPHENYLPHOSPHINE)]TUNGSTEN(0), *fac*-(ACETONE)TRICARBONYL[1,2-ETHANEDIYLBIS-(DIPHENYLPHOSPHINE)]TUNGSTEN(0), AND *mer*-TRICARBONYL[1,2-ETHANEDIYLBIS-(DIPHENYLPHOSPHINE)](PHENYLVINYLIDENE) TUNGSTEN(0)*

$$W(CO)_6 + PP \longrightarrow W(CO)_4 PP + 2CO$$
$$\mathbf{1}$$

$$W(CO)_4 PP \xrightarrow[\text{acetone}]{hv} fac\text{-}W(CO)_3 PP(\text{acetone}) + CO$$
$$\mathbf{1} \qquad\qquad\qquad\qquad \mathbf{2}$$

$$fac\text{-}W(CO)_3 PP(\text{acetone}) + PhCCH \xrightarrow{\text{THF}} mer\text{-}W(CO)_3 PP(C=CHPh)$$
$$\mathbf{2} \qquad\qquad\qquad\qquad\qquad\qquad \mathbf{3}$$

PP = 1,2-ethanediylbis[diphenylphosphine] = dppe
THF = tetrahydrofuran

Submitted by KURT R. BIRDWHISTELL†
Checked by ANNE C. DEMA,‡ XIAHOU LI,‡ C. M. LUKEHART,‡ and MARGARET D. OWEN‡

The *fac*-W(CO)$_3$(dppe)(acetone) complex (**2**) represents an important class of compounds containing labile ligands that are used as synthetic precursors. General preparative routes involve replacement of the labile ligand, acetone in complex **2**, with the ligand of choice. The following is a list of the complexes that have been prepared by room-temperature replacement of the acetone ligand of complex **2**: (1) *mer*-W(CO)$_3$(dppe)(η^2-olefin)[1], (2) *fac*-W(CO)$_3$(dppe)(η^2-alkyne)[2], (3) *mer*-W(CO)$_3$(dppe)(C = CHR)[3], (4) *mer*-W(CO)$_3$(dppe)(η^2-CS$_2$)[4], (5) *mer*-W(CO)$_3$(dppe)(η^2-SO$_2$)[5], (6) *fac*-W(CO)$_3$-(dppe)(η^1-acetylide).[6a]

Complex **2** was first reported by Schenck.[1] We found it useful to modify his procedure in order to isolate **2** and store it for future use. Complex **2**, an isolable solid, is significantly more convenient to use than labile solvent adducts such as W(CO)$_5$(thf), which must be photolytically produced each time.

---

* Chemical Abstracts Service refers to the phenylvinylidene ligand as a phenylethenylidene.
† Department of Chemistry, Loyola University, New Orleans, LA, 70118.
‡ Department of Chemistry, Vanderbilt University, Nashville, TN, 37235.

The *mer*-(CO)$_3$(dppe)W(C=CHPh), **3**, prepared as described below, contains a vinylidene ligand, a ligand that occupies a central role in the chemistry of monohapto carbon ligands in organometallic chemistry.[7] Chemical transformations relate the vinylidene ligand to acetylides, η^1-vinyls, acyls, and carbynes.[7] Low oxidation state monohapto vinylidenes of tungsten have recently been implicated as catalysts in the polymerization of terminal acetylenes.[8]

Complex **3** has been converted to a number of carbynes, disubstituted vinylidenes, and related complexes via electrophilic addition to the β- carbon of the vinylidene.[9, 3] Complex **3** has also been used to build heterodinuclear transition metal complexes.[10]

Vinylidene (**3**), as well as related vinylidenes, have been obtained in good yields via the addition of electrophiles to acetylide complexes.[6] The procedure below allows one to make these vinylidenes without the use of alkali metal acetylide reagents. Solid vinylidene **3** is not sensitive to air or water making it an attractive starting material.

A.   TETRACARBONYL[1,2-ETHANEDIYLBIS-(DIPHENYLPHOSPHINE)]TUNGSTEN(0), W(dppe)(CO)$_4$, 1

■   **Caution.**   *This reaction should be done in the hood due to the evolution of toxic carbon monoxide.*

Materials

The tungsten hexacarbonyl, dppe, and phenylacetylene were purchased from Strem Chemical Co. and Aldrich Chemical Co and used without further purification. Solvents are used as received from commercial suppliers after sparging with nitrogen. The following procedure was first reported in 1974.[11]

Procedure

A 500-mL Schlenk flask is charged with tungsten hexacarbonyl (10.56 g, 30 mmol), dppe (11.94 g, 30 mmol), 50 mL of 1,1'-oxybis[2-methoxy-ethane] (diglyme) and a magnetic stirring bar. The almost colorless mixture was sparged with nitrogen for 5 min, fitted with a reflux condenser, and heated to reflux for 4 h under nitrogen. The tungsten hexacarbonyl and dppe slowly dissolve on heating. The tungsten hexacarbonyl that sublimes out of the reaction mixture is periodically returned to the mixture by vigorously swirling the flask. The use of the large flask inhibits the tungsten hexacarbonyl from subliming into the reflux condenser. During the reaction the mixture turns a pale yellow. The solution is cooled to room temperature and

50 mL of methanol is added resulting in a pale yellow powder. The mixture is filtered in the air and the pale yellow powder is isolated. The powder is recrystallized by dissolving in chloroform, filtering, and adding methanol resulting in 14.6 g (70% yield) of very pale yellow crystals.

Properties

The $W(CO)_4$ dppe (**1**) is an air-stable pale yellow crystalline solid. Complex **1** is soluble in polar organic solvents such as dichloromethane, 1,2-dichloroethane, and tetrahydrofuran (THF). The literature melting point is 206–207°C.[11] The compound may be conveniently characterized by its infrared spectrum in the carbonyl region: $v(CO)$(1,2-dichloroethane, cm^{-1}) 2016 (s), 1912 (m), 1901 (s), 1876 (vs),[11] (KBr, cm^{-1}) 2017 (s), 1915 (s), 1886 (vs), 1872 (vs).

B. *fac*-TRICARBONYL[1,2-ETHANEDIYLBIS(DIPHENYL-PHOSPHINE)](ACETONE)TUNGSTEN(0), *fac*-$W(CO)_3$(dppe)(ACETONE), 2

■ **Caution.** *This reaction should be done in a fume hood due to the evolution of toxic carbon monoxide. The UV lamp can cause severe eye damage and should be used with adequate UV eye protection. The photolysis reactor should be hidden from sight during photolysis.*

Procedure

Complex **1** (2.78 g, 4.0 mmol) is dissolved in 300 mL of nitrogen purged acetone. Photolysis of solutions that are more concentrated than this leads to incomplete reaction and undesirable by-products. The almost colorless solution of complex **1** is transferred via stainless steel transfer tube to a submersion-type, nitrogen-flushed photochemical reactor containing a magnetic stirring bar.[12] The reactor is set up with an acetone-saturated nitrogen sparge to assist in the removal of carbon monoxide and inhibit solvent loss during photolysis. In addition to the water-cooled jacket the entire reactor is cooled in an ice bath during photolysis. This quantity of material takes 4–5 h of photolysis time. The reaction may also be followed by monitoring the solution in the CO region of the IR. The high-energy CO absorption of complex **1** at approximately 2015 cm^{-1} should be almost gone on completion of photolysis. The resulting deep yellow solution is transferred via stainless steel tubing to a Schlenk flask and the solvent removed *in vacuo*. (We use a nitrogen flushed rotary evaporator for this process.). The yellow residue is triturated and washed with 3×50 mL of a 3 : 1 petroleum ether–diethyl ether mixture yielding a yellow powder. The yellow powder is dried under vacuum at room temperature. The yield is nearly quantitative; a small amount (< 1% as determined by IR) of $W(CO)_4$(dppe) generally remains. We use this material for all subsequent syntheses.

The tetrahydrofuran adduct, [*fac*-$W(CO)_3$(dppe)(thf)], can be prepared in a similar fashion, but the THF adduct is much less stable in the solid state than the acetone adduct. Complex **2** can be quantitatively converted to the THF adduct by dissolving **2** in THF.

Properties

fac-$W(CO)_3$(dppe)(acetone) can be stored for several months in a freezer under nitrogen. Crystalline compound **2** can be transferred in the air for short periods without significant decomposition. The acetone ligand can be exchanged at room temperature by ligands such as THF, acetonitrile, olefins, alkynes, or phosphines. Complex **2** decomposes quickly in $CHCl_3$ and slowly (over several hours) in acetone or THF at room temperature, forming $W(CO)_4$dppe as the main carbonyl containing product in solution. The compound can be conveniently characterized by its IR spectrum [ν(CO), acetone, cm^{-1}]: 1921 (s), 1825 (s), 1805·(s). Checkers report some trouble in obtaining an IR of **2**. Complex **2** is very air sensitive and the solution IR needs to be obtained under rigorously air-free conditions.

C.   *mer*-TRICARBONYL[1,2-ETHANEDIYLBIS-(DIPHENYLPHOSPHINE)](PHENYLVINYLIDENE)-TUNGSTEN(0), *mer*-(CO)₃(dppe)W(C=CHPh), 3

The following procedure can be completed in 1 day, including the chromatography.

Procedure

A solution of phenylacetylene (8.00 mmol, 0.83 g) in 20 mL of tetrahydrofuran was sparged with nitrogen in a 100-mL Schlenk flask. The phenylacetylene solution is transferred via stainless steel tubing to a 250-mL Schlenk flask fitted with a rubber septum containing complex **2** (2.90 g, 4.00 mmol). After 5 min the red slurry is diluted to 100 mL with nitrogen-sparged THF. To this deep red solution of *fac*-tricarbonyl(dppe)tungsten(η^2-phenylacetylene)^2 is added approximately 0.2 mL of nitrogen-sparged water. We found that water catalyzes the alkyne to vinylidene isomerization, resulting in a higher yield of vinylidene. After stirring for 8 h at room temperature the reaction is complete and the mixture is deep green. The reaction may be monitored by following the carbonyl region of the IR. The solvent is removed *in vacuo*. After solvent removal, a green tar remains. Chromatography of the green mixture under nitrogen on alumina column (30 cm × 2–3 cm) using toluene, followed by increasing amounts of CH_2Cl_2 results in elution of a yellow band of $W(CO)_4 = $ dppe followed by a green band of **3**. Removal of the solvent in vacuo of the green fraction and washing the green powder with pentane followed by drying in vacuum, yields 2.9 g (94% yield)* of green *mer*-$W(CO)_3$(dppe)(C=CHPh), **3**.

The reddish organge methoxycarbonyl *mer*-$(CO)_3$(dppe)W[C=CH-(CO_2Me)][3], can be made in a similar fashion using methyl 2-propynoate and purification by florisil chromatography using toluene, CH_2Cl_2, and diethyl ether.

Properties

Complex **3** can be handled as a solid in the air for extended periods of time without decomposition. Solutions of this complex are air-sensitive. Complex **3** reacts with $CHCl_3$ overnight at room temperature to form a *trans*-chlorotungstencarbyne.[13] Complex **3** is soluble in CH_2Cl_2, THF, acetone, and aromatic hydrocarbons. The complex may be conveniently characterized by its IR spectrum: $v(CO)$ (THF, cm^{-1}) 2002 (m), 1940 (s), 1900 (vs) (or KBr pellet), 2000 (m), 1915 (s), 1890 (vs), and ^1H NMR ($CDCl_3$) δ: 7.5 (m, Ph), 4.99 (dd, $^4J_{PH} = 6.7$, 3.6 Hz, C=CHPh), 2.55 (m, PCH_2CH_2P). Complete spectroscopic details are given in the literature.[3]

References

1. W. A. Schenk and H. Muller, *Chem. Ber.*, **115**, 3618 (1982).
2. K. R. Birdwhistell, T. L. Tonker, and J. L. Templeton, *J. Am. Chem. Soc.*, **109**, 1401 (1987).
3. K. R. Birdwhistell, S. J. N. Burgmayer, and J. L. Templeton, *J. Am. Chem. Soc.*, **105**, 7789 (1983).
4. W. A. Schenk, T. Schwietzke, and H. Muller, *J. Organomet. Chem.*, **232**, C41–C47 (1982).
5. W. A. Schenk and F. E. Baumann, *Chem. Ber.*, **115**, 2615 (1982).
6. (a) K. R. Birdwhistell and J. L. Templeton, *Organometallics*, **4**, 2062 (1985), (b) A. Mayr, K. C. Schaefer, and E. Y. Huang, *J. Am. Chem. Soc.*, **106**, 1517 (1984).
7. (a) A. Davison and P. Selegue, *J. Am. Chem. Soc.*, **100**, 7763 (1978); (b) M. I. Bruce and A. G. Swincer, *Adv. Organomet. Chem.*, **22**, 59 (1983).
8. S. J. Landon, P. M. Shulman, and G. L. Geoffroy, *J. Am. Chem. Soc.*, **107**, 6739 (1985).
9. K. R. Birdwhistell, T. L. Tonker, and J. L. Templeton, *J. Am. Chem. Soc.*, **107**, 4474 (1985).
10. W. R. True and C. M. Lukehart, *Organometallics*, **7**, 2387 (1988).
11. S. O. Grim, W. L. Briggs, R. C. Barth, C. A. Tolman, and J. P. Jesson, *Inorg. Chem.*, **13**, 1095 (1974).

* The checkers reported a lower yield of 3, which may have been due to complex **2** being contaminated with $W(CO)_4$dppe. We have regularly obtained yields of 80–90%.

12. We use an ACE Glass # 7840 500-mL reactor with a 450-W mercury lamp and a Pyrex insert. See M. J. Therien and W. C. Trogler, *Inorg. Synth.*, **25** 151 (1989), for an example of setup.
13. K. R. Birdwhistell, Ph.D. thesis, University of North Carolina, Chapel Hill, May 1985.

35. NITRIDO AND OXO COMPLEXES OF RHENIUM(V)

Submitted by B. PATRICK SULLIVAN,* JOHN C. BREWER,† and
HARRY B. GRAY†
Checked by DOUGLAS LINEBARRIER‡ and JAMES M. MAYER‡

Rhenium compounds in high oxidation states often contain multiply bonded [Re^{V}N], [Re^{V}O], and [Re^{V}O$_2$] units. It is clearly desirable to have large-scale, convenient preparations of these compounds. Although the nitrido-rhenium complex featured here, ReNCl$_2$(PPh$_3$)$_2$, is a key starting material, a detailed preparation has not been published previously. This new, versatile preparation is based quite closely on previous preparations and chemical observations.[1-3]

The second synthesis is of the five-coordinate, *cis*-dioxo complex, ReO$_2$I(PPh$_3$)$_2$, described in 1969[4] and in 1983.[5] The complex exhibits the rare d^2 *cis*-dioxo configuration; importantly, it is a superior material for the preparation of new *trans*-dioxorhenium species. The preparation presented here is based on the previously published method involving the hydrolysis of the complex *trans*-ReOI$_2$(OCH$_2$CH$_3$)(PPh$_3$)$_2$.[4-6]

A. DICHLORONITRIDOBIS(TRIPHENYL-PHOSPHINE)RHENIUM(V)

ReOCl$_3$(PPh$_3$)$_2$ + [PhNHNH$_3$]Cl + PPh$_3$ →

$\qquad$ ReNCl$_2$(PPh$_3$)$_2$ + OPPh$_3$ + [PhNH$_3$]Cl + HCl

The compound ReNCl$_2$(PPh$_3$)$_2$ was first prepared in 1962 by reaction of hydrazine dihydrochloride with NaReO$_4$ and PPh$_3$ in aqueous ethanol, followed by distillation to remove the benzene–ethanol–water azeotrope.[1] In later papers, two new general preparative methods were reported: (1) reaction of hydrazine dihydrochloride (or with a hydrazine/hydrazine dihydrochlor-

* Department of Chemistry, University of Wyoming, Laramie, WY 82071.
† Arthur Amos Noyes Laboratory, California Institute of Technology, Pasadena, CA 91125. Contribution no. 8057.
‡ Department of Chemistry, University of Washington, Seattle, WA 98195.

ide mixture) with Re_2O_7 and PPh_3 in aqueous ethanol[2, 3] and (2) the reaction of perrhenate with PPh_3 and various substituted hydrazinium salts, for example, 4-methylbenzenesulfonic acid hydrazide hydrochloride (*p*-tosyl-hydrazine hydrochloride) and phenylhydrazine hydrochloride.[3]

It was the latter preparative method and the observation that $ReOCl_3(PPh_3)_2$ reacts with hydrazinium salts in wet ethanol[3] to give $ReNCl_2(PPh_3)_2$ that prompted the development of this synthesis. The precursor complex $ReOCl_3(PPh_3)_2$ is perfectly suited for use in this application since it is available conveniently in 10–100-g scales in 99% yield.[7]

Procedure

To a 250-mL, round-bottomed flask equipped with a reflux condenser and a nitrogen inlet are added 6.88 g (8.26 mmol) of $ReOCl_3(PPh_3)_2$,[7] 4.50 g (17.2 mmol) of PPh_3, and 1.2 g (8.3 mmol) of $[PhNHNH_3]Cl$ (Aldrich). After addition of 150 mL of absolute ethanol and 6 mL of distilled water, the mixture is heated at reflux for 2 h with vigorous magnetic stirring. The heating time at reflux is crucial, since at shorter times the isolated nitride is contaminated with unreacted $ReOCl_3(PPh_3)_2$, while at longer times the competitive formation of a dark brown, ethanol soluble material decreases the yield of product. Likewise, the addition of water to the reaction is crucial, since it shifts the ionization equilibrium of the hydrochloride toward the free hydrazine. After cooling the reaction to room temperature, the solid that has formed is filtered by suction from the red-brown filtrate. The solid is washed successively with the following solvents: water (2 × 25 mL), absolute ethanol (2 × 25 mL), acetone (3 × 25 mL), and diethyl ether (3 × 25 mL). Upon washing with acetone, a substantial amount of the red-brown impurity is removed. After air-drying, first for 1 h by suction, and then for several days at room temperature, the yield of the orange-brown microcrystalline solid is measured as 5.60 g (7.04 mmol) (85%). The material in this state is suitable for preparative purposes.

Anal. Calcd. C, 54.34%; H, 3.80%; N, 1.76%. Found. C, 54.67%; H, 3.86%; N, 1.50%. Recrystallization can be accomplished from hot benzene/ethanol mixture.

Properties

The compound $ReNCl_2(PPh_3)_2$ is an air-stable complex that is insoluble in coordinating solvents such as alcohols, acetonitrile, acetone, and diethyl ether, and only sparingly soluble in low-polarity solvents such as dichloromethane, chloroform, and benzene. The melting point is reported[2] to be

219–221°C. The characteristic infrared stretch of the ReN triple bond has not been found due to interference from Ph modes; the ReCl stretch occurs at 323 cm^{-1} (Nujol mull).[2] Despite the color of the complex in the solid state and the yellow color of a CH_2Cl_2 solution, no distinctive features appear in the visible region of the spectrum. The dipole moment of 1.6 (3) D is consistent with the X-ray crystal structure obtained elsewhere[8] that reveals a distorted square pyramid with an apical nitrido ligand and mutually trans chloride ligands in the basal plane. The triphenylphosphine ligands are distorted toward the empty rhenium coordination site.

The compound $ReNCl_2(PPh_3)_2$ is formally a 16-electron species, and despite its insolubility, it reacts smoothly and in high yield with a variety of ligands with maintenance of the robust ReN bond. In fact, the chloride and phosphine ligands can be completely stripped from the metal under appropriate conditions. Examples include reaction with $Et_2NCS_2{}^-$ to give the five-coordinate complex $ReN(Et_2NCS_2)_2$[9] and with CN^- to give six-coordinate $[ReN(CN)_5]^{3-}$ and related derivatives.[10] Other characteristic reactions include those with monodentate and bidentate phosphines to give five-coordinate (e.g., $ReNCl_2(PMePh_2)_2$)[3] and six-coordinate complexes (e.g., $ReNCl_2(PMePh_2))_3$ and $[ReNCl(diphos)_2]^+$).[3,11]

B. *trans*-ETHOXYDIIODOOXOBIS(TRIPHENYL-PHOSPHINE)RHENIUM(V)

$$KReO_4 + 3PPh_3 + 3HI + CH_3CH_2OH \rightarrow$$
$$ReOI_2(OCH_2CH_3)(PPh_3)_2 + OPPh_3 + KI + 2H_2O$$

Procedure

A mixture of 1.2 g (4.1 mmol) of $KReO_4$, 5 mL of 56% HI, and $\sim$ 30 mL absolute ethanol is heated to boiling for 10 min in an Erlenmeyer flask. After this period, 5.0 g (19 mmol) of PPh_3 is added and the sides of the flask are washed with $\sim$ 15 mL of absolute ethanol, while maintaining vigorous magnetic stirring. Boiling of the mixture is continued for $\sim$ 30 min. The light khaki-colored crystals that precipitate are filtered immediately from the hot solution, washed with absolute ethanol (3 × 50 mL) and diethyl ether (3 × 50 mL), and dried with an aspirator vacuum for $\sim$ 1 h. The yield is 3.16 g (77%).

Anal. Calcd. C, 44.50%; H, 3.44%. Found: C, 42.88%; H, 3.37%.

The material can be recrystallized from benzene–ethanol as a benzene solvate. The reaction can be run on a scale that is six times larger.

Properties

The compound *trans*-ReOI$_2$(OCH$_2$CH$_3$)(PPh$_3$)$_2$ exhibits a medium-sharp (CH$_3$CH$_2$O)-Re=O asymmetric infrared stretch at 945 cm^{-1} and a very sharp band characteristic of the ethoxy substituent (O—CH$_2$) at 910 cm^{-1}. The ^{31}P NMR spectrum shows a single resonance at -11.1 ppm (relative to 85% H$_3$PO$_4$). Proton NMR spectral shifts (CDCl$_3$ at 35°C with TMS as the internal standard) are as follows: 7.98 *m* and 7.38 *m* (C$_6$H$_5$), 1.51 *q* (OCH$_2$CH$_3$), $-0.26t$ (OCH$_2$CH$_3$).[5]

With the exception of the dehydration reaction discussed below (Section C), development of the chemistry of *trans*-ReOI$_2$(OCH$_2$CH$_3$)(PPh$_3$)$_2$ has been limited. The ethoxy group is very reactive, and easily exchanges with other free alcohols (e.g., in acetone–methanol mixtures *trans*-ReOI$_2$(OCH$_3$)(PPh$_3$)$_2$ is formed).[5] It is anticipated that the iodo ligands will undergo substitution under mild conditions so that this complex could be used as a precursor to numerous new monooxo complexes of Re(V).

C. IODODIOXOBIS(TRIPHENYLPHOSPHINE)RHENIUM(V)

$$ReOI_2(OCH_2CH_3)(PPh_3)_2 + H_2O \rightarrow ReO_2I(PPh_3)_2 + HI$$
$$+ CH_3CH_2OH$$

Procedure

In an Erlenmeyer flask, *trans*-ReOI$_2$(OCH$_2$CH$_3$)(PPh$_3$)$_2$ (1.0 g; 0.98 mmol) is suspended in a mixture of 50 mL of acetone and 2 mL of water. The suspension is magnetically stirred for 1 h, after which time a violet crystalline material is deposited. The complex ReO$_2$I(PPh$_3$)$_2$ is filtered, washed with absolute ethanol (2 × 30 mL) and diethyl ether (2 × 30 mL), and dried by suction for ∼ 1 h.

Anal. Calcd. C, 49.72%; H, 3.48%. Found: C, 49.06%; H, 3.54%.

The yield is 0.73 g (75%). This preparation has been successfully conducted on a scale that is ∼ 20 times larger.

Properties

The distinctive absorptions associated with the *cis*-ReO$_2$ grouping appear at 923 (s) and 843 (s) cm^{-1} in the infrared spectrum of the complex (Nujol mull). The ^{31}P NMR spectrum exhibits a single resonance at $+4.57$ ppm (relative to 85% H_3PO_4 as an external standard), and the ^1H NMR spectrum shows phenyl multiplets centered at 7.62 and 7.42 ppm (relative to TMS at room temperature).[5] The beautiful violet solution color arises from a characteristic visible absorption that occurs at 560 nm ($\varepsilon_{max} \approx 350\ M^{-1}\,cm^{-1}$) in CH_2Cl_2 solution. The X-ray crystal structure of the complex reveals distorted trigonal bipyramidal geometry. The phosphine ligands occupy the apical positions and are slightly bent toward the trigonal plane; the *cis*-ReO_2 core has an O—Re—O angle of 139°.[5]

$ReO_2I(PPh_3)_2$ is a coordinatively unsaturated precursor that possesses reasonably good solubility in chlorinated hydrocarbons such as CH_2Cl_2 and moderate to poor solubility in nonpolar solvents such as benzene or toluene. It is an exceedingly reactive complex and can even be used in solvents in which it has poor solubility to give complexes containing the *trans*-ReO_2 unit. For example, the complex *trans*-$[ReO_2(py)_4]I$ can be prepared in $\sim$ 90% isolated yield by gently heating $ReO_2I(PPh_3)_2$ in neat pyridine for 15 min.[4,12] A particularly interesting reaction from the chemical bonding point of view is that of $ReO_2I(PPh_3)_2$ and alkynes. This gives a novel Re(III)-oxo complex, $ReOI(alkyne)_2$.[13]

References

1. J. Chatt and G. A. Rowe, *J. Chem. Soc.*, **1962**, 4019. The compound was reported as "$ReCl_2(PPh_3)_2$".
2. J. Chatt, J. D. Garforth, N. P. Johnson, and G. A. Rowe, *J. Chem. Soc.*, **1964**, 1012.
3. J. Chatt, C. D. Falk, G. J. Leigh, and R. J. Paske, *J. Chem. Soc. (A)*, **1969**, 2288.
4. M. Freni, D. Giusto, P. Romiti, and G. Minghetti, *Gazz. Chim. Ital.*, **99**, 286 (1969).
5. G. F. Ciani, G. D'Alfonso, P. F. Romiti, A. Sironi, and M. Freni, *Inorg. Chim. Acta*, **72**, 29 (1983).
6. N. P. Johnson, C. J. L. Lock, and G. Wilkinson, *J. Chem. Soc.*, **1964**, 1054.
7. G. Parshall, *Inorg. Synth.*, **17**, 110 (1977).
8. R. J. Doedens and J. A. Ibers, *Inorg. Chem.*, **6**, 204 (1967).
9. J. F. Rowbottom and G. Wilkinson, *J. Chem. Soc. Dalton Trans.*, **1972**, 826.
10. N. P. Johnson, *J. Chem. Soc. (A)*, **1969**, 1843.
11. W. Jabs and S. Herzog, *Z. Chem.*, **12**, 268 (1972).
12. J. C. Brewer and H. B. Gray, *Inorg. Chem.*, **28**, 3334 (1989).
13. See, for example, J. M. Mayer, T. H. Tulip, J. C. Calabrese, and J. C. Valencia, *J. Am. Chem. Soc.*, **109**, 157 (1987).

36. *trans*-TRICARBONYLBIS(PHOSPHINE)IRON(0) COMPLEXES: ONE-POT SYNTHESES FROM PENTACARBONYLIRON

Submitted by J.-J. BRUNET,* F. B. KINDELA,* and D. NEIBECKER*
Checked by S. A. WANDER† and M. Y. DARENSBOURG†

$$Fe(CO)_5 + 2KOH \rightarrow K[FeH(CO)_4] + KHCO_3$$

$$K[FeH(CO)_4] + 2PR_3 + EtOH \rightarrow Fe(CO)_3(PR_3)_2 + CO + H_2 + KOEt$$

■ **Caution.** *Because of the toxic nature of pentacarbonyliron, the evolution of highly toxic carbon monoxide (a colorless and odorless gas) and highly flammable hydrogen, and the toxic nature and bad odor and risk of ignition of at least liquid phosphines, these reactions should be performed in a well-ventilated hood and gloves should be worn.*

Most of the preparations of molecular coordination compounds of iron published in *Inorganic Syntheses* or elsewhere require the readily available pentacarbonyliron or derivatives thereof. However, pentacarbonyliron is a rather stable compound that needs either thermal or photochemical activation to react. We describe here the selective, high-yield, one-pot synthesis of the title complexes from potassium tetracarbonylhydridoferrate(1 −), easily generated from pentacarbonyliron, a method that can be referred to as a "nucleophilic activation of pentacarbonyliron."

Tricarbonylbis(phosphine)iron complexes are starting materials for the synthesis of several organometallic iron complexes and are gaining attention as catalysts for the photochemical olefin hydrosilation,[1] and for the synthesis of carbamates from carbon dioxide, amines, and alkynes.[2] Two preparations that witness the interest in these compounds already appeared in *Inorganic Syntheses*. In 1966 two methods were described for the synthesis of the $Fe(CO)_3[P(C_6H_5)_3]_2$ and $Fe(CO)_4[P(C_6H_5)_3]$ complexes in moderate yields (15–34%) by a thermal reaction of dodecacarbonyltriiron or pentacarbonyliron and triphenylphosphine in tetrahydrofuran or cyclohexanol, respectively.[3] In 1989 the irradiation of $Fe(CO)_5$ in cyclohexane in the presence of several phosphines was reported to produce $Fe(CO)_3L_2$ complexes in yields ranging from 28% for the synthesis of $Fe(CO)_3[P(n\text{-}Bu)_3]_2$ to 80% for $Fe(CO)_3[P(CH_3)_3]_2$.[4] Those authors critically reviewed the known pre-

* Laboratoire de Chimie de Coordination du CNRS, unité No. 8241 liée par conventions à l'Université Paul Sabatier et à l'Institut National Polytechnique, 205 route de Narbonne, 31077 Toulouse Cedex, France.
† Department of Chemistry, Texas A&M University, College Station, TX 77843.

parative methods and emphasized the separation difficulties of the mono-
from the disubstituted product and of the phosphine from the products. In
addition to these methods, two direct syntheses of $Fe(CO)_3L_2$ complexes
should be mentioned. The first one is a time-consuming (50 h) but a fair yield
(71%) synthesis of the $Fe(CO)_3[P(C_6H_5)_3]_2$ complex from the reaction of
$Fe(CO)_5$ with $Li[AlH_4]$ in the presence of excess PPh_3 in refluxing THF.[5] In
the second one, the reaction of $Fe(CO)_5$ with $Na[BH_4]$ and excess PR_3
(3 equiv) in 1-butanol at 127°C, affords fair to high yields (47–91%) of the
$Fe(CO)_3L_2$ complexes.[6] The reaction of $Na[FeH(CO)_4]$ with PPh_3 under the
same experimental conditions is reported to give the $Fe(CO)_3[P(C_6H_5)_3]_2$
complex in 95% yield.[6]

The procedure we describe here involves the simple reaction of a stoichiometric quantity of a phosphine (2 equiv only) with $K[FeH(CO)_4]$ in EtOH to
get the $Fe(CO)_3L_2$ complexes in high yields and in a pure state without
cumbersome treatments.[7]

A. ETHANOLIC SOLUTIONS OF POTASSIUM TETRACARBONYLHYDRIDOFERRATE(1−)

$$Fe(CO)_5 + 2KOH \rightarrow K[FeH(CO)_4] + KHCO_3$$

■ **Caution.** *$Fe(CO)_5$ is a toxic liquid and should only be handled with
hands protected by gloves in a well-ventilated hood.*

Solutions of tetracarbonylhydridoferrate(1 −) salts can be prepared by
numerous methods.[8−10] The most straightforward and widely used involves
the reaction of $Fe(CO)_5$ and an excess of an alkaline base in protic media. The
conventional method to prepare ethanolic solutions of $K[FeH(CO)_4]$ is to
react $Fe(CO)_5$ and 3 equiv of KOH in EtOH at room temperature.[11] The
procedure described here uses 2KOH equivalents only and allows the rapid
and quantitative preparation of ethanolic $K[FeH(CO)_4]$ solutions, which are
stable provided that all traces of oxidizing agents are rigorously excluded.

Procedure

A 100-mL Schlenk flask containing a Teflon-coated magnetic stirring bar is
charged with KOH (content 86%, the remaining being water) (1.47 g,
22 mmol), closed with a serum cap and submitted to three vacuum–argon
cycles. Absolute ethanol (60 mL), previously degassed by bubbling argon for
0.5 h, is syringed into the Schlenk flask through the serum cap and the
mixture stirred until dissolution (∼ 0.5 h). $Fe(CO)_5$ (1.5 mL, 11 mmol) is
syringed into the resulting solution through the serum cap and stirring is

pursued for 0.5 h, during which time precipitation of $KHCO_3$ is observed. The IR spectrum of the pale pink reaction medium indicates complete reaction of $Fe(CO)_5$ (no absorption bands near 2015 and 2035 cm^{-1}) and exhibits the characteristic absorption bands of ethanolic solution of $K[FeH(CO)_4]$ near 2000 (vw), 1915 (sh), and 1890 (s) cm^{-1}.

B. *trans*-TRICARBONYLBIS(TRIPHENYLPHOSPHINE)IRON(0)

$$K[FeH(CO)_4] + 2P(C_6H_5)_3 + EtOH \rightarrow Fe(CO)_3[P(C_6H_5)_3]_2 + CO + H_2 + KOEt$$

Procedure

A 60-mL ethanolic solution of $K[FeH(CO)_4]$ (11 mmol) is prepared as described above. Solid triphenylphosphine (recrystallized from MeOH under argon and stored under argon) (5.76 g, 22 mmol) is added under a stream of argon. The Schlenk flask is fitted with a reflux condenser connected to an oil bubbler. The stream of argon is stopped and the mixture heated for 24 h at reflux by means of an oil bath. During that time, gas evolves slowly, the reaction medium becomes homogeneous and then precipitates a yellow solid. After cooling to 30–40°C under a stream of argon, the solid is filtered in air on a sintered glass, thoroughly washed with distilled water up to neutrality of the filtrate (6 × 50 mL), and finally with EtOH (3 × 50 mL). Dissolution of the yellow solid on the sintered glass with CH_2Cl_2 (3 × 50 mL) with simultaneous filtration affords a clear yellow solution. The solvent is removed on a rotatory evaporator to leave a yellow powder that is dried under reduced pressure (0.01 torr) up to a constant weight. The complex $Fe(CO)_3[P(C_6H_5)_3]_2$ requires no further purification. Yield: 7.3 g, 99%. mp 265°C (dec.).

Anal. Calcd. for $C_{39}H_{30}FeO_3P_2$: C, 70.50; H, 4.55. Found: C, 70.68; H, 4.66.

C. *trans*-TRICARBONYLBIS(TRI-*n*-BUTYLPHOSPHINE)IRON(0)

$$K[FeH(CO)_4] + 2P(n-C_4H_9)_3 + EtOH \rightarrow Fe(CO)_3[P(n-C_4H_9)_3]_2 + CO + H_2 + KOEt$$

Procedure

A 60-mL ethanolic solution of $K[FeH(CO)_4]$ (11 mmol) is prepared as described above. Tri-*n*-butylphosphine (distilled under argon and stored under argon) (4.55 g, 22 mmol) is added under a stream of argon. The Schlenk flask is fitted with a reflux condenser connected to an oil bubbler. The stream of argon is stopped and the mixture heated for 1 h at reflux by means of an oil bath. During that time, gas evolves rapidly, the reaction medium becomes homogeneous and then precipitates a pale yellow solid. After cooling to room temperature, the solvent is removed under reduced pressure. The resulting yellow slurry is treated with water distilled under argon (150 mL) to give a yellow suspension. The solid is then filtrated under argon and washed with distillated water (6 × 50 mL) up to neutrality of the filtrate and dried under reduced pressure (0.01 torr) up to a constant weight (6.05 g). Recrystallization from CH_2Cl_2 : MeOH (1 : 6) affords pure $Fe(CO)_3 [P(n-C_4H_9)_3]_2$ as a pale yellow powder. Yield: 5.6 g, 94%, mp 54–55°C.

Anal. Calcd. for $C_{27}H_{54}FeO_3P_2$: C, 59.56; H, 10.0. Found: C, 59.48; H, 10.12.

D.   *trans*-TRICARBONYLBIS(TRICYCLOHEXYLPHOSPHINE)-IRON(0)

$$K[FeH(CO)_4] + 2P(c\text{-}C_6H_{11})_3 + EtOH \rightarrow Fe(CO)_3[P(c\text{-}C_6H_{11})_3]_2$$
$$+ CO + H_2 + KOEt$$

Procedure

A 60-mL ethanolic solution of $K[FeH(CO)_4]$ (11 mmol) is prepared as described above. Solid tricyclohexylphosphine (recrystallized from MeOH under argon and stored under argon) (6.2 g, 22 mmol) is added under a stream of argon. The Schlenk flask is fitted with a reflux condenser connected to an oil bubbler. The stream of argon is stopped and the mixture heated for 8 days at reflux by means of an oil bath. During that time, gas evolves very slowly and the reaction medium becomes homogeneous and then precipitates a pale yellow solid. The reaction medium is then filtrated while hot under argon and the resulting solid is washed with distilled water up to neutrality of the filtrate (6 × 50 mL), and with degassed EtOH (3 × 50 mL). Dissolution of the yellow solid on the sintered glass with CH_2Cl_2 (6 × 50 mL) with simultaneous filtration affords a clear yellow solution. The solvent is removed on a rotatory evaporator to leave a pale yellow powder, which is dried under reduced pressure (0.01 torr) up to a constant weight. The complex $Fe(CO)_3[P(c\text{-}C_6H_{11})_3]_2$ requires no further purification. Yield: 7.2 g, 93%. mp 226°C (dec.).

TABLE I. Spectral Characteristics of Fe(CO)$_3$L$_2$ Complexes

Compound	IR^a v_{CO} (cm^{-1})	^{31}P{^{1}H}NMR^b δ (ppm)	^{13}C{^{1}H}NMR^c δ (ppm)	J_{C-P} (Hz)(mult.)
Fe(CO)$_3$[P(C$_6$H$_5$)$_3$]$_2$	1970 (w) 1887 (vs)	82.6	214.7^d	29(t)
Fe(CO)$_3$[P(n-C$_4$H$_9$)$_3$]$_2$	1958 (w) 1865 (vs)	64.4	216.0^e	28(t)
Fe(CO)$_3$[P(c-C$_6$H$_{11}$)$_3$]$_2$	1945 (w) 1855 (vs)	85.9	217.5^f	28(t)

^{a} Solution IR spectra use CH$_2$Cl$_2$ as the solvent.
^{b} ^{31}P{^{1}H}NMR spectra were recorded on a Bruker WH 90 at 36.43 MHz in CDCl$_3$.
  Chemical shifts are in parts per million (ppm) downfield from external 85% H$_3$PO$_4$.
^{c} ^{13}C{^{1}H}NMR spectra were recorded on a Bruker WM 250 at 62.89 MHz in CD$_2$Cl$_2$.
  Chemical shifts of the CO ligand are in parts per million downfield TMS, assigning the CD$_2$Cl$_2$
  resonance at 53.6 ppm.
^{d} Other signals for aromatic carbons at 136.7, 133.5, 130.3, and 128.5 ppm.
^{e} Other signals at 30.5, 26.2, 24.6, and 13.7 ppm.
^{f} Other signals at 35.0, 30.1, 28.2, and 26.8 ppm.

Anal. Calcd. for C$_{39}$H$_{66}$FeO$_3$P$_2$: C, 66.85; H, 9.49. Found: C, 66.81; H, 9.58.

Properties

The properties of the complexes Fe(CO)$_3$L$_2$ have been extensively studied and reviewed.[4,12,13] Spectroscopic data are listed in Table I.

References

1. D. K. Liu, C. G. Brindley, and M. S. Wrighton, *Organometallics*, **3**, 1449 (1984).
2. T. J. Kim, K. H. Kwon, S. C. Kwon, J. O. Baeg, S. C. Shim, and D. H. Lee, *J. Organomet. Chem.*, **389**, 205 (1990).
3. A. F. Clifford, and A. K. Mukherjee, *Inorg. Synth.*, **8**, 115 (1966).
4. M. J. Therien, and W. C. Trogler, *Inorg. Synth.*, **25**, 151 (1989).
5. W. O. Siegl, *J. Organomet. Chem.*, **169**, 191 (1975).
6. R. L. Keiter, E. A. Keiter, K. H. Hecker, and C. A. Boecker, *Organometallics*, **7**, 2466 (1988).
7. J. J. Brunet, F. B. Kindela, and D. Neibecker, *J. Organomet. Chem.*, **368**, 209 (1989).
8. J. J. Brunet, *Chem. Rev.*, **90**, 1041 (1990).
9. R. B. King and F. G. A. Stone, *Inorg. Synth.*, **7**, 193 (1963).
10. L. W. Arndt, C. J. Bischoff, and M. Y. Darensbourg, *Inorg. Synth.*, **26**, 335 (1989).

11. Y. Takegami, Y. Watanabe, H. Masada, and I. Kanaya, *Bull. Chem. Soc. Jpn.*, **40**, 1456 (1967).
12. P. N. Hawker and M. V. Twigg, *Comprehensive Coord. Chem.*, **4**, 1179 (1987).
13. D. F. Schriver and K. H. Whitmire, *Comprehensive Organomet. Chem.*, **4**, 243 (1982).

37. DICARBONYL-*cis*-DIHYDRIDO-*trans*-BIS(PHOSPHITE)IRON(II) COMPLEXES: ONE-POT SYNTHESES FROM PENTACARBONYLIRON

Submitted by J.-J. BRUNET,* F. B. KINDELA,* and D. NEIBECKER*
Checked by S. A. WANDER† and M. Y. DARENSBOURG†

$$Fe(CO)_5 + 2KOH \rightarrow K[FeH(CO)_4] + KHCO_3$$

$$K[FeH(CO)_4] + 2P(OR)_3 + H_2O \rightarrow FeH_2(CO)_2[P(OR)_3]_2 + 2CO + KOH$$

■ **Caution.** *Because of the toxic nature of pentacarbonyliron and the evolution of highly toxic carbon monoxide (a colorless and odorless gas) and the toxic nature and bad odor of liquid phosphites, these reactions should be performed in a well-ventilated hood and gloves should be worn.*

Mononuclear dicarbonyldihydridobis(phosphine) complexes of iron are scarce. The synthesis of the complex $FeH_2(CO)_2[P(C_6H_5)_3]_2$ has been claimed in the literature but without reported yield nor spectroscopic data other than the IR spectrum.[1] The complex $FeH_2(CO)_2[P(OCH_3)_3]_2$ was prepared in low overall yield (9%) by hydrogenation of the μ-dinitrogen complex $\{Fe(CO)_2[P(OCH_3)_3]_2\}_2$-$\mu$-$(N_2)$ obtained by photolysis of $Fe(CO)_3[P(OCH_3)_3]_2$ under nitrogen at $-80°C$.[2] A better synthesis of two $FeH_2(CO)_2[P(OR)_3]_2$ complexes [$P(OR)_3 = P(OC_6H_5)_3$; 89%, $P(OR)_3 = P[(OCH_2)]_3C(C_2H_5)$; 92%] is reported from the reaction of $FeH_2(CO)_4$ and the corresponding phosphite in ligroin.[3] However, these latter syntheses require the cumbersome preparation of the unstable $FeH_2(CO)_4$ just before use.

Finally, the complex $FeH_2(CO)_2(dppe)$, [dppe = 1,2-ethandiylbis-(diphenylphosphine)] was fortuitously obtained in 40% yield by the reaction of $Fe(CO)_4(H)(SiR)_3[SiR_3 = Si(C_6H_5)_3, Si(CH_3)_2C_6H_5]$ and dppe.[4]

The procedure we report here involves the very simple reaction of $K[FeH(CO)_4]$ with a phosphite ligand in protic medium. It affords the complexes $FeH_2(CO)_2[P(OR)_3]_2$ in high yields.[5]

* Laboratoire de Chimie de Coordination du CNRS, unité No. 8241 liée par conventions à l'Université Paul Sabatier et à l'Institut National Polytechnique, 205 route de Narbonne, 31077 Toulouse Cedex, France.
† Department of Chemistry, Texas A&M University, College Station, TX 77843.

A. WATER–THF SOLUTIONS OF POTASSIUM TETRACARBONYLHYDRIDOFERRATE(1−)

$$Fe(CO)_5 + 2KOH \rightarrow K[FeH(CO)_4] + KHCO_3$$

■ **Caution.** *$Fe(CO)_5$ is a toxic liquid and should be handled with hands protected by gloves in a well-ventilated hood.*

Solutions of tetracarbonylhydridoferrate(1−) salts can be prepared by numerous methods.[6−8] The most straightforward and widely used involves the reaction of $Fe(CO)_5$ and an excess of an alkaline base in protic media. The procedure described here uses two KOH equivalents only and allows the rapid and quantitative preparation of $K[FeH(CO)_4]$ solutions in H_2O/THF solvent mixtures, which are stable provided that all traces of oxidizing agents are rigorously excluded.

Procedure

A 250-mL Schlenk flask containing a Teflon-coated magnetic stirring bar is charged with KOH (content 86%, the remaining being water) (1.47 g, 22 mmol), closed with a serum cap and submitted to three vacuum–argon cycles. A H_2O : THF solvent mixture (50 : 30 mL), previously degassed by bubbling argon for 0.5 h, is syringed into the Schlenk flask through the serum cap and the mixture stirred until dissolution (∼ 0.5 h). $Fe(CO)_5$ (1.5 mL, 11 mmol) is syringed into the resulting solution through the serum cap and stirring is pursued for 0.5 h. The IR spectrum of the pale pink reaction medium indicates complete reaction of $Fe(CO)_5$ (no absorption bands near 2015 and 2035 cm^{-1}) and exhibits the characteristic absorption bands of solutions of $K[FeH(CO)_4]$ in H_2O/THF solvent mixtures at 2000 (w), 1911 (sh), and 1882 (s) cm^{-1}.

B. DICARBONYL-*cis*-DIHYDRIDO-*trans*-BIS(TRIMETHYL PHOSPHITE)IRON(II)

$$K[FeH(CO)_4] + 2P(OCH_3)_3 + H_2O \rightarrow FeH_2(CO)_2[P(OCH_3)_3]_2$$
$$+ 2CO + KOH$$

Procedure

A 80-mL H_2O : THF (50 : 30 mL) solution of $K[FeH(CO)_4]$ (11 mmol) is prepared under argon as described above in a 250-mL Schlenk flask stoppered with a serum cap. The stopcock of the Schlenk flask is closed and connected to a gas buret. Trimethyl phosphite (distilled under argon and stored under argon) (2.73 g, 22 mmol) is syringed dropwise, through the serum cap, into the vigorously stirred reaction medium. The stopcock of the Schlenk flask is opened immediately after the first drop of trimethyl phosphite is added. A vigourous evolution of CO gas is observed. The reaction is complete when gas evolution ceases (~ 1 h) that is, when the volume of CO gas evolved equals the theoretical amount (~ 520 mL, 22 mmol at room temperature and 760 torr). THF is then evaporated under reduced pressure (**Caution:** *The reaction product sublimes!*) up to the appearance of a milk-like precipitate. Extraction of the product from the remaining aqueous phase is performed, under argon, with pentane (6 × 30 mL). Careful concentration of the pentane solution to 50 mL and cooling to − 78°C for 3h affords a white solid and a light green pentane phase. The latter is removed with a syringe and the product is dried at − 20°C under reduced pressure (0.1 torr) up to a constant weight. The isolated $FeH_2(CO)_2[P(OCH_3)_3]_2$ is a colorless, sometimes light green, oil at room temperature. Yield: 3.6 g, 90%.

Anal. Calcd. for $C_8H_{20}FeO_8P_2$: C, 26.54; H, 5.57 . Found: C, 26.32; H, 5.60.

C. DICARBONYL-*cis*-DIHYDRIDO-*trans*-BIS(TRIETHYL PHOSPHITE)IRON(II)

$$K[FeH(CO)_4] + 2P(OC_2H_5)_3 + H_2O \rightarrow FeH_2(CO)_2[P(OC_2H_5)_3]_2$$
$$+ 2CO + KOH$$

Procedure

The same procedure as above is followed, using triethyl phosphite (3.66 g, 22 mmol). The isolated $FeH_2(CO)_2[P(OC_2H_5)_3]_2$ is a colorless, sometimes light green, oil at room temperature. Yield: 4.45 g, 91%.

Anal. Calcd. for $C_{14}H_{32}FeO_8P_2$: C, 37.69; H, 7.23. Found: C, 37.75; H, 7.41.

D. DICARBONYL-*cis*-DIHYDRIDO-*trans*-BIS(TRIPHENYL PHOSPHITE)IRON(II)

$$K[FeH(CO)_4] + 2P(OC_6H_5)_3 + H_2O \rightarrow FeH_2(CO)_2[P(OC_6H_5)_3]_2$$
$$+ 2CO + KOH$$

Procedure

The same procedure as above is followed, using triphenyl phosphite (6.83 g, 22 mmol). At the end of the CO gas evolution (24 h), evaporation of THF

TABLE I. Spectral Characteristics of $FeH_2(CO)_2[P(OR)_3]_2$ Complexes

Compound	IR $\nu_{CO}(cm^{-1})$	¹H NMR[c] δ (ppm)[d] J_{H-P}(Hz)(mult.)		¹³C{¹H}NMR[c] δ (ppm)[d] J_{C-P}(Hz)(mult.)	³¹P NMR[c] δ(ppm)(mult.)[e]
$FeH_2(CO)_2[P(OCH_3)_3]_2$	2020 (s)[a] 1975 (s)	3.57 (OCH_3) −11.2 (Fe—H)	60 (t)	211.9 (CO)[f] 18 (t)	188.0 (t)
$FeH_2(CO)_2[P(OC_2H_5)_3]_2$	2020 (s)[a] 1970 (s)	1.25 (CH_3) 3.93 (OCH_2) −11.0 (Fe—H)	60 (t)	212.5 (CO)[g] 18 (t)	182.2 (t)
$FeH_2(CO)_2[P(OC_6H_5)_3]_2$	2040 (s)[b] 1990 (s)	7.2 (OC_6H_5) −11.3 (Fe—H)	62 (t)	209.0 (CO)[h] 16 (t)	175.1 (t)

[a] Pentane solution.
[b] KBr pellet.
[c] NMR spectra were recorded on a Bruker WH 90 in $CDCl_3$ at 90, 36.43, and 22.63 MHz for ¹H, ³¹P, and ¹³C NMR, respectively.
[d] Relative to TMS.
[e] Relative to H_3PO_4.
[f] Other signal at 51.5 ppm (s) (OCH_3).
[g] Other signals at 60.3 (s) (OCH_2) and 15.8 ppm (t) (CH_3).
[h] Other signals for aromatic carbons (s) at 151.5 (C—O), 129.4 (C ortho), 124.4 (C para) and 121.6 ppm (C meta).

under reduced pressure leads to an aqueous phase and a white precipitate. Water is removed by means of a syringe and the white crystals washed with water distilled under argon (3×30 mL). Extraction with THF (2×20 mL) under argon and filtration into another Schlenk flask affords a clear solution that is concentrated to 10 mL under reduced pressure. Addition of pentane (50 mL) and cooling at $-20°C$ overnight precipitates white crystals, which are separated, washed with cold pentane ($-20°C$), and dried at $-20°C$ under reduced pressure (0.1 torr) up to a constant weight. $FeH_2(CO)_2[P(OC_6H_5)_3]_2$ is a white solid that melts at 84–85°C. Yield: 7.6 g, 94%.

Anal. Calcd. for $C_{38}H_{32}FeO_8P_2$: C, 62.12; H, 4.39. Found: C, 62.32; H, 4.31.

Properties

$FeH_2(CO)_2[P(OC_6H_5)_3]_2$ is stable at room temperature and can be handled in air. On the contrary, $FeH_2(CO)_2[P(OCH_3)_3]_2$ and $FeH_2(CO)_2$ $[P(OC_2H_5)_3]_2$ are moderately stable at room temperature. They decompose with hydrogen evolution to give complex mixtures of $Fe(CO)_4[P(OR)_3]$, and $Fe(CO)_3[P(OR)_3]_2$ (as evidenced by IR analysis) and unidentified red iron complexes. However, they can be stored with little, if any, decomposition, at $-20°C$, in the dark, under argon. All the complexes are insoluble in water and soluble in common organic solvents at room temperature (pentane, hexane, diethyl ether, THF, acetone, dichloromethane and chloroform). Spectral characteristics are listed in Table I.

References

1. S. Cenini, F. Porta, and M. Pizzotti, *Inorg. Chim. Acta.*, **20**, 119 (1976); F. Porta, S. Cenini, S. Giordano, and M. Pizzotti, *J. Organomet. Chem.*, **150**, 261 (1978).
2. H. Berke, W. Bankhardt, G. Huttner, J. von Seyerl, and L. Zsolnai, *Chem. Ber.*, **114**, 2754 (1981).
3. H. Berke, G. Huttner, and L. Zsolnai, *Chem. Ber.*, **114**, 3549 (1981).
4. U. Schubert and M. Knorr, *Inorg. Chem.*, **28**, 1765 (1989).
5. J. J. Brunet, F. B. Kindela, D. Labroue, and D. Neibecker, *Inorg. Chem.*, **29**, 4152 (1990).
6. J. J. Brunet, *Chem. Rev.*, **90**, 1041 (1990).
7. R. B. King and F. G. A. Stone, *Inorg. Synth.*, **7**, 193 (1963).
8. L. W. Arndt, C. J. Bischoff, and M. Y. Darensbourg, *Inorg. Synth.*, **26**, 335 (1989).

38. THIONITROSYL COMPLEXES OF RUTHENIUM(II)

Submitted by K. C. JAIN* and U. C. AGARWALA*
Checked by T. W. DEKLEVA,† P. LEGZDINS‡ and J. C. OXLEY†

The chemistry of transition metal nitrosyl complexes has grown during the past three decades. The thionitrosyl complexes have, however, eluded synthesis until recently, apparently as a result of a lack of suitable thionitrosylating agents; the literature contains syntheses of only a few thionitrosyl complexes. The first exploratory work in this area was carried out by Chatt, et al.,[1] who prepared thionitrosyl complexes of osmium, rhenium, and molybdenum in poor yield via reactions of respective metal nitrides with elemental sulfur, propenesulfide-, or disulfur dichloride. An easier alternative approach has been reported[2a,b] that utilizes the reaction of $(NSCl)_3$ with transition metal complexes. The syntheses of two thionitrosyl derivatives of ruthenium are given here.

A. TRICHLORO(THIONITROSYL)BIS(TRIPHENYL-PHOSPHINE)-RUTHENIUM(II)

$$3RuCl_2(PPh_3)_3 + 2(NSCl)_3 \xrightarrow{\text{THF/CH}_2\text{Cl}_2} Ru(NS)Cl_3(PPh_3)_2 + \ldots$$

Trichloro(thionitrosyl)bis(triphenylphosphine)ruthenium(II)was previously prepared by the reaction of trithiazyl trichloride in THF with $RuCl_3 \cdot xH_2O$ and triphenylphosphine in absolute ethanol.[3] The product crystallizes over a period of several hours, and the yield is very poor. The same complex can be prepared from dichlorotris(triphenylphosphine)ruthenium(II)[4a] in CH_2Cl_2 and trithiazyl trichloride[5] in tetrahydrofuran (THF) in a better yield and in less time.

Procedure

After 2.87 g (3 mmol) of $RuCl_3(PPh_3)_2$ is placed in a 250-mL round-bottomed flask, 75 mL of CH_2Cl_2 is added, followed by the addition of a solution of $(NSCl)_3$ (0.5 g, 2 mmol) in 30 mL of dry THF. The mixture is stirred for 30 min and 100 mL of methanol is added. $Ru(NS)Cl_3(PPh_3)_2$ precipitates as shining brown crystals. The precipitation is completed within 10–15 min. The

* Department of Chemistry, Indian Institute of Technology, Kanpur-208 016, Uttar Pradesh, India.
† Department of Chemistry, University of British Columbia, Vancouver, B.C., V6T1Y6.

crude product is filtered, washed with methanol (4×10 mL) and ether (6×10 mL) and dried *in vacuo* (1 h). The crude complex is recrystallized by dissolving it in boiling CH_2Cl_2 (200 mL), evaporated to small volume (~ 20 mL) and adding 50 mL of methanol, after which shining brown crystals appear. These are centrifuged or filtered through a sintered-glass filtered funnel and washed with methanol (2×10 mL) and ether (4×10 mL) and dried *in vacuo* (2 h). Yield, 0.6 g ($\sim 25\%$ based on ruthenium complex).

Anal. Calcd. for $C_{36}H_{30}Cl_3NSP_2Ru$: C, 55.6; H, 3.8; Cl, 13.6; S, 4.1; N, 1.8; P, 8.0. Found: C, 54.22; H, 3.84; Cl, 13.7; S, 4.0; N, 1.8; P, 7.8.

Properties

Trichlorothionitrosylbis(triphenylphosphine)ruthenium(II) is a reddish brown crystalline substance, which melts at 176–178° in air. The IR spectrum of the compound shows a strong absorption both in KBr pellet as well as in Nujol mull band due to $v(NS)$ at 1310 cm^{-1}. The compound is sparingly soluble in dichloromethane, chloroform, and benzene but insoluble in alcohol, ether, and hexane. The compound is stable in air for an indefinite period. The X-ray diffraction pattern of powdered $Ru(NS)Cl_3(PPh_3)_2$ shows it to be isomorphic with $Ru(NO)Cl_3(PPh_3)_2$.[6]

B.   TRICHLORO(THIONITROSYL)BIS(TRIPHENYLARSINE)-RUTHENIUM(II)

$$RuCl_3(AsPh_3)_2 \cdot MeOH + 2(NSCl)_3 + 2(AsPh_3) \xrightarrow{\text{THF/CH}_2\text{Cl}_2}$$

$$Ru(NS)Cl_3(AsPh_3)_2 + \ldots$$

Trichloro(thionitrosyl)bis(triphenylarsine)ruthenium(II) was prepared earlier in a poor yield by keeping overnight a mixture of $RuCl_3 \cdot xH_2O$ and triphenylarsine in ethanol and trithiazyl trichloride in THF. The procedure described below is based on the reaction of $(NSCl)_3$ in THF with $RuCl_3(AsPh_3)_2 \cdot MeOH$[4b] in a slightly better yield and in considerably less time.

Procedure

Solutions of 1.7 g (2 mmol) of trichlorobis(triphenylarsine) monomethanol ruthenium(III) in 40 mL of CH_2Cl_2 and of 1.0 g (4 mmol) of trithiazyl trichloride in 30 mL of dry THF are mixed together rapidly. The initial mixing of the two solutions results in the formation of a precipitate. The mixture is stirred under nitrogen atmosphere for 30 min and 100 mL of methanol is added to it. The mixture was filtered and the filtrate is kept on the water bath for 10 min to remove dichloromethane and a solution of 1.2 g (4 mmol) of triphenylarsine in 20 mL of boiling methanol is added, which causes immediate precipitate formation.* After cooling the mixture to room-temperature orange-brown glistening plates are separated by centrifuging the mixture or by filtering it through a sintered-glass crucible. These are washed with methanol (5 × 10 mL) and ether (2 × 10 mL). The orange-brown complex is extracted in boiling dichloromethane (5 × 50 mL) and the solution is evaporated to 25 mL. Then 100 mL of hot methanol is added to it and allowed to crystallize at room temperature for 10 min, after which orange brown crystals appear. These are washed with methanol (2 × 10 mL) and ether (2 × 10 mL) and dried *in vacuo* (2 h). Yield: 0.6 g [~ 40% based on $RuCl_3(AsPh_3)_2 \cdot MeOH$].

Anal. Calcd. for $C_{36}H_{30}Cl_3NSAs_2Ru$: C, 49.9; H, 3.5; Cl, 12.2; S, 3.7; N, 1.6; As, 17.3. Found: C, 49.8; H, 3.5; Cl, 12.4; S, 3.5; N, 1.7; As, 17.1.

Properties

Trichloro(thionitrosyl)bis(triphenylarsine)ruthenium(II) forms orange-brown glistening plates that do not melt upto 260°C in air. The infrared spectrum of the compound in KBr shows a strong absorption band at $1300 \, cm^{-1}$ assigned to $\nu(NS)$. The complex is moderately soluble in dichloromethane, chloroform, and benzene but insoluble in alcohol, ether, and hexane. The compound is stable in air for an indefinite period. X-ray diffraction pattern of powdered $Ru(NS)Cl_3(AsPh_3)_2$ shows it to be isomorphic with $Ru(NO)Cl_3(PPh_3)_2$.[6]

References

1. M. W. Bishop, J. Chatt and J. R. Dilworth, *J. Chem. Soc. Dalton Trans.* **20**, 1979, 1.
2. (a) K. K. Pandey, D. K. M. Raju, H. L. Nigam, and U. C. Agarwala, "Proceedings of the Indian National Science Academy," *Proc. Indian Nat. Sci. Acad.*, **48**, 16–64 (1982); (b) T. J. Greenhough, B. W. S. Kolthammer, P. Legzdins, and J. Trotter, *Inorg. Chem.*, **18**, 3548 (1979).
3. K. K. Pandey and U. C. Agarwala, *Z. Anorg. Allgem. Chem.*, **461**(2), 231 (1980).

*Although 1.2 g of $AsPh_3$ does not dissolve in 20 mL of MeOH at room temperature, it dissolves in boiling methanol. One may add instead a $AsPh_3$ slurry in methanol prepared at room temperature to the warm concentrated reaction mixture. It will have practically no effect on the yield of the complex.

4. (a) P. S. Hallmar, T. A. Stephenson, and G. Wilkinson, *Inorg. Synth.*, **12**, 237 (1970); (b) T. A. Stephenson and G. Wilkinson, *J. Inorg. Nucl. Chem.*, **28**, 945 (1966).
5. W. L. Jolly and K. D. Maguire, *Inorg. Synth.*, **9**, 102 (1967).
6. B. L. Haymore and J. A. Ibers, *Inorg. Chem.*, **14**, 3060 (1975).

39. *trans*-DICHLORO TETRAMINE COMPLEXES OF RUTHENIUM(III)

Submitted by CHI-MING CHE,* TAI-CHU LAU,* and
CHUNG-KWONG POON*
Checked by JILL GRANTHAM† and THOMAS G. RICHMOND†

Linear and cyclic polyamine complexes of ruthenium(III) have been little studied relative to those of first-row transition elements such as cobalt(III) and nickel(II), presumably due to a lack of general routes of syntheses. A convenient and reproducible method for the syntheses of *trans*-[RuLCl$_2$],[+] where L is any multidentate aliphatic ligand with a four-amine donor combination, has been developed.[1,2] Procedures for the preparation of *trans*-[RuLCl$_2$]$^+$ [L = (en)$_2$,‡ 2,3,2,-tet,‡ and cyclam‡] are described here. The literature method is closely followed except that the much safer hexafluorophosphate(1 −) anion is employed in place of perchlorate.

A. *trans*-DICHLOROBIS(1,2-ETHANEDIAMINE)RUTHENIUM(III) HEXAFLUOROPHOSPHATE

$$K_2[RuCl_5(OH_2)] + 2en + NH_4[PF_6] \rightarrow trans\text{-}[Ru(en)_2Cl_2][PF_6] + 2KCl$$
$$+ NH_4Cl + H_2O$$

Procedure

Potassium aquapentachlororuthenate(III), K$_2$[RuCl$_5$(OH$_2$)] [Johnson Matthey*] (1.0 g, 2.7 mmol), suspended in methanol (100 mL, A.R. grade), is placed in a 500 conical flask fitted with a reflux condenser. This mixture is heated at reflux with vigorous stirring for 15 min. 1,2-Ethanediamine (0.33 g, 5.5 mmol) dissolved in methanol (150 mL) is added dropwise (about 1 drop

---

* Department of Chemistry, University of Hong Kong, Pokfulam Road, Hong Kong.
† Department of Chemistry, University of Utah, Salt Lake City, UT 84112.
‡ Abbreviations: en = 1,2-ethanediamine; 2,3,2-tet = *N*,*N*′-bis(2-aminoethyl)-1,3-propane-diamine; cyclam = 1,4,8,11-tetraazacyclotetradecane.

per second) through the condenser while the stirring and refluxing are maintained.

■ **Caution.** *1,2-Ethanediamine is toxic and an irritant. Manipulation should be performed in a fume hood.*

The addition process takes about 3 h for completion. Heating at reflux is continued (12–16 h). The yellowish brown solution is allowed to cool and then filtered to remove any unreacted $K_2[RuCl_5(OH_2)]$. The filtrate is rotary-evaporated to dryness. The brown solid is dissolved in a minimum amount of 0.1 M hydrochloric acid (~ 30 mL) and the suspension is then filtered. Ammonium hexafluorophosphate(1 −) (~ 4 g) is added to this solution, immediately precipitating the orange-yellow product. The solid is removed by filtering through a medium porosity frit, washed with ice-cold water (~ 10 mL), and dried *in vacuo.* The crude product is recrystallized by dissolving it in hot 0.1 M hydrochloric acid (~ 40 mL). Golden-yellow platelets appear on cooling overnight at ~ 4° in a refrigerator and are isolated as noted above. Yield: 0.6 g (51%).*

Anal. Calcd. for $C_4H_{16}N_4Cl_2PF_6Ru$: C, 10.99; H, 3.66; N, 12.82. Found: C, 11.02; H, 3.62; N, 13.24.

B. *trans*-[*R,S*)-*N,N'*-BIS(2-AMINOETHYL)-1,3-PROPANEDIAMINE]DICHLORORUTHENIUM(III) HEXAFLUOROPHOSPHATE

$$K_2[RuCl_5(OH_2)] + 2,3,2\text{-tet} + NH_4[PF_6] \rightarrow trans\text{-}[Ru(2,3,2\text{-tet})Cl_2][PF_6]$$
$$+ 2KCl + NH_4Cl + H_2O$$

Procedure

This compound is prepared from $K_2[RuCl_5(OH_2)]$ (1.0 g, 2.7 mmol) and 2,3,2-tet (0.43 g, 2.7 mmol) by the method described above for the bis(1,2-ethanediamine) analog. Yield: 0.7 g (54%).†

Anal. Calcd. for $C_7H_{20}N_4Cl_2PF_6Ru$: C, 17.61; H, 4.19; N, 11.74. Found: C, 17.76; H, 4.15; N, 11.96.

---

* Using $K_2[RuCl_5(OH_2)]$ from other sources may result in poorer product yields.

† The checkers performed these reactions at 50% of the scale reported and found the product yields for *trans*-[Ru(en)$_2$Cl$_2$]PF$_6$ and *trans*-[Ru(2,3,2-tet)Cl$_2$]PF$_6$ were 40 and 26%, respectively.

C. *trans*-DICHLORO(1,4,8,11-TETRAAZACYCLO-
TETRADECANE)RUTHENIUM(III) CHLORIDE
DIHYDRATE

$$K_2[RuCl_5(OH_2)] + cyclam \rightarrow trans\text{-}[Ru(cyclam)Cl_2]Cl + 2KCl + H_2O$$

Procedure

This compound is prepared by essentially the same method as that for the
bis(1,2-ethanediamine)   analog   described   in   Section   A.   Potassium
aquapentachlororuthenate(III) [Johnson Matthey] (1.0 g, 2.7 mmol), suspended in methanol (100 mL, A.R. grade), is placed in a 500-mL conical flask
fitted with a reflux condenser. The mixture is heated at reflux with vigorous
stirring for 15 min. 1,4,8,11-Tetraazacyclotetradecane (cyclam, prepared by
the method of Barefield et al.,[3] 0.53 g, 2.7 mmol) dissolved in methanol
(150 mL, A.R. grade) is added dropwise through the condenser at about
1 drop per second. The addition process takes about 3 h for completion.
Heating at reflux is continued overnight (12–16 h). The yellowish brown
solution is cooled and filtered to remove any unreacted $K_2[RuCl_5(OH_2)]$.
The filtrate is rotary-evaporated to dryness and the solid redissolved in a
minimum amount ($\sim$ 40 mL) of 0.1 M boiling HCl. The solution is filtered
while hot, and 12 M. HCl (2 mL) is added to the filtrate, which is cooled
overnight at $\sim 4°C$ in a refrigerator. Light brown leaflets of the product as
the dihydrate slowly crystallize. The solids are removed by filtering off,
washed with ethanol ($\sim$ 10 mL) and diethyl ether ($\sim$ 15 mL), and dried *in
vacuo* at room temperature. The complex was characterized as the dihydrate
with   characteristic   water   bands   observed   at   about $3300\ cm^{-1}$ and
$1630\ cm^{-1}$. Yield: 0.46 g (42%).

Anal. Calcd. for $C_{10}H_{24}N_4Cl_3Ru\cdot 2H_2O$: C, 27.06; H, 6.36; N, 12.62. Found:
C, 27.2; H, 6.20; N, 12.50.

Properties

All the prepared complexes are highly colored and give well-formed crystals
that appear to be stable indefinitely in the solid state. They are low-spin
monomeric species (μ_{eff} lies between 2.1 and 2.2 Bm and Λ between 100 and
$110\ \Omega^{-1}\ cm^2\ mol^{-1}$ in water at room temperature). All these complexes are
characterized by the presence of an intense ligand-to-metal charge transfer
band. In CH_3CN, *trans*-$[Ru(en)_2Cl_2][PF_6]$ absorbs at 346 (ε 5740
$M^{-1}\ cm^{-1}$), *trans*-$[Ru(2,3,2\text{-tet})Cl_2][PF_6]$ absorbs at 353 (ε 4480) and
trans-$[Ru(cyclam)Cl_2]Cl$ absorbs at 357 nm (ε 2260).

References

1. C. K. Poon and C. M. Che, *J. Chem. Soc. Dalton Trans.*, **1988**, 756.
2. C. K. Poon and C. M. Che, *Inorg. Chem.*, **20**, 1640 (1981).
3. E. K. Barefield, F. Wagner, A. W. Herlinger, and A. R. Dahl, *Inorg. Synth.*, **16**, 220 (1976).

40. *trans*-DICHLOROBIS(1,2-ETHANEDIAMINE)COBALT(III) CHLORIDE

Submitted by W. GREGORY JACKSON*
Checked by STANLEY KIRSCHNER,† THADDEUS J. GISH,† and STEVEN BARBER†

$$4\text{-}trans\text{-}[Co(en)_2(NO_2)_2]NO_3 + 16HCl + 5CH_2O + 4(NH_2)_2CO \rightarrow$$
$$4\text{-}trans\text{-}[Co(en)_2Cl_2]Cl\cdot HCl\cdot 2H_2O + 6N_2 + 9CO_2 + 11H_2O$$

A preparation of the *trans*-$[Co(en)_2Cl_2]^+$ ion has appeared in *Inorganic Syntheses* on three occasions.[1-3] The last[3] of these is the most economical with respect to materials and especially time, but the very low solubility of the *trans*-$[Co(en)_2Cl_2]NO_3$ product precludes its general usefulness. However, the modification detailed below leads to the well-known *trans*-$[Co(en)_2Cl_2]\cdot Cl\cdot HCl\cdot 2H_2O$ (or $[H(OH_2)_2]\cdot[trans\text{-}Co(en)_2Cl_2]\cdot Cl_2$) salt[4] (en = 1,2-ethanediamine), which on drying[1] affords the useful and very soluble *trans*-$[Co(en)_2Cl_2]Cl$.

The synthesis employs unrecrystallized *trans*-$[Co(en)_2(NO_2)_2]NO_3$, obtained[5] in 90% yield within 2.5 h from basic reagents‡. The product, washed with methanol and then ether and air-dried, is sufficiently pure for the next step.

Urea is well known as a scavenging reagent for HNO_2, and in the following modified synthesis it may be included to reduce toxic fumes (NO_2):

$$trans\text{-}[Co(en)_2(NO_2)_2]^+ + 2HCl \rightarrow trans\text{-}[Co(en)_2Cl_2]^+ + 2HNO_2$$
$$2HNO_2 + (NH_2)_2CO \rightarrow 2N_2 + CO_2 + H_2O$$

---

* University College (N.S.W.), Canberra, A.C.T., Australia 2600.
† Department of Chemistry, Wayne State University, Detroit, MI 48202.
‡ In that preparation it is advisable to dilute each of the 1,2-ethanediamine and nitric acid with twice its mass of ice prior to their careful mixing.

Less well known is the action of formaldehyde, which is used to destroy NO_3^- *without* reduction of cobalt(III) to cobalt(II):

$$4HNO_3 + 5H_2CO \rightarrow N_2 + 5CO_2 + 7H_2O$$

The complete reaction requires ~ 3 h and gives a yield of 57–69% based on *trans*-[Co(en)$_2$(NO$_2$)$_2$]NO$_3$, or 46–64% based on $CoCl_2 \cdot 6H_2O$. It is also possible to convert *trans*-[Co(en)$_2$Cl$_2$]NO$_3^3$ to *trans*-[Co(en)$_2$Cl$_2$] Cl·HCl·2H$_2$O (87–94% yield) using the procedure below (but omitting urea).

Procedure

A mixture of *trans*-[Co(en)$_2$(NO$_2$)$_2$]NO$_3$ (100 g, 0.30 mol) in HCl (32%; 350 mL) containing urea (20 g; 0.33 mol) and formalin (40% w/v HCHO, 50 mL; 0.67 mol) in a 2-L flask is heated with occasional stirring on a steam bath. Within minutes, there is copious frothing, as the initially orange solution rapidly darkens to a red-brown. This quickly subsides, and then the solution slowly changes to black-green. After 2.5 h, the product mixture is cooled on ice, and swirled occasionally to induce crystallization.* After 0.5 h, the fine green plates of *trans*-[Co(en)$_2$Cl$_2$]Cl·HCl·2H$_2$O are collected by filtration.† Dull green metamorphs result on thorough washing with ethanol (3 × 100 mL) and diethyl ether (1 × 100 mL), which removes the lattice water and hydrogen chloride; this process is completed by oven drying at 110°C. Typical yield: 49 g (57%).

Anal. Calcd. for [Co(C$_4$H$_{16}$N$_4$)Cl$_2$]Cl: C, 16.83; H, 5.65; N, 19.63; Cl, 37.26. Found: C, 17.0; H, 5.6; N, 19.6; Cl, 37.4%. The product shows a single-line ^{13}C NMR spectrum (D$_2$O), and its visible absorption spectrum (ε_{620}(max) 37.5) in 0.01 M HCl is in close agreement with the literature value.[2]

References

1. J. C. Bailar, Jr., *Inorg. Synth.*, **2**, 222 (1946).
2. J. Springborg and C. E. Schäffer, *Inorg. Synth.*, **14**, 63 (1974).
3. J. Zektzer, *Inorg. Synth.*, **18**, 73 (1978).
4. J. Williams, *Inorg. Synth.*, **13**, 232 (1972).
5. H. F. Holtzclaw, D. P. Sheetz, and B. D. McCarty, *Inorg. Synth.*, **4**, 177 (1953).

* If the mixture is stirred continuously, or left too long, some *cis*-[Co(en)$_2$Cl$_2$]Cl crystallizes with the trans isomer.

† The addition of nitric acid (60 mL) to the filtrate results in the deposition of *cis*-[Co(en)$_2$Cl$_2$]NO$_3$ (18.0 g). It contains a little *trans*-[Co(en)$_2$Cl$_2$]NO$_3$, which is difficult to remove by recrystallization.

41. RESOLUTION OF THE DODECAAMMINEHEXA-
μ-HYDROXO-TETRACOBALT(III) ION

Submitted by TAKAJI YASUI,* TOMOHARU AMA,* and
GEORGE B. KAUFFMAN†
Checked by DUŠAN J. RADANOVIĆ‡ and MILOŠ I. DJURAN‡

Dodecaamminehexa-μ-hydroxo-tetracobalt(III) salts (ammine-hexol salts), first prepared by Jørgensen,[1] are analogous to the corresponding tris(1,2-ethanediamine) compounds with the 1,2-ethanediamine (ethylenediamine) molecule replaced by the bidentate ligand

$$\left[\begin{array}{c} H \\ O \\ \diagdown \\ \diagup \\ O \\ H \end{array} Co(NH_3)_4 \right]^{+}$$

Because of the then prevalent view that optical activity was always connected with carbon atoms, a number of Alfred Werner's contemporaries argued that the optical activity of all the coordination compounds that he had resolved, beginning in 1911, was somehow due to the 1,2-ethanediamine or bipyridyl molecules or to the oxalate ions that they contained, even though these symmetric ligands are themselves optically inactive. Therefore, in 1914 Werner successfully resolved the ammine–hexol chlorides or bromides, which do not contain carbon, using silver (+)-3-bromocamphor-9-sulfonate (silver[(1R)-(*endo,anti*)]-3-bromo-1,7-dimethyl-2-oxobicyclo[2.2.1]heptane-7-methanesulfonate) as a resolving agent.[2] This resolution silenced even Werner's most skeptical opponents, unequivocally proved his postulated octahedral configuration for the cobalt atom, and confirmed his long-held view that no fundamental difference exists between organic and inorganic compounds.

However, Werner's classic resolution involves a complicated procedure that often gives isomers of low optical purity, as he himself admitted.[2] Mason and Wood[3] also resolved ammine–hexol salts, but they did not give a detailed procedure. Kudo and Shimura[4] modified Werner's resolution

* Department of Chemistry, Faculty of Science, Kochi University, Akebono-cho, Kochi 780, Japan.
† Department of Chemistry, California State University, Fresno, CA 93740.
‡ Department of Chemistry, Faculty of Science, Svetozar Markovic University, Kragujevac 34000, Yugoslavia.

method by using sodium bis(μ-d-tartrato)diantimonate(III) as a resolving reagent. They obtained the diastereomer as four fractions, but only the first and second crops indicated relatively high optical purity. The convenient procedure described below, which is a modification of Kudo and Shimura's method, gives an optically pure product of high optical activity.

Racemic ammine–hexol iodide is obtained by converting the sulfate, which is prepared according to the procedure of Kauffman and Pinnell,[5] to the chloride with barium chloride solution and then by metathesis with sodium iodide solution. The racemic ammine–hexol iodide is stirred with sodium bis(μ-d-tartrato)diantimonate(III) in a weakly acidic solution. The precipitated (+)-diastereomer is converted to the (+)-bromide with sodium bromide. The (−)-enantiomer is isolated as the sparingly soluble sulfate by adding ammonium sulfate to the filtrate from the (+)-diastereomer precipitation, and then converted into the chloride by adding barium chloride, and the (−)-enantiomer is finally isolated as the bromide by metathesis with sodium bromide.

A. (+)-DODECAAMMINEHEXA-μ-HYDROXO-TETRACOBALT(III) DI[BIS(μ-d-TARTRATO)-DIANTIMONATE(III)] IODIDE DODECAHYDRATE

$$[Co\{(OH)_2Co(NH_3)_4\}_3](SO_4)_3 + 3BaCl_2 \rightarrow$$
$$[Co\{(OH)_2Co(NH_3)_4\}_3]Cl_6 + 3BaSO_4\downarrow$$

$$[Co\{(OH)_2Co(NH_3)_4\}_3]Cl_6 + 6NaI + 4H_2O \rightarrow$$
$$[Co\{(OH)_2Co(NH_3)_4\}_3]I_6 \cdot 4H_2O + 6NaCl$$

$$2(\pm)\text{-}[Co\{(OH)_2Co(NH_3)_4\}_3]I_6 \cdot 4H_2O + 2Na_2[Sb_2(d\text{-}C_4H_2O_6)_2]\cdot 5H_2O \rightarrow$$
$$(+)\text{-}[Co\{(OH)_2Co(NH_3)_4\}_3][Sb_2(d\text{-}C_4H_2O_6)_2]_2I_2 \cdot 12H_2O\downarrow$$
$$+ (-)\text{-}[Co\{(OH)_2Co(NH_3)_4\}_3]I_6 + 4NaI + 6H_2O$$

Procedure

To a solution containing 9.0 g of barium chloride dihydrate in 200 mL of cold water 8.0 g of hexol–ammine sulfate[5] is added, and the suspension is stirrred for about 5 min. The white precipitate of barium sulfate is removed by suction filtration, and a solution containing 30 g of sodium iodide in 40 mL of cold water is added to the filtrate with stirring. After the solution has been cooled in an ice bath for about 15 min, the fine crystals of the hexol iodide are filtered, washed with a solution containing 2 g of sodium iodide in 10 mL of cold water, and then with 20 mL of ethanol, and air-dried. The yield is 10.2 g. The hexol iodide is used for the resolution without further recrystallization.

Anal. Calcd. for $[Co\{(OH)_2Co(NH_3)_4\}_3]I_6 \cdot 4H_2O$ $(Co_4H_{50}O_{10}N_{12}I_6)$: C, 0.00; H, 3.66; N, 12.22. Found: C, 0.11; H, 3.79; N, 12.21.

Racemic ammine–hexol iodide tetrahydrate (8.25 g, 0.006 mol) is dissolved in 300 mL of 0.001 M hydrochloric acid. A solution containing sodium bis-(μ-d-tartrato)diantimonate(III) pentahydrate (4.2 g, 0.00625 mol) in 100 mL of water (this solution is acidic[6]) is added to the first solution. A white-brown precipitate appears immediately in the solution, but it is not the desired diastereomer. (This precipitate may contain other compounds, e.g., ($\pm$)-$[Co\{(OH)_2Co(NH_3)_4\}_3][Sb_2(d\text{-}C_4H_2O_6)_2]_3$). The desired diastereomer in the form of a pale brown crystalline powder is formed by stirring the solution at 20° for 30 min and is accompanied by disappearance of the white-brown precipitates. The crystalline powder is collected by suction filtration and suspended again in 200 mL of cold 0.001 M hydrochloric acid. The suspension is stirred for 4–5 min. The powder is collected by suction filtration on a sintered-glass filter and is used as described in Section B without being dried completely. Therefore yield was not calculated.

Anal. (of Sample of Air-Dried Product). Calcd. for $[Co\{(OH)_2Co(NH_3)_4\}_3]$ $[Sb_2(d\text{-}C_4H_2O_6)_2]_2I_2 \cdot 12H_2O$ $(Sb_4Co_4C_{16}H_{74}O_{42}N_{12}I_2)$: C, 9.22; H, 3.58; N, 8.07. Found: C, 9.32; H, 3.45; N, 7.94.

The filtrate may be stored in an ice bath to obtain the $(-)_{589}$-enantiomer of the hexol bromide, but it should be used as described in Section C as soon as possible to minimize racemization.

B. (+)-DODECAAMMINEHEXA-μ-HYDROXO-TETRACOBALT(III) BROMIDE DIHYDRATE

$$(+)\text{-}[Co\{(OH)_2Co(NH_3)_4\}_3][Sb_2(d\text{-}C_4H_2O_6)_2]_2I_2 \cdot 12H_2O + 6NaBr \rightarrow$$
$$(+)\text{-}[Co\{(OH)_2Co(NH_3)_4\}_3]Br_6 \cdot 2H_2O\downarrow + 2Na_2[Sb_2(d\text{-}C_4H_2O_6)_2]$$
$$+ 2NaI + 10H_2O$$

Procedure

The diastereomer powder obtained as described above is suspended in 30 mL of cold 0.001 M hydrochloric acid, saturated with sodium bromide, and the suspension is vigorously stirred for 3 min in an ice bath. The resulting brown crystals of the crude (+)-hexol bromide are collected by suction filtration, and washed with 5 mL of cold water containing 1.0 g of sodium bromide, and then with 10 mL of cold methanol.

The crude crystals are purified by recrystallization as follows. They are placed in 60 mL of cold 0.001 M hydrochloric acid, the mixture is stirred for

2–3 min, and insoluble materials are removed by suction filtration. To the filtrate 25 mL of cold 0.001 M hydrochloric acid, saturated with sodium bromide, is added, and the solution is cooled in an ice bath for 10 min. The resulting crystalline deposit is collected by suction filtration, washed with 20 mL of methanol, and air-dried. The yield is 2.7 g (85%, based on the assumption of equal amounts of each antipode in the racemic hexol iodide tetrahydrate).*

The specific and molecular rotations, $[\alpha]_{589} = +2640°$ and $[M]_{589} = +27,906°$ (in 0.001 M hydrochloric acid), and $[\alpha]_{589} = +2380°$ and $[M]_{589} = +25,156°$ ($c = 0.05\%$ in a 1 : 1 acetone : water mixture at 18°).† (For 0.05% solutions Werner reported $[\alpha]_{589.3} = +2620°$ and $[M]_{589.3} = +27,713°$ or $[\alpha]_{560} = +4446°$ and $[M]_{560} = +47,038°$ in 1 : 1 aqueous acetone.[2]) $\Delta\varepsilon_{614} = -13.20$ and $\Delta\varepsilon_{507} = +11.15$ in 0.01 M hydrochloric acid at 18°. (Kudo and Shimura reported $\Delta\varepsilon_{614} = -13.5$ and $\Delta\varepsilon_{505} = +11.3$.[4]) An additional recrystallization does not increase the optical purity.

C. (−)-DODECAAMMINEHEXA-μ-HYDROXO-TETRACOBALT(III) BROMIDE DIHYDRATE

$(-)\text{-}[\text{Co}\{(\text{OH})_2\text{Co}(\text{NH}_3)_4\}_3]\text{I}_6 + 3(\text{NH}_4)_2\text{SO}_4 \rightarrow$
$(-)\text{-}[\text{Co}\{(\text{OH})_2\text{Co}(\text{NH}_3)_4\}_3](\text{SO}_4)_3\downarrow + 6\text{NH}_4\text{I}$

$(-)\text{-}[\text{Co}\{(\text{OH})_2\text{Co}(\text{NH}_3)_4\}_3](\text{SO}_4)_3 + 3\text{BaCl}_2\cdot2\text{H}_2\text{O} \rightarrow$
$(-)\text{-}[\text{Co}[(\text{OH})_2\text{Co}(\text{NH}_3)_4\}_3]\text{Cl}_6 + 3\text{BaSO}_4\downarrow + 6\text{H}_2\text{O}$

$(-)\text{-}[\text{Co}\{(\text{OH})_2\text{Co}(\text{NH}_3)_4\}_3]\text{Cl}_6 + 6\text{NaBr} + 2\text{H}_2\text{O} \rightarrow$
$(-)\text{-}[\text{Co}\{(\text{OH})_2\text{Co}(\text{NH}_3)_4\}_3]\text{Br}_6\cdot2\text{H}_2\text{O}\downarrow + 6\text{NaCl}$

Procedure

To the filtrate from the (+)-diastereomer precipitation (Section A) a solution containing 1.5 g (0.00223 mol) of sodium bis(μ-d-tartrato)diantimonate(III) pentahydrate in 10 mL of water is added, and the solution is stirred at 20° for 5 min. The resulting white-brown precipitate‡ is removed by suction filtra-

---

* If the resolution is carried out on a larger scale, the optical purity of both antipodes is decreased.

† In 50% acetone–water the optical rotatory power of the optically active hexol–ammine bromide decreases with time. The half-life period of the intensity loss at 22° is about 55 mins.

‡ This precipitate is probably a mixture of (±)-diastereomer with (−)-enantiomer in excess, and (−)-hexol bromide obtained from it has low optical purity. If the procedure in these two sentences is omitted, the (−)-hexol bromide obtained is of low optical purity.

tion, and a solution containing 5.0 g of ammonium sulfate in 30 mL of water is added to the filtrate with stirring. Violet-brown crystalline flakes of the (−)-hexol sulfate appear immediately. After the solution is cooled in an ice bath for 30 min, the crystals are collected by suction filtration, washed with 10 mL of cold water, and then with 20 mL of methanol, and air-dried. The yield is 1.1 g.

Inasmuch as the hexol sulfate is sparingly soluble in water, it is converted into the chloride by metathesis with barium chloride solution and then isolated as the bromide as follows. The sulfate (1.1 g) is added to a solution containing 1.6 g of barium chloride dihydrate in 50 mL of cold 0.001 M hydrochloric acid, and the suspension is stirred for about 3 min. The white precipitate of barium sulfate is removed by suction filtration, and 15 mL of cold 0.001 M hydrochloric acid, saturated with sodium bromide, is added to the filtrate. The fine, dark brown crystalline deposit of (−)-hexol bromide dihydrate is filtered, washed with 10 mL of methanol, and air-dried. The yield is 1.0 g (31.5%, based on the assumption of equal amounts of each antipode in the racemic hexol iodide tetrahydrate).

Anal. Calcd. for $[Co\{(OH)_2Co(NH_3)_4\}_3]Br_6 \cdot 2H_2O(Co_4H_{46}O_8N_{12}Br_6)$: C, 0.00; H, 4.38; N, 15.89. Found: C, 0.04; H, 4.53; N, 15.96.

The specific rotation $[\alpha]_{589} = -4350°$. The molecular rotation $[M]_{589} = -46,005°$. $\Delta\varepsilon_{614} = +20.75$, and $\Delta\varepsilon_{507} = -17.86$. These values were obtained in 0.01 M hydrochloric acid at 18°.

Properties

The structure of the dodecaamminehexa-μ-hydroxo-tetracobalt(III) ion has been established by an X-ray crystal structure analysis of the racemic chloride,[7] and several investigations of the optical activity of the resolved isomers and related hexa-μ-hydroxo-tetracobalt(III)-type complex ions have been made.[2, 3, 8−10]

Both enantiomers of dodecaamminehexa-μ-hydroxo-tetracobalt(III) bromide are obtained as dihydrates. They are very soluble in water and rapidly undergo racemization. The highest specific rotation values that Werner[2] reported for 0.05% solutions are $[\alpha]_{589.3} = +2620°$ or $[\alpha]_{560} = +4446°$ for the (+)-bromide and $[\alpha]_{560} = -4500°$ for the (−)-bromide in 1 : 1 aqueous acetone.

The active bromides exhibit strongly anomalous rotatory dispersion.[2, 11, 12] Kudo and Shimura[4] reported that the circular dichroism (CD) intensity of the optically active ammine–hexol is affected by the anion coexisting in the solution, that the antipodes racemize readily in neutral and basic aqueous solutions, and that the racemization rate in acidic aqueous solution is slow enough for measurement of the CD spectra. On the basis of their detailed CD measurements they determined the absolute configuration of the $(+)_{589}$-antipode as Δ, based on a negative sign of the E_a component of the CoO_6 chromophore. Previously, two different assignments for the absolute configuration of this type of complex had been made.[3, 10]

References

1. S. M. Jørgensen, *Z. anorg. Chem.*, **16**, 184 (1898).
2. A. Werner, *Ber.*, **47**, 3087 (1914); for an annotated English translation, see G. B. Kauffman, *Classics in Coordination Chemistry, Part 1: The Selected Papers of Alfred Werner*, Dover Publications, New York, 1968, pp. 177–184.
3. S. F. Mason and J. W. Wood, *Chem. Commun.*, **1967**, 209.
4. T. Kudo and Y. Shimura, *Bull. Chem. Soc. Jpn.*, **52**, 1648 (1979).
5. G. B. Kauffman and R. P. Pinnell, *Inorg. Synth.*, **6**, 176 (1960).
6. Y. Yoshikawa and K. Yamasaki, *Coord. Chem. Rev.*, **28**, 205 (1979).
7. I. Sotöfte and E. Bang, *Acta Chem. Scand.*, **25**, 1164 (1971).
8. H. A. Goodwin, E. C. Gyarfas, and D. P. Mellor, *Austr. J. Chem.*, **11**, 426 (1958).
9. R. D. Kern and R. A. D. Wentworth, *Inorg. Chem.*, **6**, 1018 (1967).
10. I. Masuda and B. E. Douglas, *J. Coord. Chem.*, **1**, 189 (1971).
11. A. Werner, *Compt. rend.*, **159**, 426 (1914).
12. I. Lifschitz, *Z. physik. Chem.*, **105**, 46 (1923).

42. TRICARBONYL PHOSPHINE, PHOSPHITE, AND ARSINE DERIVATIVES OF COBALT(I)

Submitted by CHRISTA LOUBSER* and SIMON LOTZ*
Checked by JOHN E. ELLIS†

Pentacoordinated complexes of cobalt(I) may be readily prepared by the reductive cleavage of the cobalt–cobalt bond in $Co_2(CO)_8$ or its derivatives $[Co(CO)_3L]_2$:

$$[Co(CO)_3L]_2 \rightarrow 2[Co(CO)_3L]^-$$

where L may be a phosphine, phosphite, or arsine.

We found successful reduction to largely depend on the purity of the dimer $[Co(CO)_3L]_2$. Because of their low solubilities in nonpolar solvents (thereby facilitating purification), $[Co(CO)_3L]_2$ (L = PPh_3, $P(OPh)_3$ or $AsPh_3$), in

* Department of Chemistry, University of Pretoria, Pretoria 0002, South Africa.
† Department of Chemistry, University of Minnesota, Minneapolis, MN 55455.

particular, lend themselves to effective reduction with high yields of the anions. Further, depending on the electrophilic reagent employed for the stabilisation of $[Co(CO)_3L]^-$, it is possible to prepare cobalt(I) complexes of vastly varying stabilities. In the syntheses described here, chlorotrimethyl-stannane is used, affording compounds of high stability (decreasing in the order $PPh_3 > AsPh_3 > P(OPh)_3$). This enables them to be used as starting materials for a whole range of pentacoordinated organocobalt compounds, e.g., the analogous $[Co(Ph_3Sn)(CO)_3(PPh_3)]$, which proved a useful starting compound for stable cobalt(I) carbenes.[1]

The syntheses can be represented by the following set of general equations:

$$Co_2(CO)_8 + 2L \rightarrow [Co(CO)_3L]_2 + 2CO$$

$$[Co(CO)_3L]_2 + Na/Hg \rightarrow 2Na[Co(CO)_3L]$$

$$Na[Co(CO)_3L] + Me_3SnCl \rightarrow [Co(Me_3Sn)(CO)_3L] + NaCl$$

$$(L = PPh_3, P(OPh)_3, AsPh_3)$$

A. TRICARBONYL(TRIMETHYLSTANNYL) (TRIPHENYLPHOSPHINE) COBALT(I)

Procedure

■ **Caution.** *Organotin compounds and benzene are highly toxic and should be handled with extreme care. Drying tetrahydrofuran (THF) is hazardous and should be carried out with due caution.[2] Carbon monoxide (which is released during reactions) is a colorless, odorless gas of extreme toxicity. All reactions should be carried out in a well-ventilated hood.*

All reactions and sample preparations are carried out under inert atmosphere. Solvents are distilled under nitrogen atmosphere and dried in the following manner. Benzene is distilled from sodium. *n*-Pentane and *n*-hexane are distilled from sodium–lead alloy. Tetrahydrofuran is predried over calcium hydride and refluxed with sodium, whereafter benzophenone is added. After the solution turns blue, the tetrahydrofuran is distilled. Dichloromethane is distilled from phosphorous pentoxide. Commercial octacarbonyldicobalt* is recrystallized from hexane prior to use. Commercial triphenylphosphine† is used without further purification.

Glassware is oven-dried at 110°C for 24 h and left to cool under inert atmosphere. Glassware attached to the vacuum line should be heavy-walled and able to withstand pressures as low as 0.1 mmHg.

---

* Strem Chemicals, P.O. Box 212, Danvers, MA 01923.
† E. Merck, Darmstadt, Germany.

The method described for the preparation of $[Co(CO)_3(PPh_3)]_2$ is based on that of Manning.[3] $[Co(CO)_3(PPh_3)]_2$ is poorly soluble in most common solvents. It is therefore convenient to prepare the required amount immediately prior to use and to convert the entire batch.

In a 500-mL, two-necked, round-bottomed flask equipped with a magnetic stirring bar and a nitrogen inlet, a solution of 3.42 g (10 mmol) $Co_2(CO)_8$ in 120 mL of benzene is prepared while a stream of nitrogen is passed through. Then, 5.25 g (20 mmol) of triphenylphosphine is dissolved in 30 mL of benzene and added slowly to the contents of the flask while vigorous stirring is maintained. The flask is fitted with a condenser. A T junction, placed on top of the condenser, is connected to a supply of nitrogen and a mineral-oil bubbler. The mixture is refluxed on an oil bath at 80°C for 1 h. The reaction is complete when the initial orange-brown color has changed to red-brown. The heat is removed and the product allowed to cool and settle out. The benzene layer is carefully decanted. The flask is stoppered with a rubber stopper and the other inlet connected to the vacuum pump via a liquid nitrogen trap (− 196°C). The remaining solvent is then removed at reduced pressure (30°C, 0.5 mm). The fine red-brown powder is finally washed with 100 mL of pentane, the last traces of which are also subsequently removed *in vacuo*. The flask is stoppered and kept for the next preparation. Yield > 95%; IR: $v(CO)$ = 1950 cm^{-1} (s), 1977 cm^{-1} (w).[2] ($[Co(CO)_3(PPh_3)]_2$ is stable indefinitely at room temperature in the solid form, but gradually decomposes in solution.)*

A 1-L, two-necked, round-bottomed flask is placed under nitrogen atmosphere. The stopper is removed and under a stream of nitrogen, 500 g of mercury is introduced. A 1% sodium amalgam is prepared by carefully stirring 5 g of sodium into the mercury by adding 0.2-g portions at a time.[4] (**Warning:** *The formation of the amalgam is highly exothermic, and care should be taken to prevent overheating.*) After complete addition the amalgam is left to cool to room temperature, the $[Co(CO)_3(PPh_3)]_2$ is suspended in 300 mL of THF and added to the amalgam. The reaction mixture is stirred vigorously at room temperature with a mechanical stirrer. After 16 h, stirring is ceased and the flask left to stand for another 3 h to allow settling of the fine gray metal particles. The dark yellow solution is carefully decanted and filtered through a fine-porosity glass frit fitted on top of a 500-mL Schlenk tube[5] (Fig. 1).†

Filtration is aided by allowing a slight buildup of nitrogen pressure above the solution and by opening the bottom outlet. Further amounts of

---

* The checker prepared this and the other dimeric products in about the same yield in toluene, which is a much less hazardous solvent than benzene. Instead of refluxing the solvent, the reaction mixture in toluene was vigorously stirred at about 80°C.

† The checker used dry filter aid (Celite) to facilitate the filtration.

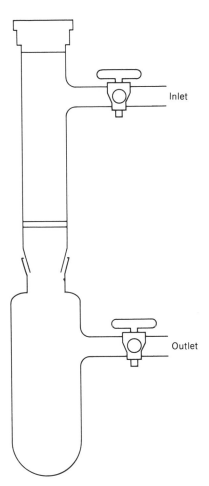

Fig. 1. Apparatus for filtration.

Na[Co(CO)$_3$(PPh$_3$)] are retrieved by carefully admitting 3×50 mL THF to the amalgam and filtering as before.

Then 3.2 g (16 mmol) chlorotrimethylstannane* is dissolved in 20 mL of THF and added to the Na[Co(CO)$_3$(PPh$_3$)] solution while stirring rapidly at room temperature. After 30 min stirring is ceased and the mixture left to allow settling of the formed NaCl. After filtering through a medium-porosity glass frit (using a similar setup as before), the NaCl is washed with 20 mL of dichloromethane. This washing is then filtered into the THF solution. The

* E. Merck, Darmstadt, Germany.

Schlenk tube is stoppered with a rubber stopper and the solution is evaporated under reduced pressure. The residue is taken up in a minimum volume (5–10 mL) dichloromethane : hexane (1 : 1) and introduced to an alumina* (neutral, activity III) column (15 cm × 3 cm) prepared in hexane. The product is eluated as a yellow band with hexane (100 mL). Subsequent removal of the solvent under vacuum and further drying (30°C, 0.5 mm, 1 h) gives typically 4.4–5.0 g (40–45% based on $[Co(CO)_3(PPh_3)]_2$) of orange yellow crystals (mp 173–175°C (dec)).

Anal. Calcd. for $C_{24}CoH_{24}O_3PSn$: C, 50.66; H, 4.25. Found: C, 50.86; H, 4.30. The product is sufficiently pure to be used as starting material in subsequent syntheses, but further purification may be effected by crystallization from hexane solution.

The experiment may be successfully scaled down (enabling the yield to be increased to 70%), but scaling up more than twice is not recommended because of the large volumes of THF required. Also, no concomitant increase in yield is obtained.

Properties

$[Co(Me_3Sn)(CO)_3(PPh_3)]$ crystallizes as oblong translucent yellow crystals. An infrared spectrum in CCl_4 solution consists of two bands in the carbonyl region: 2013 cm^{-1} (w) and 1936 cm^{-1} (vs). The ^1H NMR spectrum (benzene-d_6) shows a sharp singlet at δ 0.72 ppm (9H) due to the methyl protons and two multiplets centered at δ 6.97 ppm (9H) and δ 7.60 ppm (6H) due to the phenyl protons. The $^{31}P\{^1H\}$NMR spectrum (benzene-d_6, relative to 85% H_3PO_4) shows an extremely broad singlet at δ 53.4 ppm, while the $^{119}Sn\{^1H\}$NMR spectrum (benzene-d_6, relative to Me_4Sn) has a doublet at δ 117.3 ppm $[^2J(PSn) = 240.8$ Hz].

$[Co(Me_3Sn)(CO)_3(PPh_3)]$ is stable indefinitely at room temperature in the solid form under inert atmosphere. It dissolves readily in polar solvents and sparingly in nonpolar solvents.

B.   TRICARBONYL(TRIMETHYLSTANNYL) (TRIPHENYLPHOSPHITE) COBALT(I)

Procedure

■   **Caution.**   *Triphenyl phosphite is a toxic liquid.*
All reactions and sample preparations are carried out under inert atmosphere. Solvents are distilled under nitrogen atmosphere and dried as de-

---

* M. Woelm, Eschwege, Germany.

scribed in Section A. Commercial triphenyl phosphite* is used without further purification.

The method for the preparation of $[Co(CO)_3\{P(OPh)_3\}]_2$ is based on that described in Section A for $[Co(CO)_3(PPh_3)]_2$.

A solution of 5 mL (19 mmol) of triphenyl phosphite in 10 mL of benzene is added slowly to a rapidly stirred solution of 3.42 g (10 mmol) of octacarbonyldicobalt in a 250-mL round-bottomed flask. After the initial effervescence ceases, a condenser is placed on the flask and the reaction mixture stirred at 60°C for 2 h until the dark red color persists. The flask is cooled to room temperature and left to stand, thereby allowing the product to settle out. The benzene layer is decanted and the dimer washed repeatedly with 50-mL portions of warm hexane until the latter remains colorless.

The last traces of hexane are removed under vacuum to yield typically 7.8–8.0 g (86–88%) $[Co(CO)_3\{P(OPh)_3\}]_2$. IR: $v(CO) = 1966$ cm^{-1} (vs), 1989 cm^{-1} (sh)[2].

A 500-mL, two-necked, round-bottomed flask is equipped with a nitrogen inlet and flushed. Then 300 g of mercury is introduced under a stream of nitrogen. By adding 3 g of sodium (taking the necessary precautions as before), a 1% amalgam is prepared and allowed to cool to room temperature. The $[Co(CO)_3\{P(OPh)_3\}]_2$ is suspended in 100 mL of THF and added to the amalgam. The reaction mixture is stirred vigorously with a mechanical stirrer (300 rpm) for 10–15 min. The color changes from dark red to brown to gray.† The yellow solution, still containing the suspended metal particles, is quickly decanted‡ and filtered through a glass frit (fine porosity). Any remaining sodium cobaltate salt can be retrieved from the amalgam by rinsing twice with 30 mL THF.

Chlorotrimethylstannane 2.32 g (11.6 mmol) is dissolved in 20 mL of THF and added to the $Na[Co(CO)_3\{P(OPh)_3\}]$ solution while stirring vigorously. After 30 min the THF is completely removed under vacuum (-20°C, 0.5 mm). The residue is then repeatedly extracted with 20-mL portions of hexane until the latter becomes colorless. The orange-yellow hexane solution is filtered and reduced to dryness under vacuum to leave a dark yellow solid. If necessary, further purification can be accomplished by repeated dissolution of the product in a minimum volume hexane, followed by rapid cooling to -30°C in a Dry Ice–acetone bath.§ The crystals are dried *in vacuo* for 2 h

---

* Aldrich Chemical Company Inc., P.O. Box 2060, Milwaukee, WI 53201.

† The yellow solution appears gray as a result of the suspension of the fine metal particles therein.

‡ $Na[Co(CO)_3\{P(OPh)_3\}]$ is extremely unstable in solution (even at subzero temperatures), and manipulation thereof should be swift.

§ As $[Co(Me_3Sn)(CO)_3\{P(OPh)_3\}]$ appears to decompose on alumina, this means of purification is not recommended.

(0°C, 0.1 mm). The yield is typically 5.1–5.6 g (48–53%) based on [Co(CO)$_3$-{P(OPh)$_3$}]$_2$ (mp 91–92°C).

Anal. Calcd. for C$_{24}$H$_{24}$CoO$_6$PSn: C, 46.72; H, 3.92. Found: C, 46.73; H, 3.89.

An increase in yield can be obtained on scaling down the experiment.

Properties

[Co(Me$_3$Sn)(CO)$_3${P(OPh)$_3$}] crystallizes as pale yellow crystals. An infrared spectrum in hexane solution consists of two bands: v(CO) = 2030 cm^{-1} (w, sh) and 1961 cm^{-1} (vs). The ^{1}H NMR spectrum (benzene-d_6) shows a sharp singlet at δ 0.45 ppm (9H) due to the methyl protons and a triplet at δ 6.87 ppm (3H), a triplet at δ 7.03 ppm (6H), and a broad doublet at δ 7.31 ppm (6H) due to the phenyl protons. The phosphorus atom gives rise to an extremely broad singlet at δ 130.6 ppm on the ^{31}P{^{1}H} NMR spectrum (benzene-d_6, relative to 85% H$_3$PO$_4$). The ^{119}Sn{^{1}H}NMR spectrum shows a doublet [^2J(PSn) = 418.7 Hz] at δ 136.2 ppm (benzene-d_6, relative to Me$_4$Sn). In the solid form, [Co(Me$_3$Sn)(CO)$_3${P(OPh)$_3$}] can be exposed to air for very short periods of time, but it should not be kept above -20°C for longer than a few minutes.

C. TRICARBONYL(TRIMETHYLSTANNYL) (TRIPHENYLARSINE) COBALT(I)

Procedure

■ **Caution.** *Triphenylarsine is extremely toxic and should be handled with the greatest of care.*

Commercial triphenylarsine* is used as purchased.

In a 250-mL, two-necked, round-bottomed flask (flushed with nitrogen), a solution of 1.5 g (5 mmol) of octacarbonyldicobalt in 50 mL of benzene is prepared. Then 2.7 g (10 mmol) of triphenylarsine is dissolved in 20 mL of benzene and slowly added to the former solution while thorough stirring is maintained. After the initial gas evolution has subsided, the flask is fitted with a condenser. The solution is then stirred at 60°C. After 1 h the heat is removed and the contents of the flask allowed to cool to room temperature, where-upon the dimer settles as a fine purple-red solid. The benzene is carefully decanted and the product washed with 50-mL portions of warm hexane until the latter remains colorless. Subsequent *in vacuo* drying yields typically 3.2 g

---

* E. Merck, Darmstadt, Germany.

(71%) $[Co(CO)_3(AsPh_3)]_2$. An infrared spectrum in Nujol shows strong absorption at 1949 cm^{-1} in the carbonyl region.

In a 500-mL, two-necked, round-bottomed flask, a 1% sodium amalgam is prepared (see Section A) using 2.5 g of sodium and 250 g of mercury and left to cool to room temperature. The dimer is suspended in 100 mL of THF and added to the amalgam. Using a mechanical stirrer, the suspension is stirred vigorously for 20 min, after which the dark red color has vanished completely. The suspension is decanted quickly* and to it is added 1 g of chlorotrimethyl-stannane dissolved in 10 mL of THF. After stirring for 30 min at room temperature, the THF is removed under reduced pressure. The residue is repeatedly extracted with hexane (20–30-mL portions) until the solvent no longer becomes yellow. The hexane solution is filtered through a fine-porosity glass frit and reduced to complete dryness with the aid of the vacuum pump (0°C, 0.5 mm). The dark yellow product is taken up in 4 mL dichloromethane : hexane (1 : 1) and introduced to an alumina (neutral, activity IV) column (20 cm × 1 cm) prepared in hexane under inert atmosphere. The product is eluted as a yellow band with hexane, which is subsequently removed under vacuum. Further drying (25°C, 0.1 mm, 2 h) yields 2.9–3.2 g (66–73%) of primrose yellow crystals [mp 122–124°C (dec)].

Anal. Calcd. for $C_{24}H_{24}AsCoO_3Sn$: C, 47.03; H, 3.95. Found: C, 47.08; H, 3.88.

Properties

$[Co(Me_3Sn)(CO)_3(AsPh_3)]$ is sparingly soluble in hexane and soluble in benzene and more polar solvents. Dissolution in acetone results in rapid decomposition. The infrared spectrum in hexane shows $\nu(CO) = 1946$ cm^{-1} (vs) and 2015 cm^{-1} (w). The ^{1}H NMR spectrum (C_6D_6) exhibits a singlet at δ 0.73 ppm (9H) due to the methyl protons and multiplets centered at δ 6.98 ppm (9H) and δ -7.58 ppm (6H) due to the phenyl protons. The $^{119}Sn\{^1H\}$NMR spectrum is typified by a singlet at δ 143.6 ppm (benzene-d_6, relative to Me_4Sn).

$[Co(Me_3Sn)(CO)_3(AsPh_3)]$ should be stored below 0°C under inert conditions, but may be exposed to air for very short periods of time.

References

1. A. C. Filippou, E. Herdtweck, and H. G. Alt, *J. Organomet. Chem.*, **355**, 437 (1988).
2. R. W. Parry, *Inorg. Synth.*, **12**, 317 (1970).

* $Na[Co(CO)_3(AsPh_3)]$ is unstable and should be reacted with the chlorotrimethylstannane as soon as possible.

3. A. R. Manning, *J. Chem. Soc. (A)*, **1968**, 1135.
4. R. J. Angelici, *Synthesis and Technique in Inorganic Chemistry*, Saunders, Philadelphia, 1977, p. 153.
5. D. F. Shriver, *The Manipulation of Air-Sensitive Compounds*, McGraw-Hill, New York, 1969.

43. DECAKIS(ACETONITRILE)DIRHODIUM(II) TETRAFLUOROBORATE

Submitted by KIM R. DUNBAR* and LAURA E. PENCE*
Checked by JOANNA CZUCHAJOWSKA† and F. ALBERT COTTON†

While the existence of fully solvated transition metal monomers is fairly common,[1] binuclear compounds of this type are rare.[2-5] The advantages of solvated systems are numerous, allowing for designed syntheses of substitution products whose formation may be inhibited or otherwise influenced by the presence of strong donor ligands. Metal dimers that are stabilized solely by labile solvent ligands are expected to be valuable synthons for a variety of applications. Attempts to access these types of complexes have typically involved the action of strong acids on binuclear metal carboxylates.[4-6] The acidic medium, however, complicates the isolation of the desired solvated products and promotes decomposition. By taking advantage of the demonstrated use of trialkyloxonium reagents to esterify and remove bridging carboxylate ligands,[7,8] we have prepared the first example of an easily accessible solvated binuclear compound whose synthesis and crystallization are presented herein.

■ **Caution.** *Triethyloxonium tetrafluoroborate* $(1-)$ *is toxic, corrosive, a neurologic hazard, a suspected carcinogen, and moisture-sensitive, and should be stored under nitrogen in a freezer.*

$$Rh_2(O_2CCH_3)_4(MeOH)_2 + (Et_3O)[BF_4] \xrightarrow[\Delta]{CH_3CN/CH_2Cl_2}$$

$$[Rh_2(NCCH_3)_{10}][BF_4]_4$$

Procedure

Using standard Schlenk-line techniques, 200 mg of $Rh_2(O_2CCH_3)_4$-$(MeOH)_2{}^9$ (0.395 mmol), 5 mL of a 1 M solution of $(Et_3O)[BF_4]$ in CH_2Cl_2 (5.0 mmol, Aldrich), and 10 mL of CH_3CN (freshly distilled from calcium hydride) are refluxed in a 100-mL, three-necked, round-bottomed flask at

* Department of Chemistry, Michigan State University, East Lansing, MI 48824.
† Department of Chemistry, Texas A & M University, College Station, TX 77843.

atmospheric pressure for 7 days. The reaction is performed under N_2 without stirring; the reaction vessel is fitted with a 12-in West condenser. The methanol adduct of rhodium acetate is preferred over the anhydrous form; the latter undergoes much more sluggish reactions with triethyloxonium tetrafluoroborate (1 −). All glass joints should be lubricated with Apiezon H grease and well secured. The color of the reaction solution changes from purple to red within 2 h and gradually becomes deep orange over a period of several days. The initial purple color is due to the presence of $Rh_2(O_2CCH_3)_4(NCCH_3)_2$, which undergoes a facile reaction with Et_3O^+ to give the red-purple cation $[Rh_2(O_2CCH_3)_2(NCCH_3)_6]^{2+}$ and EtO_2CCH_3.[7] The conversion of $[Rh_2(O_2CCH_3)_2(NCCH_3)_6]^{2+}$ to $[Rh_2(NCCH_3)_{10}]^{4+}$ proceeds much more slowly and vigorous reflux conditions are required.

Between the period of several days and one week of reflux time, large orange block-shaped crystals of $[Rh_2(NCCH_3)_{10}][BF_4]_4$ form on the walls and bottom of the flask; these are collected by filtration under an inert atmosphere, washed three times with 5 mL amounts of CH_2Cl_2 followed by 10 mL of diethyl ether, and dried *in vacuo*. These crystals do not require any further purification. The filtrate from the reaction solution is promptly pumped to a low volume (~ 10 mL) and chilled at − 20° for 2 h. A crop of orange microcrystals admixed with orange solid is obtained in this manner. Reduction of the volume and subsequent chilling is repeated several times until a minimal volume of 1 mL is achieved. Additional product is obtained by adding 5–18 mL of diethyl ether to the above concentrated solution, chilling for 2 h, and decanting the solution from the solid. At this point, if the product is decidedly red instead of dark orange, indicating that some unreacted intermediate remains, the entire amount should be redissolved in acetonitrile and made to react further react with triethyloxonium tetrafluoroborate(11). On grinding, a crystalline sample should yield an orange rather than a red-purple powder.

The mixture of solid and microcrystalline material obtained in the aforementioned procedure is treated with three 5-mL portions of CH_2Cl_2 to remove excess triethyloxonium reagent and residues of vacuum grease. (Prolonged exposure to CH_2Cl_2 should be avoided as it will eventually decompose the compound.) This crude form of the product may be purified by dissolving ~ 200 mg in 10–20 mL of warm acetonitrile, filtering, and alternating cycles of reduction of the volume and chilling to − 20°C to produce microcrystals. The final amount of solvent (~ 1 mL) should be decanted in order to ensure that one obtains a clean product free of sticky residues. Combined yields are typically ~ 260 mg (70%). Single crystals may be grown by slow diffusion of diethyl ether or CH_2Cl_2 into a solution of the compound in acetonitrile. Highly concentrated solutions of the dimer coupled with an excessive amount of precipitating solvent should be avoided as this invariably results in an oiling of the product.

The 200-mg reaction scale, while small, has resulted in the highest overall yield. Doubling the scale not only significantly increases the reaction time, but tends to yield large amounts of $[Rh_2(O_2CCH_3)_2(NCCH_3)_6]^{2+}$ with only small quantities of $[Rh_2(NCCH_3)_{10}]^{4+}$. The purity of a given sample is easily established by ^1H NMR and infrared spectroscopy.

Properties

The compound $[Rh_2(NCCH_3)_{10}][BF_4]_4$ is fairly stable in air, both in solution and in the solid state, but exposure to light should be minimized because of its photochemical lability.[10] The title compound is hygroscopic, being prone to formation of the pink axial water adduct $[Rh_2(NCCH_3)_8(OH_2)_2]^{4+}$. The water is easily removed by redissolving the pink solid in acetonitrile. The compound exhibits two absorptions in the electronic spectrum in CH_3CN at $\lambda_{max}(nm) = 468$ ($\varepsilon = 3.90 \times 10^2\ M^{-1}\ cm^{-1}$) and 277 ($2.2 \times 10^4$) in the absence of air and minimum exposure to light. The salient features of the infrared spectrum (Nujol mull, CsI, cm^{-1}) are the three $C\equiv N$ stretches at 2342 (m), 2317 (m), and 2300 (w) as well as the strong $v(B—F)$ at 1024 (vs) and 1062 (vs). The ^1H NMR spectrum in CD_3CN performed under anaerobic conditions shows only one resonance at $+1.95$ ppm(s) due to rapid exchange of both axial and equatorial CH_3CN ligands with the deuteriated solvent. Cyclic voltammetric studies reveal a single irreversible reduction for $[Rh_2(NCCH_3)_{10}]\ [BF_4]_4$ at $E_{p,c} = -0.05$ V versus Ag/AgCl.

Anal. Calcd. for $C_{20}H_{33}F_{16}N_{10}B_4Rh_2$: C, 24.93; H, 3.13; F, 31.55. Found: C, 25.44; H, 3.58; F, 31.37.

Acknowledgment

The authors would like to thank Steven C. Haefner for his assistance.

References

1. F. A. Cotton and G. Wilkinson, *Advanced Inorganic Chemistry*, 5th ed, Wiley, New York, 1988.
2. K. R. Dunbar, *J. Am. Chem. Soc.*, **110**, 8247 (1988).
3. (a) F. Maspero and H. Taube, *J. Am. Chem. Soc.*, **90**, 7361 (1968); (b) C. R. Wilson, H. Taube, *Inorg. Chem.*, **14**, 2276 (1975).
4. J. M. Mayer and E. H. Abbott, *Inorg. Chem.*, **22**, 2774 (1983).
5. A. R. Bowen and H. Taube, *Inorg. Chem.*, **13**, 2245 (1974).

6. (a) P. Legzdins, G. L. Rempel, and G. Wilkinson, *Chem. Commun.*, **1969**, 825; (b) P. Legzdins, R. W. Mitchell, G. L. Rempel, J. D. Ruddick, and G. Wilkinson, *J. Chem. Soc. A*, **1970**, 3322.
7. G. Plimbett, C. D. Garner, and W. Clegg, *J. Chem. Soc. Dalton Trans.*, **1986**, 1257.
8. F. A. Cotton, A. H. Reid, Jr., and W. Schwotzer, *Inorg. Chem.*, **24**, 3965 (1985).
9. G. A. Rempel, P. Legzdins, H. Smith, and G. Wilkinson, *Inorg. Synth.*, **13**, 87 (1972).
10. D. E. Morris, K. R. Dunbar, C. A. Arrington, S. K. Doorn, L. E. Pence, and W. H. Woodruff, *J. Am. Chem. Soc.* (in press).

44. (2,2'-BIPYRIDINE)DICHLOROPALLADIUM(II) AND [N'-(2-DIAMINOETHYL)-1,2-ETHANEDIAMINE]CHLOROPALLADIUM(II) CHLORIDE

Submitted by F. L. WIMMER*, S. WIMMER,† and P. CASTAN†
Checked by RICHARD J. PUDDEPHATT‡

The synthesis of $[Pd(bpy)(L)]^{2+}$ (L = chelating ligand) usually involves the reaction of $Pd(bpy)Cl_2$ with L. Although the preparation of $Pd(bpy)Cl_2$ has been described in *Inorganic Syntheses*,[1] this method is generally tedious as the reactants $K_2[PdCl_4]$ and 2,2'-bipyridine in water initially form the insoluble salt $[Pd(bpy)_2][PdCl_4]$. Conversion of this complex to $Pd(bpy)Cl_2$ involves treatment with HCl, but the completeness of this reaction is not evident, as both $[Pd(bpy)_2][PdCl_4]$ and $Pd(bpy)Cl_2$ are insoluble and have similar colors. An alternate approach is to react bpy with $K_2[PdCl_4]$ under acidic conditions. However, again there is a possibility of formation of the salt, and contamination by this is not obvious from elemental analyses. These methods generally give a powdery product.

$[Pd(dien)Cl]Cl$§ is employed extensively to study substitution reactions because of its relatively simple chemistry and as a model to understand the behavior of the antitumor drug *cis*-$Pt(NH_3)Cl_2$ (*cis*-platinum) toward nucleotides and nucleosides.[2,3] The same problem of formation of the salt $[Pd(dien)Cl]_2[PdCl_4]$ arises on reaction of dien with $K_2[PdCl_4]$. To prevent this, $PdCl_2$ is normally allowed to react with dien in water, but the eventual workup to isolate $[Pd(dien)Cl]Cl$ is irksome with yields of only 40–50%.[4]

* Laboratoire de Pharmacologie et de Toxicologie Fondamentales du CNRS.
† Laboratoire de Chimie Inorganique, 205, route de Norbonne, 31077 Toulouse Cedex, France.
‡ Department of Chemistry, University of Western Ontario, London, Ontario N6A5B7, Canada.
§ Diethylenetriamine (dien): *N*-(2-aminoethyl)-1,2-ethanediamine.

We present here a simplified synthesis of pure crystalline Pd(bpy)Cl$_2$ and [Pd(dien)Cl]Cl with almost quantitative yields. The method works equally well for the complexes of 4,4'-dimethyl-2,2'-bipyridine, 1,10-phenanthroline, 5-nitrophenanthroline, 4,7-dimethylphenanthroline, and 3,4,7,8-tetramethylphenanthroline.

A. (2,2'-BIPYRIDINE)DICHLOROPALLADIUM(II)

$$PdCl_2 + 2CH_3CN \rightarrow Pd(CH_3CN)_2Cl_2$$

$$Pd(CH_3CN)_2Cl_2 + bpy \rightarrow Pd(bpy)Cl_2 + 2CH_3CN$$

This method is essentially that of Newkome and coworkers.[5] Preparation and isolation of the complex requires ~ 4 h.

■ **Caution.** *Acetonitrile is toxic and flammable, and 2,2'-bipyridine is a skin irritant. Any contact with skin for both chemicals should be avoided and the manipulations should be performed in a well-ventilated fume hood.*

Dichloropalladium(II) (PdCl$_2$) (0.404 g, 2.275 mmol) is dissolved in boiling acetonitrile (120 mL), and the mixture is then filtered to give a red solution. The dissolution is facilitated by extracting the PdCl$_2$ with small portions of hot acetonitrile. 2,2'-Bipyridine (0.3905 g, 2503 mmol) in acetonitrile (25 mL) is added dropwise to the hot stirred solution of Pd(CH$_3$CN)$_2$Cl$_2$ in acetonitrile contained in a 250-mL Erlenmeyer flask. Orange needles of Pd(bpy)Cl$_2$ rapidly separate. The mixture is digested at room temperature for 3 h, filtered and the solid washed with acetonitrile, acetone, and diethyl ether and dried *in vacuo* (silica gel). The yield is 0.69–0.71 g (91–94%, based on PdCl$_2$).

Anal. Calcd. for Pd(bpy)Cl$_2$: C, 35.98; H, 2.40; N. 8.40. Found: C. 35.89; H, 2.32; N, 8.30.

Properties

The compound Pd(bpy)Cl$_2$ is obtained as fine light orange needles. It is insoluble in water and organic solvents, but soluble (with solvolysis) in polar solvents such as dimethyl sulfoxide (DMSO) and *N,N*-dimethylformamide (DMF). It dissolves in 2 *M* sodium hydroxide solution with hydrolysis of the chloro groups. The Pd—Cl stretching vibrations occur at 340 and 347 cm^{-1}.

B. [N-(2-AMINOETHYL)-1,2-ETHANEDIAMINE]-CHLOROPALLADIUM(II) CHLORIDE

$$PdCl_2 + 2CH_3CN \rightarrow Pd(CH_3CN)_2Cl_2$$

$$Pd(CH_3CN)_2Cl_2 + dien \rightarrow [Pd(dien)Cl]Cl + 2CH_3CN$$

■ **Caution.** *Acetronitrile is toxic and flammable. Dien is toxic and a strong irritant. Any contact with skin for both chemicals should be avoided and the manipulations should be performed in a well-ventilated fume hood.*

Preparation of the complex requires ~ 2 h. Dien (0.3 mL, 2.80 mmol) in acetonitrile (20 mL) is added dropwise to the hot stirred solution of $PdCl_2$ (0.487 g, 2.75 mmol) in acetonitrile (100 mL) contained in a 250-mL round-bottomed flask. Undissolved $PdCl_2$ should be removed before the dien is added. A creamy yellow precipitate separates instantaneously. The mixture is heated under reflux with stirring for 5 min, whereupon the solid gradually becomes yellow. The solid is collected, washed several times with acetonitrile and then ethanol and dried *in vacuo* (silica gel). The yield is 0.694–0.730 g (90–95%, based on $PdCl_2$).

Anal. Calcd. for [Pd(dien)Cl]Cl: C, 17.11; H, 4.63; N, 15.00. Found: C, 17.21; H, 4.68; N, 15.01.

Properties

The compound [Pd(dien)Cl]Cl separates as a light yellow microcrystalline solid. It is very soluble in water, but only slightly soluble in polar organic solvents such as acetone. The electronic spectrum depends on the chloride–ion concentration. Thus, the complex has an absorption maximum in water at 330 nm ($\varepsilon = 462$ cm^{-1} dm^3 mol^{-1})[6] and in 0.1 M NaCl at 334 nm ($\varepsilon = 480$ cm^{-1} dm^3 mol^{-1}). The Pd—Cl stretching vibration is situated as a strong band at 331 cm^{-1}.

References

1. B. J. McCormick, E. N. Jaynes, Jr. and R. I. Kaplan, *Inorg. Synth.*, **13**, 216 (1972).
2. E. L. J. Breet and R. van Eldik, *Inorg. Chem.*, **23**, 1865 (1984).
3. R. B. Martin, in *Platinum, Gold and Other Metal Chemotherapeutic Agents*, ACS Symposium Series 209, American Chemical Society, Washington DC, 1983, p. 231.
4. W. H. Baddley and F. Basolo, *J. Am. Chem. Soc.*, **88**, 2944 (1966).
5. G. R. Newkome, V. K. Gupta, H. C. R. Taylor, and F. R. Fronczek, *Organometallics*, **3**, 1549 (1984).
6. E. L. J. Breet, R. van Eldik, and H. Kelm; *Polyhedron*, **2**, 1181, (1983).

45. *trans*-DICARBONYL-DI-μ-IODO-DIIODODIPLATINUM(II)

Submitted by BIANCA PATRIZIA ANDREINI,* DANIELA BELLI
DELL'AMICO,* FAUSTO CALDERAZZO,* and NICOLA PASQUALETTI*
Checked by ALLEN L. SELIGSON† and WILLIAM C. TROGLER†

$$PtI_2(s) + 2CO \rightarrow PtI_2(CO)_2 \tag{1}$$

$$2PtI_2(CO)_2 \rightleftharpoons Pt_2I_4(CO)_2 + 2CO \tag{2}$$

Platinum(II) iodo carbonyl derivatives were first prepared by carbonylation of platinum iodides at high pressure and temperature.[1]

A simple preparation of the *trans*-dicarbonyl-di-μ-iododiiododi-platinum(II) is reported here, which is carried out at subatmospheric pressure of CO starting from PtI_2 with the intermediate formation of $PtI_2(CO)_2$ according to equilibrium (2).

Procedure

Platinum(II) iodide is prepared in substantially quantitative yields from an aqueous solution of $[PtCl_4]^{2-}$ obtained from dihydrogen hexachloro-platinate(IV) and hydrazine hydrochloride[2] by addition of the stoichiometric amount of aqueous KI^3. The platinum iodide so obtained has a lower platinum content (35.8%) than expected (43.5%) after drying *in vacuo* at room temperature, due to the presence of unknown impurities not containing platinum. The yields of the presently described preparation are based on the platinum content of the starting platinum(II) iodide.

A quantity of PtI_2 (2.327 g) corresponding to 5.18 mmol of platinum is suspended in 100 mL of toluene under an atmosphere of CO at room temperature ($\sim 20°C$). After stirring for 12 h, the suspension is filtered to eliminate a small amount of a black residue. Under these conditions the solution obtained contains the species $PtI_2(CO)_2$, an equilibrium mixture of the trans and the cis isomers, with prevalence of the former [$\sim 95\%$, $\tilde{\nu}_{CO}$, 2120 (s) cm^{-1}]. Only one band [2147 (w) cm^{-1}] is observable for the cis form, probably as a result of overlapping of the second band with that of the trans isomer. The solution of the $PtI_2(CO)_2$ isomers is evaporated to dryness and a violet-black residue is obtained, consisting mostly of $Pt_2I_4(CO)_2$.

The violet-black residue is suspended in pentane (150 mL), and the suspension is stirred for 12 h under CO at 24°C. The temperature of the

* Dipartimento di Chimica e Chimica Industriale, Sezione di Chimica Inorganica, via Risorgimento 35, 56100 Pisa, Italy.
† Department of Chemistry, University of California, San Diego, CA 92093.

carbonylation must be carefully controlled with a thermostat; 24°C and pentane as a solvent appear to be the best choice for optimizing the yield of the dimer under the relatively low partial pressure of CO existing under these conditions. The violet dimer $Pt_2I_4(CO)_2$ is collected by filtration, dried *in vacuo* for $\frac{1}{2}$h, and then sealed in vials under N_2 (1.70 g, 68.8% yield).

Anal. Calcd. for $Pt_2C_2I_4O_2$: Pt, 40.9; Found: Pt, 40.6. CO (gasvolumetric)[4] Calcd.: 5.9. Found: 5.5.

Properties

trans-Dicarbonyl-di-μ-iodo-diiododiplatinum(II) is sensitive to moisture over a long exposure to air and moderately soluble in the common organic solvents. It is characterized by a carbonyl stretching vibration at 2104 cm^{-1} in toluene. The product decomposes without melting at 120–140°C in air, and sublimes *in vacuo* (10^{-2} mmHg) at 110°C with extensive decomposition. It darkens *in vacuo*, and the product so obtained is no longer completely soluble in hydrocarbons unless it is treated under a CO atmosphere. It reacts with NBu_4I in CH_2Cl_2 to give[4] $[PtI_3(CO)]^-$. The crystal and molecular structure of the compound has been solved by X-ray diffraction methods. The complex is a planar centrosymmetric iodo bridged dimer.[4]

References

1. L. Malatesta and L. Naldini, *Gazz. Chim. Ital.*, **90**, 1505 (1960).
2. W. E. Cooley and D. H. Busch, *Inorg. Synth.*, **5**, 208 (1957).
3. L. Ramberg, *Z. Anorg. Allgem. Chem.*, **83**, 36 (1913).
4. B. P. Andreini, D. Belli Dell'Amico, F. Calderazzo, M. G. Venturi, G. Pelizzi, and A. Segre, *J. Organomet. Chem.*, **354**, 357 (1988).

46. CHLOROHYDRIDOBIS(TRIALKYLPHOSPHINE)-PLATINUM(II) COMPLEXES

Submitted by JULIA R. PHILLIPS* and WILLIAM C. TROGLER*
Checked by MICK BRAMMER† and DIANNE L. PACKETT†

Chlorohydridobis(trialkylphosphine)platinum(II) complexes are useful as precursors to a variety of platinum hydrides[1-5] and platinum alkyls,[1,6-7] as

* Department of Chemistry, University of California at San Diego, La Jolla, California 92093.
† Research and Development Department, Union Carbide Chemicals and Plastics Co., Inc., South Charleston, WV 25303.

well as catalysts or catalyst precursors.[8,9] Previously reported methods of synthesis of *trans*-chlorohydridobis(trimethylphosphine)platinum(II) use hydrazine hydrate,[1] (naphthalene)sodium-hydrogen,[4] or sodium tetrahydroborate(1 −) in methanol[1,11] as the reductant. The former methods give an unstable product of questionable purity, while the latter procedure gives a low yield. The problem may arise from the heterogeneous nature of the reaction, as the *cis*-dichlorobis(trimethylphosphine)platinum(II) starting material exhibits a very low solubility in most solvents. Addition of diethylamine to a methanol suspension of *cis*-dichlorobis(trimethylphosphine)platinum(II) solubilizes the starting material, so the sodium tetrahydroborate(1 −) reduction results in a good yield of product. A $^{31}P\{^{1}H\}NMR$ spectrum was recorded to investigate the nature of the species in the methanol–diethylamine solution. This showed a doublet of doublets (δ − 23.41, $^{1}J_{Pt-P}$ 3003 Hz; δ − 24.19, $^{1}J_{Pt-P}$ 3527 Hz; $^{2}J_{P-P}$ 21 Hz) consistent with displacement of chloride by diethylamine, to form a *cis*-chloromonoamine complex *in situ*, which is then reduced by tetrahydroborate (1 −). After reduction, the volatile diethylamine can be removed easily. This method can also be used to synthesize the triethylphosphine derivative, as an alternative to the hydrazine hydrate reduction.[1,12]

A. *trans*-CHLOROHYDRIDOBIS(TRIMETHYLPHOSPHINE)-PLATINUM(II)

$$cis\text{-}PtCl_2[P(CH_3)_3]_2 + Na[BH_4] \xrightarrow[\text{CH}_3\text{OH}]{\text{HN(C}_2\text{H}_5)_2} trans\text{-}Pt(H)Cl[P(CH_3)_3]_2$$

Procedure

cis-Dichlorobis(trimethylphosphine)platinum(II) is prepared by adding slowly 1.6 mL of trimethylphosphine to a stirred solution of 3.2 g of *cis*-dichlorobis(diethyl sulfide)platinum(II)[13] dissolved in a minimum amount of benzene[4] under an atmosphere of nitrogen. The white precipitate is collected in the air on a fritted funnel, washed with diethyl ether to remove excess trimethylphosphine, and recrystallized from a minimum amount of hot *N,N'*-dimethylformamide. Yield: 2.4 g, 70%.

Into a 50-mL Schlenk flask is weighed 0.25 g of the recrystallized *cis*-dichlorobis(trimethylphosphine)platinum(II). Under a nitrogen atmosphere, the solid is suspended in 10 mL of methanol (refluxed under nitrogen and distilled from magnesium). To the suspension is added 3 mL of diethylamine (distilled and stored under nitrogen), and the suspension is stirred until the solid dissolves. Then 1 equiv (0.022 g) of sodium tetrahydroborate(1 −) is added and stirring continued for about 5 min until gas evolution ceases. The solvent is removed from the light yellow solution *in vacuo*. The residue is taken up in 15 mL of benzene (refluxed under nitrogen and distilled from potassium-benzophenone) and filtered into a 25-mL Schlenk flask through a piece of filter paper that had been folded and wired to the tip of a cannula.[14] The benzene is removed in vacuo and the resulting white residue is dissolved in 10 mL of toluene (refluxed under nitrogen and distilled from sodium benzoylbiphenyl) and filtered, again using filter paper wired to a cannula, into another 25-mL Schlenk flask. The resulting clear solution is again filtered into a 25-mL Schlenk flask, the volume reduced *in vacuo*, and the solution is cooled ($-10°$) overnight. White crystals of the product are isolated by removal of the supernatant by cannulation and drying under vacuum. The crystals are not rinsed because of their extreme solubility. The supernatant is concentrated and cooled to produce a second crop of crystals. Yield: 0.17 g, 75%.

Anal. Calcd. for $C_6H_{19}P_2ClPt$: C, 18.78; H, 4.99. Found (Galbraith Laboratories): C, 18.99; H, 5.10.

Properties

trans-Chlorohydridobis(trimethylphosphine)platinum(II) is air-sensitive and must be stored and manipulated under an inert atmosphere to avoid decomposition. It is soluble in benzene, toluene, acetone, and methanol. Chlorinated solvents should be avoided. The ^{1}H NMR spectrum in C_6D_6 shows the hydride resonance at $\delta - 16.13$ ($^1J_{Pt-H}$ 1311 Hz, $^2J_{P-H}$ 16.6 Hz), and the $^{31}P\{^1H\}$NMR spectrum exhibits a single resonance at $\delta - 14.26$ with ^{195}Pt satellites ($^1J_{Pt-P}$ 2627 Hz).

B. *trans*-CHLOROHYDRIDOBIS(TRIETHYLPHOSPHINE)-PLATINUM(II)

$$cis\text{-PtCl}_2[P(C_2H_5)_3]_2 + Na[BH_4] \xrightarrow[\text{CH}_3\text{OH}]{\text{HN}(C_2H_5)_2} trans\text{-Pt(H)Cl}[P(C_2H_5)_3]_2$$

Procedure

cis-Dichlorobis(triethylphosphine)platinum(II) is synthesized from 3.2 g of *cis*-dichlorobis(diethyl sulfide)platinum(II)[13] and 2.4 mL of triethylphosphine using the same method as given above for the synthesis of *cis*-dichlorobis-(trimethylphosphine)platinum(II), with recrystallization from boiling acetonitrile. Yield: 2.9 g, 80%.

The procedure given for the trimethylphosphinechlorohydrido derivative is then followed, starting with 0.26 g of the recrystallized *cis*-dichlorobis(triethylphosphine)platinum(II) and 0.019 g of sodium tetrahydroborate(1 −). When gas evolution ceases, after addition of the sodium tetrahydroborate-(1 −), the solvent is removed from the yellow solution *in vacuo*. The residue is extracted using pentane (refluxed under nitrogen and distilled from sodium-benzophenone) and filtered through filter paper wired to a cannula as described above. The solvent is removed *in vacuo* and the residue is re-dissolved in more dry pentane. The solution is treated with activated charcoal, filtered as before, and the solvent is removed from the resultant colorless solution *in vacuo*. The white product is recrystallized in the air from boiling hexanes. The supernatant remaining after isolation of the first crop of crystals is concentrated and a second crop is obtained. Yield: 0.14 g, 57%.

Anal. Calcd. for $C_{12}H_{31}P_2ClPt$: C, 30.81; H, 6.68. Found (Galbraith Laboratories): C, 30.73; H, 6.74.

Properties

trans-Chlorohydridobis(triethylphosphine)platinum(II) is air stable for an indefinite period of time. It is soluble in most organic solvents, including pentane, but chlorinated solvents should be avoided. The ^1H NMR spectrum in C_6D_6 shows the hydride resonance at δ − 16.84 ($^1J_{Pt-H}$ 1273 Hz, $^2J_{P-H}$ 14 Hz), while the $^{31}P\{^1H\}$ NMR spectrum has a single resonance at δ 23.54 with ^{195}Pt satellites ($^1J_{Pt-P}$ 2721 Hz).

References

1. J. Chatt and B. L. Shaw, *J. Chem. Soc.*, **1962**, 5075.
2. R. G. Goel and R. C. Srivastava, *Can. J. Chem.*, **61**, 1352 (1983).
3. B. L. Shaw and M. F. Uttley, *J. Chem. Soc. Chem. Commun.*, **1974**, 918.
4. D. L. Packett, C. M. Jensen, R. L. Cowan, C. E. Strouse, and W. C. Trogler, *Inorg. Chem.*, **24**, 3578 (1985).
5. T. G. Attig and H. C. Clark, *Can. J. Chem.*, **53**, 3466 (1975).
6. H. C. Clark and H. Kurosawa, *J. Chem. Soc. Chem. Commun.*, **1971**, 957.
7. H. C. Clark, P. L. Fiess, and C. S. Wong, *Can. J. Chem.*, **55**, 177 (1977).
8. C. M. Jensen and W. C. Trogler, *J. Am. Chem. Soc.*, **108**, 723 (1986).
9. H. C. Clark, C. Billard, and C. S. Wong, *J. Organomet. Chem.*, **173**, 341 (1979).
10. H. C. Clark and H. Kurosawa, *J. Organomet. Chem.*, **36**, 399 (1972).
11. D. Ramprasad, H. J. Yue, and J. A. Marsella, *Inorg. Chem.*, **27**, 3151 (1988).
12. G. W. Parshall, *Inorg. Synth.*, **12**, 26 (1970).
13. G. B. Kauffman and D. O. Cowan, *Inorg. Synth.*, **6**, 211 (1960).
14. J. P. McNally, V. S. Leong, and N. J. Cooper, *Experimental Organometallic Chemistry*, ACS Symposium Series 357, American Chemical Society, Washington, DC, 1987, p. 6.

Chapter Four

TRANSITION METAL ORGANOMETALLICS AND LIGANDS

47. LARGE-SCALE SYNTHESIS OF 1,2,3,4,5-PENTA-METHYLCYCLOPENTADIENE

Submitted by CAROL M. FENDRICK,* LARRY D. SCHERTZ,*
ERIC A. MINTZ,* and TOBIN J. MARKS†
Checked by T. E. BITTERWOLF,† P. A. HORINE,† T. L. HUBLER,†
J. A. SHELDON,† and D. D. BELIN†

The pentamethylcyclopentadienyl ligand has proved to be of great utility in main group, d-element, and f-element organometallic chemistry. In comparison to the unsubstituted cyclopentadienyl ligand, the pentamethylcyclopentadienyl ligand is a greater donor of electron density, and generally imparts enhanced solubility and crystallizability to a variety of metal complexes. As an ancillary ligand, it increases thermal stability in a number of metal hydrocarbyl systems. Furthermore, the greater steric demands of the permethyl ligand prevent oligomer formation and accompanying coordinative saturation in several organo-f-element systems.

King and Bisnette reported the first extensive study of transition metal pentamethylcyclopentadienyl chemistry.[1] The ligand was prepared using the laborious (six steps) and expensive procedure of de Vries.[2] This approach was later improved somewhat by substituting CrO_3/pyridine for MnO_2 in the step involving oxidation of di-sec-2-butenylcarbinol.[3] A method that begins with hexamethyl Dewar benzene is more straightforward but not economical.[4] The three-step procedure of Feitler and Whitesides beginning with

*Department of Chemistry, Northwestern University, Evanston, IL 60208. Research Supported by NSF Grant CHE-8800813.
†Department of Chemistry, University of Idaho, Moscow, ID 83843.

2-butanol and tiglic acid represents a significant, further improvement.[5] Several years ago, we reported a large-scale pentamethylcyclopentadiene synthesis[6] based on a refinement of a procedure originally developed by Sorensen et al.[7] and subsequently modified by Threlkel and Bercaw.[8] This procedure is relatively economical and is amenable to scales of several hundred grams. Nevertheless, it has the drawback of necessarily handling large quantities of noxious bromine and 2-bromo-2-butenes, as well as efficiently converting the latter to the corresponding lithium reagents. We report here a synthesis[9,10] that is even more economical and that, in our laboratory, has been carried out on scales of 500 g or more. Such a scale ensures that a typical organometallic research group need prepare this valuable ligand only at infrequent intervals. Moreover, the intermediate 2,3,4,5-tetramethylcyclopent-2-enone is a precursor for a host of other useful ligands.[9b] The present approach to pentamethylcyclopentadiene is a refinement and/or simplification of a reaction sequence originally reported by Burger et al.[4] and is shown below. The overall yield of pentamethylcyclopentadiene is 41%.

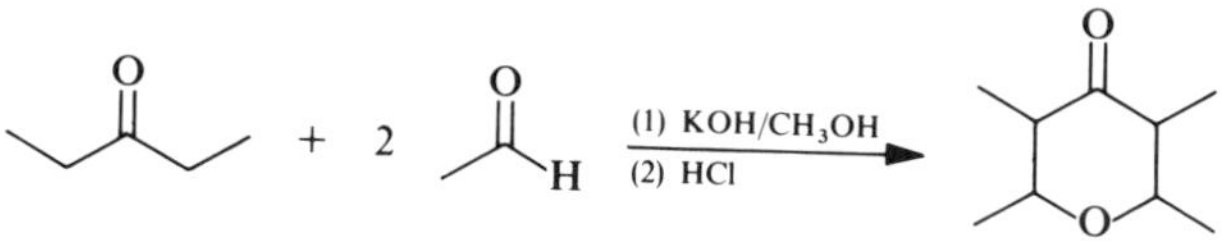

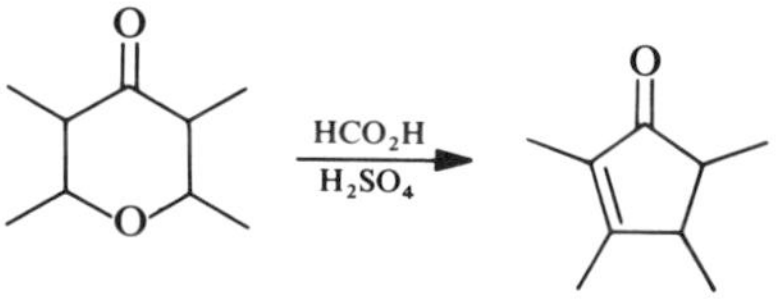

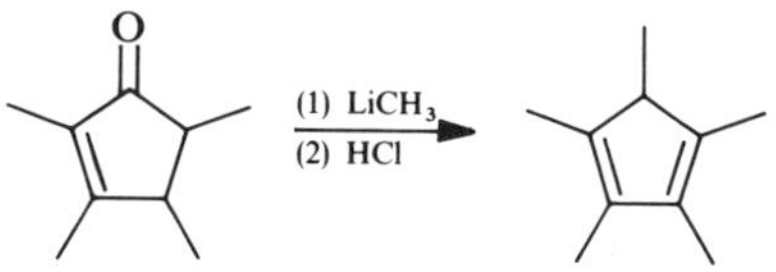

Procedure for Reaction 1

■ **Caution.** *Acetaldehyde is toxic, and inhalation should be avoided. Adequate ventilation and gloves are recommended.*

2,3,5,6-Tetrahydro-2,3,5,6-tetramethyl-γ-pyrone (1). A 12-L, three-necked, round-bottomed flask fitted with an efficient mechanical stirrer and a gas inlet is flushed with dinitrogen. Under a dinitrogen flush, 2360 mL of methanol and 380 g (6.77 mol) of KOH are added to the flask. The solution is then cooled to 0°C in an ice bath and 2000 mL (18.9 mol) 3-pentanone (Aldrich) is added under a dinitrogen flush. The solution is maintained at 0°C, and 4245 mL (75.5 mol) of acetaldehyde (Fluka, puriss) is added dropwise with stirring over a period of 36 h. During the addition of acetaldehyde, the color of the reaction mixture changes to a deep red, then to a dark brownish red color. After acetaldehyde addition is complete, the reaction solution is stirred an additional 12 h at 0°C.

With rapid stirring, cold concentrated HCl (600 mL) is next added dropwise to the reaction mixture at 0°C over a period of 4 h. Using the 12-L flask, rapid mechanical stirring, and glass tube–nitrogen pressure siphoning techniques, the aqueous and organic layers are then separated, and the organic layer is washed three times with 1 L of 2 N aqueous HCl solution. The aqueous phase is then backextracted with two 2.5-L portions of ether. After the ether is removed by rotary evaporation, the organic layers are combined and washed once with 1 L of 2 N NaCl solution. The remaining methanol is next removed by rotary evaporation, leaving a viscous red oil. Fractional distillation through a 600-mm Vigreux column at 70–83°, 15 torr, affords 1780 g (60% yield) of a clear yellow liquid, which is predominantly 2,3,5,6-tetrahydro-2,3,5,6-tetramethyl-γ-pyrone (1), but contains a small quantity (∼ 5%) of 3-methylhex-2-en-4-one. The 3-methylhex-2-ene-4-one does not interfere with the subsequent reaction, and this product is not subjected to additional purification.

Properties

The 2,3,5,6-tetrahydro-2,3,5,6-tetramethyl-γ-pyrone obtained is a clear yellow liquid. The ^{1}H NMR (60 MHz, CCl$_4$) exhibits signals at δ 1.45 (d, 6H, $J = 7$ Hz), 1.80 (d, 6H, $J = 7$ Hz), 2.60 (m, 2H), 3.80 (m, 2H).

Procedure for Reaction 2

2,3,4,5-Tetramethylcyclopent-2-enone (2). A 12-L, three-necked, round-bottomed reaction flask is cooled in an ice bath. It is charged with 6500 mL of 95–97% HCOOH (Aldrich) and 2250 mL of concentrated H$_2$SO$_4$ (exothermic), and the solution is allowed to cool to room temperature. The 2,3,5,6-tetrahydro-2,3,5,6-tetramethyl-γ-pyrone, 1580 g (1, 10.1 mol), is then added with rapid stirring to the acid solution at a fast trickle. After addition is complete (4 h), the reaction solution is heated to 50°C and stirred for 24 h. The resulting dark brown mixture is next slowly poured in 500-mL aliquots with swirling into 2-L Erlenmeyer flasks, each containing 1000 g of ice. Two or three flasks can be used repetitively in this procedure. Rapid temperature rises should be avoided to prevent frothing. Ether (500 mL) is added to each flask, and each aqueous phase is separated and backextracted twice with 250 mL of ether using a 2-L separatory funnel. The combined ether layers of each aliquot are then washed twice with 500 mL 2 *M* NaCl solution and with 250-mL aliquots of 10% aqueous NaOH solution until the aqueous layer becomes yellow in color. This indicates approximate neutrality, which should be additionally verified using pH paper. If necessary, further washing should be carried out until the aqueous phase is alkaline. Failure to remove all HCOOH from the organic layer at this stage results in product decomposition during the subsequent distillation. The organic layer is finally washed with two 500-mL portions of aqueous 2 *N* NaCl solution, dried over 300 g of anhydrous Na_2SO_4, filtered, and the ether removed by rotary evaporation. The combined crude product mixture is then purified by fractional distillation through a 600-mm Vigreux column. The colorless to yellow liquid boiling at 77–90°C at 15 torr is collected. Yield: 1045 g (75%).

Properties

The 2,3,4,5-tetramethylcyclopenta-2-enone is obtained as a yellow liquid. If purified by an additional fractional distillation, it is a colorless liquid. The ^1H NMR spectrum (400 MHz, $CDCl_3$, major isomer) exhibits resonances at δ 1.16 (d, 3H, $J = 7.4$ Hz), 1.18 (d, 3H, $J = 7.2$ Hz), 1.69 (s, br, 3H), 1.89 (qd, 1H, $J = 7.4$, 2.5 Hz), 1.98 (s, 3H), 2.25 (q, br, 1H).

Procedure for Reaction 3

1,2,3,4,5-Pentamethylcyclopentadiene (3). A 2-L three-necked flask containing a large magnetic stirring bar is fitted with a reflux condenser and purged with dinitrogen. Methyllithium (Aldrich, 1714 mL, 1.4 *M* in ether, 2.40 mol) is transferred into the flask by cannula. Behind a safety shield, the solution volume is reduced to approximately 1200 mL by vacuum transfer. The solution is then cooled in an ice bath and with rapid stirring, 2,3,4,5-tetramethylcyclopent-2-enone (**2**, 276 g, 298 mL, 2.00 mol) is added dropwise over a period of 2 h. The solution is next allowed to warm to room temperature and stirred overnight. After this time, any remaining methyllithium is hydrolyzed by slow, dropwise (exothermic reaction) addition of methanol (80 mL) followed by water (400 mL). The reaction mixture is next divided into two portions, and each is washed with a solution prepared from ammonium chloride (240 g), concentrated hydrochloric acid (200 mL), and water (1 L), until the aqueous phase becomes acidic. The combined aqueous phases are then backextracted twice with 500 mL of ether. The ether fractions are next combined and reduced to a total volume of 700 mL on the rotary evaporator. At this point, water elimination from the intermediate cyclopentenol is nearly complete (as judged by ^{1}H NMR and gas chromatography using an OV-101 column at 130°C). As a precaution, the ether concentrate is stirred for 1 h with 2 mL of 6 *N* HCl solution, then washed three times with 200 mL of 5% NaHCO$_3$ solution and dried over 100 g of anhydrous K$_2$CO$_3$. After filtration, the remaining ether is removed on the rotary evaporator and the product purified by careful fractional distillation using a 600-mm Vigreux column. The fraction boiling at 60–65°C at 15 torr affords 245 g (90% yield) of pentamethylcyclopentadiene. Gas chromatography (OV-101 column at 130°) indicates the product to be greater than 97% pure.

Properties

1,2,3,4,5-Pentamethylcyclopentadiene is a colorless to pale yellow liquid with a sweet olefinic odor. It can be stored indefinitely under dinitrogen in a − 15°C freezer. The ^{1}H NMR spectrum (60 MHz, CCl$_4$) exhibits resonances at δ 0.95 (d, 3H, $J = 8$ Hz), 1.75 (s, br, 12H), and 2.40 (m, 1H). The infrared spectrum (neat liquid) exhibits transitions at 2960 (vs), 2915 (vs), 2855 (vs), 2735 (w); 1660 (m), 1640 (w), 1390 (s), 1355 (m), 1150 (w), 1105 (mw), 1048 (2), 840 mw, and 668 (w) cm^{-1}.

References and Note

1. R. B. King and M. B. Bisnette, *J. Organometal. Chem.*, **8**, 287 (1967).
2. L. deVries, *J. Org. Chem.*, **25**, 1838 (1960).
3. J. E. Bercaw, R. H. Marvich, L. G. Bell, and H. H. Brintzinger, *J. Am. Chem. Soc.*, **94**, 1219 (1972).
4. U. Burger, A. Delay, and F. Mazenod, *Helv. Chim. Acta*, **57**, 2106 (1974).
5. D. Feitler and G. M. Whitesides, *Inorg. Chem.*, **15**, 466 (1976).
6. (a) J. M. Manriquez, P. J. Fagan, L. D. Schertz, and T. J. Marks, *Inorg. Synth.*, **21**, 181 (1982) and **28**, 317 (1990); (b) for another version of this procedure, see R. S. Threlkel, J. E. Bercaw, P. F. Seidler, and R. G. Bergman, *Org. Synth.*, **65**, 42 (1987).
7. P. H. Campbell, N. W. K. Chiu, K. Deugan, I. J. Miller, and T. S. Sorensen, *J. Am. Chem. Soc.*, **91**, 6404 (1969).
8. R. S. Threlkel and J. E. Bercaw, *J. Organometal. Chem.*, **136**, 1 (1977).
9. (a) C. M. Fendrick, E. A. Mintz, L. D. Schertz, T. J. Marks, and V. W. Day, *Organometallics*, **3**, 819 (1984); (b) C. M. Fendrick, L. D. Schertz, V. W. Day, and T. J. Marks, *Organometallics*, **7**, 1828 (1988).

10. Another refinement of the Burger, Delay, Mazenod procedure utilizes LiCl or LiBr to promote the cyclization in Eq. 1 and CH_3MgI in Eq. 3. The overall yield (34%) is slightly lower than in the present procedure. See F. X. Kohl and P. Jutzi, *J. Organometal. Chem.*, **343**, 119 (1983).

48. DICHLOROBIS(η^5-CHLOROCYCLOPENTADIENYL)-TITANIUM, $(ClC_5H_4)_2TiCl_2$

Submitted by ERIC A. ANSLYN* and ROBERT H. GRUBBS*
Checked by CHRISTIAN FELTEN† and DIETER REHDER†

Methyl-substituted cyclopentadienyl ligands have played a major role in organometallic chemistry. Corresponding cyclopentadienyl ligands with electron-withdrawing groups are rare with chlorocyclopentadienyl[1] and trifluoromethylcyclopentadiene[2] as the major exceptions. Transition metal complexes of these ligands are prepared through their thallium derivatives. An important complex that is needed in multigram quantities for substituent effect and other studies is dichlorobis(η^5-chlorocyclopentadienyl)titanium.[3] The literature preparation of this complex requires the use of chlorocyclopentadienyl thallium and yields only 180 mg of product. Increasing the scale of the literature preparation by several fold proved laborious and often resulted in very poor yields. Because of the toxicity of thallium,[3] its high price on a multigram scale and the small quantity of material produced per run, another synthetic pathway was developed. A route (Eq. 1) was developed that yields 8–10 g of dichlorobis(η^5-chlorocyclopentadientyl)titanium in an overall 40% yield from readily available (1α,3aα,4α,7α,7aα)-3a,4,7,7a-tetrahydro-4,7-methano-1*H*-inden-1-ol.[5] This quantity of product would require approximately 420 g of CpTl and 192 g of TlOEt if synthesized by the current procedure.[1]

Procedure

■ **Caution.** *All reactions are performed under Ar using Schlenk-tube techniques. All solvents are dried and then distilled from sodium benzophenone ketyl. The vacuum flash pyrolysis must be done at less than 0.005 torr. If the vacuum is not maintained, the reaction should be stopped and repeated. Carbon tetrachloride is a carcinogen and must be handled with care in a hood.*

* Contribution No. 8034 from the Arnold and Mabel Beckman Laboratories of Chemical Synthesis, 164-30, California Institute of Technology, Pasadena, CA 91125.
† Department of Chemistry, University of Hamburg, D-2000, Hamburg 13, Germany.

$$(1)$$

$$(2)$$

$$(3)$$

$$(4)$$

$$(5)$$

A.   (1α,3aβ,4β,7β,7aβ)-1-CHLORO-3a,4,7,7a-TETRAHYDRO-4,7-METHANO-1H-INDENE

A 100 mL, three-necked, round-bottomed flask equipped with an argon inlet is charged with 10 g (67.5 mmol) of α-1-hydroxy-endo-dicyclopentadiene.[4] The flask is purged with Ar, and 30 mL of CCl_4 (distilled from P_4O_{10}) is added via syringe. Triphenyl phosphine (19.7 g, 75.2 mmol) is added to the mixture, which is then stirred under Ar at 60°C for 9 h. The clear solution forms a white precipitate of Ph_3PO. The reacton mixture is cooled to room temperature and 50 mL pentane is added to precipitate the remaining Ph_3PO. The suspension is filtered through a 7-cm coarse frit covered with a 1-cm layer of silica gel. The precipitate is then washed with 3×40 mL of diethyl ether. The filtrate is concentrated to a light yellow oil, which is distilled at 35°C at 5 torr to yield 8.5 g (75%) of product. ^1H NMR ($CDCl_3$)

(400 MHz JEOL GX-400) δ 5.92, 5.84, 5.74, 5.51 (each 1H, m olefinic), 1.36, 1.54 (each 1H, d, CH_2 bridge), 2.82, 3.07 (each 1H, s, bridgeheads), 2.97, 3.44 (each 1H, m, fused bridgeheads), 4.27 (1H, s, CHCl); ^{13}C NMR (CDCl$_3$) δ 137.2, 137.8, 133.3, 131.3, 66.1, 54.8, 54.2, 51.1, 45.1, 45.0.

Anal. Calcd. C, 72.07: H, 6.65. Found: C, 72.32: H, 6.48.

B.   (CHLOROCYCLOPENTADIENYL)LITHIUM

Two medium Schlenk tubes are connected by a Teflon stopcock (plug type) to opposite ends of a glass-bead-filled 2-cm × 50-cm heat-resistant, glass tube equipped with a vacuum attachment. Then (1α,3aβ,4β,7β,7aβ)-1-chloro-3a,4,7,7a-tetrahydro-4,7-methano-1H-indene (18.7 g, 112.6 mmol) is transferred by syringe into one of the Schlenk tubes, degassed twice, and frozen. The entire apparatus is then evacuated to 0.002 torr. The tube is heated under vacuum to 450°C in an oven. The empty Schlenk tube is cooled to − 78°C and the flask containing the chlorodicyclopentadiene is thawed and heated to 90°C. The pyrolysis requires about 1 h. The Schlenk tube containing the product is filled with Ar and warmed to 0°C. The reaction product is kept at 0°C and used immediately or stored at − 50°C (only slight decomposition is observed over week periods at this temperature). The resulting 18 g of product, containing cyclopentadiene and chlorocyclopentadiene, is added by a cannula to 180 mL of THF (distilled from Na/benzophenone) at 0°C. To this solution, 44.4 mL (0.9 equiv., 97.2 mmol) of 2.2 *M* *n*-BuLi in hexane is added dropwise over a 5-min period. The reaction mixture is then stirred at room temperature for $\frac{1}{2}$ h and one-half the volume of solvent is evaporated under vacuum (room temperature). The THF solution is then transferred by cannula into 500 mL of vigorously stirring pentane at − 78°C. The resulting pink precipitate is collected by filtration at − 78°C, washed twice with 30 mL of pentane at − 78°C, dried *in vacuo*, and transferred in a dry box. (**Caution**: Extremely air-sensitive.) Yield 9.5 g (80% based on chloro-*endo*-dicyclopentadiene) ^1H NMR D^8 THF δ 5.47 s; ^{13}C NMR, 102.5 and 102.0.

C.   DICHLOROBIS(η^5-CHLOROCYCLOPENTADIENYL)-TITANIUM

A 500-mL round-bottomed Schlenk flask is charged with (chlorocyclopentadienyl)lithium (8.12 g, 76.6 mmol) in a dry box. Diethyl ether, 200 ml, is added via syringe and the salt suspended by stirring. The reaction mixture is cooled to 0°C, and 4.2 mL (38.3 mmol) TiCl$_4$ is added via syringe. The resulting dark red mixture is stirred for $\frac{1}{2}$ h at room temperature. Aqueous HCl (40 mL, 3 *M*) is added by syringe and the reaction mixture is vigorously stirred for 5 min to yield a red precipitate. The mixture is filtered in air through a coarse frit, and the brick-red solid is washed with 4×20 mL of EtOH, then 4×20 mL of ether. The precipitate is dried under vacuum to yield 8.2 g (67%) of product. An analytic sample can be prepared by sublimation at 220°C at (0.005 torr). The ^{1}H NMR was identical to that reported in the literature.[1]

Anal. Calcd. C, 37.78: H, 2.54. Found: C, 37.44: H, 2.51.

Properties

The dichloride bis(η^5-chlorocyclopentadienly)titanium is a brick-red solid that is stable in air and light. It is sparingly soluble in THF and insoluble in hydrocarbons.

References

1. B. G. Conway and M. D. Rausch, *Organometallics*, **4**, 688 (1985).
2. P. G. Gassman and C. H. Winter, *J. Am. Chem. Soc.*, **108**, 4228 (1986).
3. H. Heydlauf, *Eur. J. Pharmacol.*, **6**, 340 (1969).
4. W. C. Finch, E. V. Anslyn, and R. H. Grubbs, *J. Am. Chem. Soc.*, **110**, 2406 (1988).
5. R. B. Woodward and T. J. Katz, *Tetrahedron*, **5**, 79 (1959); M. Rosenblum, *J. Am. Chem. Soc.*, **79**, 3179 (1957).

49. η^3-ALLYLHYDRIDOBIS[ETHYLENEBIS (DIPHENYLPHOSPHINE)]MOLYBDENUM(II)

$$MoCl_5 + 2Ph_2PCH_2CH_2PPh_2 + C_3H_6 \xrightarrow[\text{Na/Hg}]{\text{THF}}$$
$$MoH(\eta^3\text{-}C_3H_5)\,[Ph_2PCH_2CH_2PPh_2]_2 + 5\,NaCl$$

Submitted by CHRISTOPHER J. BISCHOFF,* JOHN M. HANCKEL,*
MARIA M. LUDVIG,* and MARCETTA Y. DARENSBOURG*
Checked by M. MACIEJEWSKI† and M. RAKOWSKI DUBOIS†

The title compound is of interest because its method of preparation represents one of the first definite examples of C—H bond activation.[1] Furthermore, the temperature-dependent ^{1}H NMR spectrum indicates the reversible

* Department of Chemistry, Texas A&M University, College Station, TX 77843.
† Department of Chemistry, University of Colorado, Boulder, CO 80309.

abstraction of hydrogen from a metal-bound olefin, again providing a bona fide example of an important organometallic process, olefin isomerization.

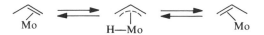

The reported procedures for the preparation of $HMo(\eta^3\text{-}C_3H_5)(diphos)_2$ (diphos = $(C_6H_5)_2PCH_2CH_2P(C_6H_5)_2$) are two-step syntheses beginning with $MoCl_5$. One method uses isolated $MoCl_4(diphos)\cdot C_6H_6$ as intermediate,[1] and the other uses $MoCl_3$ (tetrahydrofuran)$_3$.[2] Both intermediates are reduced further by sodium amalgam in the presence of 1,2-bis(diphenylphosphino)ethane and propene, with overall yields, calculated from $MoCl_5$, of 23 and 31%, respectively.[1] The synthetic route described below is a "one-pot" reaction of $MoCl_5$ and diphos over a Na/Hg amalgam in the presence of propylene. This approach, which is identical to the synthesis of $trans\text{-}Mo(N_2)_2(diphos)_2$,[3] both simplifies the synthesis and increases the yield to 50–60%.

Procedure

■ **Caution.** *$MoCl_5$ is very moisture-sensitive and reacts violently with water to form HCl.*

The first portion of the synthetic procedure must be performed under an argon atmosphere. A 500-mL three-necked flask is equipped with a gas adapter, glass stopper, rubber septum, and magnetic stirring bar. A 1% w/w Na/Hg amalgam is prepared in the flask, utilizing 14 mL of Hg and 2.0 g of Na. A solution of $(C_6H_5)_2PCH_2CH_2P(C_6H_5)_2$ (4.4 g) in 125 mL of tetrahydrofuran (THF) freshly distilled from sodium benzophenone ketyl is prepared in a 250-mL round-bottomed flask and is added to the Na/Hg amalgam through a stainless steel cannula. A solid addition tube equipped with gas inlet is filled with $MoCl_5$ (1.25 g). Using Schlenk-line techniques, with a strong argon flow, the addition tube is fitted to the reaction flask. The gray mixture is vacuum-degassed and the flask is charged with propene before the slow addition of $MoCl_5$. As the $MoCl_5$ is added, a white vapor is evolved and the solution immediately turns light green-brown. The solution is stirred for 24 h, under a propene atmosphere with an oil bubbler, to allow for a very slow flow of gas. During this time the solution changes to a dark yellow-brown.

The remainder of the synthesis may be carried out under N_2, although argon is preferable. The solution is removed from the Na/Hg amalgam, filtered through Celite, and collected in a 250-mL Schlenk flask. The solvent is removed under vacuum, and the resulting orange-brown solid is redissolved in toluene (75 mL) to give an orange-brown solution. The solution is filtered through Celite, and the volume reduced by half. Methanol (75 mL) is added to precipitate the product, which is collected by filtration on a fine frit. A second recrystallization may be necessary to remove all uncomplexed diphos. Yield after a second recrystallization is 2.076 g (48.5%).

Anal. Calcd. for $C_{55}H_{54}P_4Mo$: C, 70.66; H, 5.82; P, 13.25. Found: C, 70.39; H, 5.74; P, 12.91.

Properties

The product is a deep orange-red crystalline solid, soluble in THF, benzene, and toluene. The allyl–hydride is moderately air-stable; however, it must be kept under argon or nitrogen when in solution. The solid decomposes at temperatures $> 142°C$. The ^1H NMR spectrum in d_6-benzene at $+ 5°C$ shows a quintet centered at δ $- 2.6$ ppm (Mo–H) ($J_{P-H} = 39$ Hz), with signals at δ 0.38, 1.16, and 3.72 ppm for the η^3-C_3H_5. The frequencies listed here agree with the data acquired by Osborn and coworkers.[4] The hydride quintet coalesces as the temperature is increased, as the hydride is undergoing rapid site exchange with the terminal hydrogen atoms on the allyl group.

The free diphos ligand at room temperature shows proton NMR resonances in d_6-benzene centered at δ 7.32 and 7.02 ppm for the phenyl hydrogens, and a triplet centered at δ 2.21 ppm for the ethylene hydrogens. In the case of the bound diphos the phenyl resonances remain unchanged, while the triplet transforms into two very broad peaks shifted upfield to δ 2.10 and 1.58 ppm. The ^{31}P NMR signal for free diphos is a singlet at δ $- 13.4$ ppm in C_6H_6 and the bound is shifted downfield to δ 84.7 ppm, relative to an external standard of H_3PO_4.

The allyl–hydride ligands cannot be displaced by dinitrogen, nor will the Mo(O) complex *trans*-$Mo(N_2)_2(diphos)_2$[5] react with propene to give the allyl–hydride, even at elevated temperatures. Propene can be liberated on reaction with CO (1 atm), yielding the known *trans*-$Mo(CO)_2(diphos)_2$ [$\nu(CO) = 1823$ cm^{-1}, THF solution], which slowly (48 h) isomerizes to *cis*-$Mo(CO)_2(diphos)_2$ [$\nu(CO) = 1861$ and 1797 cm^{-1}].[6]

References

1. J. W. Byrne, Ph.D. dissertation, Harvard University, 1976.
2. M. W. Anker, J. Chatt, G. J. Leigh, and A. G. Wedd, *J. Chem. Soc. Dalton Trans.*, **1975**, 2639.
3. T. A. George and M. E. Noble, *Inorg. Chem.*, **17**, 1678 (1978).
4. J. W. Byrne, H. U. Blaser, and J. A. Osborn, *J. Am. Chem. Soc.*, **97**, 3871 (1975).
5. T. A. George and C. D. Siebold, *Inorg. Chem.*, **12**, 2548 (1973).
6. B. Dezube, J. K. Kouba, and S. S. Wreford *J. Am. Chem. Soc.*, **100**, 4404 (1978).

50.   BIS(η^5-CYCLOPENTADIENYL)MOLYBDENUM(IV) COMPLEXES

Submitted by NED D. SILAVWE,* MICHAEL P. CASTELLANI,* and DAVID R. TYLER*
Checked by MARK A. BECK,† JOSEPH D. LICHTENHAN,† and NANCY M. DOHERTY†

Dihydridobis(η^5-cyclopentadienyl)molybdenum(IV), Cp_2MoH_2, was first reported in 1959 from the reaction between $MoCl_5$, NaCp ($Cp = \eta^5\text{-}C_5H_5$), and $NaBH_4$.[1] Since then, several reports have appeared that have refined this basic procedure.[2] The procedure reported here incorporates the use of solid NaCp·DME, which simplifies the product work-up. An alternate method to prepare Cp_2MoH_2 by metal-vapor synthesis in moderate yields has been described, but this preparation is not practical for synthesis on the gram scale.[3] Dihydridobis(η^5-methylcyclopentadienyl)molybdenum(IV) has been prepared by the reaction of $MoCl_5$, $Na(C_5H_4Me)$, and $NaBH_4$ in a manner analogous to that described for the unsubstituted compound.[4, 5] Photolysis of Cp_2MoH_2 provides a convenient method for the *in situ* generation of molybdocene.[6, 7]

The synthesis of dichlorobis(η^5-cyclopentadienyl)molybdenum(IV), Cp_2MoCl_2, was first reported in 1964.[8] This synthesis involved the reaction between Cp_2MoH_2 and $CHCl_3$. A second procedure developed by the same investigators involves the reaction of $Mo_2(O_2CCH_3)_4$, NaCp, and PPh_3 followed by the addition of HCl_{aq}.[9] This reaction provides the dichloride in comparable overall yields to the following procedure, but requires significantly less time to complete. If Cp_2MoH_2 is not needed, then this synthesis should be considered. Dichlorobis(η^5-methylcyclopentadienyl)molybdenum(IV) has been prepared by an analogous method.[5] The chloride ligands on these complexes are easily metathesized, providing a convenient source of the Cp_2Mo^{2+} moiety.[10, 11] Reduction of Cp_2MoCl_2 with Na/Hg amalgam provides a source of molybdocene.[6]

The study of organometallic oxo complexes has been hindered by the limited number of pure compounds available. One of these complexes, oxobis(η^5-cyclopentadienyl)molybdenum(IV), Cp_2MoO, was first reported in 1972.[11] That preparation involved the reaction of dichlorobis(η^5-cyclopentadienyl)molybdenum(IV) with aqueous NaOH. Unfortunately, the product obtained by this method is difficult to purify. The procedure that follows includes a new and reproducible method of purification. A second,

---

* Department of Chemistry, University of Oregon, Eugene, OR 97403.

† Department of Chemistry, University of Washington, Seattle, WA 98195.

less convenient, procedure involving the Na/Hg amalgam reduction of Cp_2MoCl_2 in the presence of epoxides has also been reported.[6] The synthesis of oxobis(η^5-methylcyclopentadienyl)molybdenum(IV) has been reported.[12]

All manipulations should be carried out under a nitrogen atmosphere, either in a glove box or by employing standard Schlenk techniques. Tetrahydrofuran (THF), pentane, hexanes, benzene, toluene, and 1,2-dimethoxyethane (DME) should be distilled under nitrogen from sodium benzophenone. Dichloromethane and chloroform should be distilled from CaH_2 under a nitrogen atmosphere. The 6 M HCl_{aq} and 12 M $NaOH_{aq}$ solutions can be sufficiently deoxygenated by placing them under dynamic vacuum and then freezing the solution with liquid nitrogen. The acetone and water should be deoxygenated by three freeze–pump–thaw cycles.

A. DIHYDRIDOBIS(η^5-CYCLOPENTADIENYL) MOLYBDENUM(IV)

$$MoCl_5 + 3Na(C_5H_5) \cdot DME + 2NaBH_4 \rightarrow Cp_2MoH_2 + 5NaCl + \tfrac{1}{2}Cp_2 + B_2H_6$$

Procedure

This procedure requires 5–6 days to complete. Inside a glove box, combine 28.0 g (0.102 mol) of $MoCl_5$, 90.0 g (0.505 mol) of $NaCp \cdot DME$,[13] and 18.0 g (0.476 mol) of $NaBH_4$ in a 2-L three-necked flask equipped with a vacuum adapter,[14] a 500-ml equal-pressure addition funnel stoppered with a septum cap, and a mechanical stirrer. On a Schlenk line, transfer 200-mL hexanes to the flask via cannula or syringe. Cool the flask, with stirring, in a dry ice–acetone bath. Fill the addition funnel with 700 mL of THF (in portions) via cannula from a flask containing the distilled solvent and slowly drip it into the slurry (addition time is $\sim$ 90 min). Remove the dry ice–acetone bath and allow the reaction mixture to warm to room temperature. Replace the addition funnel with a reflux condenser equipped with a vacuum adapter. Next heat the reaction mixture to reflux for 5 h with a heating mantle. After cooling the mixture to room temperature, add a magnetic stirring bar to the flask and replace the mechanical stirrer and the reflux condenser with ground-glass stoppers. Remove the solvent under vacuum with stirring. This last step will require several hours.

Replace one of the stoppers with a 500-mL addition funnel, and add 6 M HCl_{aq} to the addition funnel with a cannula. (A $\tfrac{1}{8}$-in.-i.d. polyethylene cannula is best used for this transfer. Use a cork borer to put holes in two septa and insert the cannula into the septa before inserting the septa into the ground-glass joints of the addition funnel and the storage flask initially containing the degassed HCl_{aq}.) Cool the flask containing the reaction mixture in an ice bath and add the HCl_{aq} dropwise with occasional swirling or stirring of the solution. A gas is evolved, accompanied by the evolution of white fumes. The latter should be kept to a minimum by controlling the addition rate of the acid. After adding sufficient HCl_{aq} to completely react with the solid in the reaction flask, the gas evolution will cease. Filter the brown mixture by transferring the slurry to a separate Schlenk frit (attached to a 2-L Schlenk flask to be used as a receiving vessel) with a polyethylene cannula. Wash the solid that remains on the frit with 150–200-mL portions of 6 *M* HCl until the HCl solution becomes pale brown to colorless. (The HCl_{aq} that is used to wash the solid is conveniently added to the frit with a cannula. The volume of the combined filtrate and washing solutions is 900–1000 mL.) Remove the Schlenk frit and replace it with a ground-glass stopper. Cool the solution in an ice bath with stirring. Slowly add 12 *M* $NaOH_{aq}$ to this solution until the solution becomes basic to litmus paper. As the solution approaches the endpoint, a yellow precipitate will form. Transfer this basic slurry in portions to a Schlenk frit with a $\frac{1}{8}$-in. polyethylene cannula and filter it. (This process takes several hours.) Dry the yellow solid* under vacuum for several hours. The product is purified by extracting it into acetone,† followed by evaporating the extract to dryness under vacuum. Dry the resulting yellow solid at room temperature under vacuum for at least 12 h. The product can be further purified by sublimation at 100–110°C at 10^{-3} torr employing a liquid nitrogen-filled cold finger. Be sure the solid is dry before subliming it. The yield of yellow powder is 9.37 g (41.1 mmol, 40%).‡

B. DIHYDRIDOBIS(η^5-METHYLCYCLOPENTADIENYL) MOLYBDENUM(IV)

$$MoCl_5 + 3Na(MeCp) + 2NaBH_4 \rightarrow (MeCp)_2MoH_2 + 5NaCl + \tfrac{1}{2}(MeCp)_2 + B_2H_6$$

Procedure

This procedure requires 5–6 days to complete. This procedure is the same as above except for the following changes. Reflux 90 mL (0.529 mol) of freshly cracked methylcyclopentadiene dimer with 20.0 g (0.870 mol) of sodium

---

*The checkers report that the color of this solid is sometimes gray-green.

† The checkers report that benzene is more efficient at extracting the product from the crude material.

‡ The checkers report a 33% yield.

sand in 350 mL of dimethoxyethane for 12 h to prepare the Na(MeCp) (MeCp = $C_5H_4CH_3$).

■ **Caution.** *Do only in a well-ventilated hood because H_2 is liberated in this reaction.*

In the three-necked, 2-L, round-bottomed flask combine 36.0 g (0.132 mol) of $MoCl_5$ and 27.0 g (0.714 mol) of $NaBH_4$, cover the solids with 150 mL of toluene, and cool the flask in a Dry Ice–Acetone bath. Transfer the Na(MeCp) solution to the addition funnel via cannula and slowly drip it into the reaction mixture. The addition of the Na(MeCp) solution will cause the production of a white gas that should be kept to a minimum by controlling the rate of addition. While still cold, slowly add 500 mL of THF, then allow the reaction mixture to warm to room temperature and follow by heating to reflux for 4–5 h. The reaction mixture may become viscous, presumably from the polymerization of THF. Proceed to completion according to the pre-paration of Cp_2MoH_2 described previously. If the sublimate is tinted slightly green, the product may be purified by eluting it through a NaOH-treated alumina column (see the preparation of Cp_2MoO) with hexanes. The yield of golden yellow $(MeCp)_2MoH_2$ is 14.5 g (56.7 mmol, 43%).*

Properties

Both compounds are oxygen- and light-sensitive and thermally sensitive and should be stored in the freezer in the dark. Exposure to room light at room temperature for a day or so causes little decomposition. Both of these compounds are soluble in common aromatic hydrocarbons, CH_2Cl_2, THF, and acetone. The methylcyclopentadienyl analogue is also soluble in ali-phatic hydrocarbons, diethyl ether, methanol, and acetonitrile, while the unsubstituted derivative is only slightly soluble in these solvents. The methyl-cyclopentadienyl derivative is generally more soluble in a given solvent than the cyclopentadienyl derivative. They react with $CHCl_3$ (slowly) and $CHBr_3$ (violently) at room temperature to form the corresponding Cp_2MoX_2 (X = Cl or Br) compounds (see next synthesis). The ^1H NMR spectrum of Cp_2MoH_2 in benzene-d_6 consists of singlets from the Cp rings at 4.35 ppm and from the hydrides at − 8.80 ppm. The ^1H NMR spectrum of $(MeCp)_2MoH_2$ in benzene-d_6 consists of two broad singlets at 4.45 and 4.20 ppm (Cp protons), a sharp singlet at 1.86 ppm (Me protons), and a singlet from the hydrides at − 8.25 ppm. The Mo—H stretch in the infrared spectrum of Cp_2MoH_2 is observed at 1826 cm^{-1} in CH_2Cl_2 solution, while v(Mo—H) in the MeCp complex is observed at 1835 cm^{-1} in CH_2Cl_2. In Nujol mulls, v(Mo—H) are observed at 1847 cm^{-1} for Cp_2MoH_2[2a] and at 1840 cm^{-1} for $(MeCp)_2MoH_2$.[4]

*The checkers report a 36% yield.

C. DICHLOROBIS(η^5-CYCLOPENTADIENYL)MOLYBDENUM(IV)

$$Cp_2MoH_2 + 2CHCl_3 \rightarrow Cp_2MoCl_2 + 2CH_2Cl_2$$

Procedure

Inside a glove box, dissolve 5.00 g (21.9 mmol) of Cp_2MoH_2 in 125 mL of $CHCl_3$ in a 500-mL Schlenk flask equipped with a magnetic stirring bar. A yellow-brown solution is obtained. Attach a reflux condenser equipped with a vacuum adapter[14] to the flask and transfer the apparatus to a Schlenk line. Reflux the solution for 10 h and then allow it to cool to room temperature for several hours. A green precipitate will separate out. Replace the reflux condenser with a Schlenk frit, filter the solution, and dry the olive green solid under vacuum. The yield is 5.95 g (20.0 mmol, 91%).

D. DICHLOROBIS(η^5-METHYLCYCLOPENTADIENYL) MOLYBDENUM(IV)

$$(MeCp)_2MoH_2 + 2CHCl_3 \rightarrow (MeCp)_2MoCl_2 + 2CH_2Cl_2$$

Procedure

This preparation is the same as above except that (1) 6.64 g (25.9 mmol) of $(MeCp)_2MoH_2$ is dissolved in 150 mL $CHCl_3$ and (2) after the refluxed solution cools to room temperature, add 50 mL of pentane with stirring. Stir this solution for 2 h, after which time filtration followed by drying *in vacuo* yields 6.86 g (21.1 mmol, 81%) of a gray-green solid. Adding another 50 mL of pentane to the filtrate followed by cooling to $-25°C$ produces an additional 0.48 g (26.6 mmol, 87% total) of product.

Properties

Both Cp_2MoCl_2 and $(MeCp)_2MoCl_2$ are somewhat light-sensitive in the solid state, although the latter compound is much less so. Both should be stored in the dark. While both compounds are oxygen-sensitive, they may be briefly exposed to the air without significant decomposition. The cyclopentadienyl complex is soluble in dimethyl sulfoxide (DMSO) and $CHCl_3$ and is insoluble in most other common organic solvents. The methylcyclopentadienyl derivative is soluble in DMSO, CH_2Cl_2, and $CHCl_3$ and is slightly soluble in methanol and acetonitrile. The ^1H NMR of Cp_2MoCl_2 in DMSO-d_6 consists of a singlet at 5.64 ppm (Cp protons). The ^1H NMR of

$(MeCp)_2MoCl_2$ in DMSO-d_6 consists of a sharp singlet at 1.92 ppm (Me protons) and two broad singlets at 5.23 and 5.30 ppm (Cp protons).

E. OXOBIS(η⁵-CYCLOPENTADIENYL)MOLYBDENUM(IV)

$$Cp_2MoCl_2 + 2NaOH \rightarrow Cp_2MoO + H_2O + 2NaCl$$

Procedure

In a glove box, combine 2.01 g (6.77 mmol) of Cp_2MoCl_2 and 9.0 g (0.22 mol) of NaOH in a 100-ml Schlenk flask equipped with a magnetic stirring bar. With a syringe, add 40 mL of water with stirring and cool the solution in an ice bath until it is approximately room temperature, then leave it stirring at room temperature with the Schlenk flask wrapped in aluminum foil. After 10 h, filter the solution through a Schlenk frit and dry the emerald green precipitate under high vacuum for at least 12 h. Extract the brown filtrate with dichloromethane ($\sim$ 25 mL per extraction) until the aqueous layer is pale brown. (The CH_2Cl_2 layers are emerald green.) Evaporate the combined extracts to dryness under vacuum, then further dry them on a high-vacuum line for 12 h to remove any residual water. The yield of emerald green solid is 0.8 g (3.5 mmol, 52%).*

The product can be purified by eluting it through a specially treated alumina column. In the air, treat 500 g of basic alumina with 500 mL of saturated alcoholic NaOH in a 1-L Erlenmeyer flask. Shake the mixture occasionally over the next 2 h. Filter the mixture through a Büchner funnel, transfer the wet solid to a 1-L Schlenk flask, and remove the residual ethanol *in vacuo*. (A substantial amount of the solution will remain with the alumina after filtering.) Next heat the alumina in a silicon oil bath at 140°C under high vacuum for 4–5 h. The alumina should be stored in a glove box.

Prepare a 6-in-long column (25 mm i.d.) with this alumina and pentane as the solvent. Load a solution of 0.50 g (2.1 mmol) of crude Cp_2MoO in 10 mL of benzene onto the column. Elution of the complex with pentane ($\sim$ 75 mL using flash chromatography) removes any yellow Cp_2MoH_2 that may be present; elution with benzene ($\sim$ 100 mL) removes a blue band;† and a 1 : 1 mixture of benzene/THF ($\sim$ 150 mL) elutes the blue-green Cp_2MoO band. Removal of the solvent *in vacuo* yields 0.25 g (1.0 mmol, 50%) of Cp_2MoO. The overall yield is 26% of emerald green Cp_2MoO. The product is pure by ^{1}H NMR spectroscopy.

---

* The checkers report a 70% yield.
† The checkers report that this band is sometimes green.

F. OXOBIS(η^5-METHYLCYCLOPENTADIENYL)-MOLYBDENUM(IV)

$$(MeCp)_2MoCl_2 + 2NaOH \rightarrow (MeCp)_2MoO + H_2O + 2NaCl$$

This procedure is the same as above except that 3.10 g (9.54 mmol) of Cp_2MoCl_2 and 12.4 g (0.310 mol) of NaOH are stirred in 120 mL of water. The yield is 2.10 g (7.78 mmol, 82%) of crude green product.* Purifying the product as above yields $(MeCp)_2MoO$ as a green solid in 40% overall yield. **Note:** *With this complex the second band eluted from the column is purple.*

Properties

Both Cp_2MoO and $(MeCp)_2MoO$ are oxygen-, light-, and temperature-sensitive and should be stored in the dark and in the freezer. Both compounds are soluble in benzene, THF, CH_2Cl_2, acetone, and acetonitrile. The MeCp complex is also soluble in diethyl ether, but the Cp complex is only slightly soluble in this solvent. Both complexes react rapidly with methanol. Both complexes react with $CHCl_3$ to form the corresponding dichloro complex. The ^1H NMR of Cp_2MoO in benzene-d_6 consists of a singlet at 5.11 ppm (Cp protons). The ^1H NMR of $(MeCp)_2MoO$ in benzene-d_6 consists of a singlet at 1.67 ppm (Me protons) and two triplets at 4.20 ppm (J_{H-H} = 2.2 Hz) and 5.90 ppm (J_{H-H} = 2.2 Hz) (Cp protons).

References

1. M. L. H. Green, C. N. Street, and G. Wilkinson, *Z. Naturforsch., B: Anorg. Chem. Org. Chem.*, **14**, 738 (1959).
2. (a) E. O. Fischer and Y. Hristidu, *Z. Naturforsch., B: Anorg. Chem. Org. Chem.*, **15**, 135 (1960); (b) M. L. H. Green, J. A. McCleverty, L. Pratt, and G. Wilkinson, *J. Chem. Soc.*, **1961**, 4854; (c) R. B. King, in *Organometallic Syntheses*, J. J. Eisch and R. B. King (eds)., Academic Press, London, 1965, p. 79; (d) M. L. H. Green and P. J. Knowles, *J. Chem. Soc. Perkins Trans. 1*, **1973**, 989.
3. (a) M. J. D'Aniello, Jr. and E. K. Barefield, *J. Organomet. Chem.*, **76**, C50 (1974); (b) E. M. Van Dam, W. N. Brent, and M. P. Silvon, *J. Am. Chem. Soc.*, **97**, 465 (1975).
4. A. Nakamura and S. Otsuka, *J. Mol. Catal.*, **1**, 285 (1976).
5. G. J. S. Adam and M. L. H. Green, *J. Organomet. Chem.*, **208**, 299 (1981).
6. M. Berry, S. G. Davies, and M. L. H. Green, *J. Chem. Soc. Chem. Commun.*, **1978**, 99.
7. (a) G. L. Geoffroy and M. G. Bradley, *J. Organomet. Chem.*, **134**, C27 (1977); (b) G. L. Geoffroy and M. G. Bradley, *Inorg. Chem.*, **17**, 2410 (1978); (c) M. Berry, N. J. Cooper, M. L. H. Green, and S. J. Simpson, *J. Chem. Soc. Dalton Trans.*, **1980**, 29.
8. (a) R. L. Cooper and M. L. H. Green, *Z. Naturforsch., B: Anorg. Chem. Org. Chem.*, **19**, 652 (1964); (b) R. L. Cooper and M. L. H. Green, *J. Chem. Soc. (A)*, **1967**, 1155.

* The checkers report a 76% yield.

9. (a) M. L. H. Green, M. L. Poveda, J. Bashkin, and K. Prout, *J. Chem. Soc. Chem. Commun.*, **1982**, 30; (b) J. Bashkin, M. L. H. Green, M. L. Poveda, and K. Prout, *J. Chem. Soc. Dalton Trans.*, **1982**, 2485.
10. A. R. Dias and C. C. Romao, *J. Organomet. Chem.*, **233**, 223 (1982).
11. M. L. H. Green, A. H. Lynch, and M. G. Swanwick, *J. Chem. Soc. Dalton Trans.*, **1972**, 1445.
12. N. D. Silavwe, M. Y. Chiang, and D. R. Tyler, *Inorg. Chem.*, **24**, 4219 (1985).
13. J. Smart and C. J. Curtis, *Inorg. Chem.*, **16**, 1788 (1977).
14. Also called a *nitrogen flow adapter*. (See Kontes Glassware cat. no. K-211300.)

51. CYCLOPENTADIENYL RHENIUM COMPLEXES

Submitted by FRANCINE AGBOSSOU,* EDWARD J. O'CONNOR,*
CHARLES M. GARNER,* N. QUIRÓS MÉNDEZ,* JESÚS M.
FERNÁNDEZ,* ALAN T. PATTON,* JAMES A. RAMSDEN,* and
J. A. GLADYSZ*
Checked by JOSEPH M. O'CONNOR† and TRACY TAJIMA† (Sections A–G),
and KEVIN P. GABLE‡ (Section H)

Chiral, pseudotetrahedral cyclopentadienyl rhenium complexes of formulas $Re(\eta^5-C_5H_5)(NO)(PPh_3)(X)$ and $[Re(\eta^5-C_5H_5)(NO)(PPh_3)(L)]^+X^-$ have proved to be of broad utility in synthetic and mechanistic organometallic chemistry.[1-41] As described in the procedures that follow, they are readily available in optically active form. This provides a valuable stereochemical probe for mechanistic studies.[2,5,7,8,11,12,14,17,19,21,22,26,28-32,34,35,40,41] Furthermore, the $Re(\eta^5-C_5H_5)(NO)(PPh_3)$ moiety can function as an efficient chiral auxiliary for the stereospecific introduction of new ligand-based chiral centers.[3,8,12,16,18,24,29,31-33,35] Organic compounds of high enantiomeric purity can be prepared subsequently. The complexes described here have also proved useful in the development of C_1 ligand chemistry.[1,4,5,11,13,15]

A. (η^5-CYCLOPENTADIENYL)(TRICARBONYL)RHENIUM

* Department of Chemistry, University of Utah, Salt Lake City, UT 84112.
† Department of Chemistry, University of Nevada-Reno, Reno, Nevada 89577
‡ Department of Chemistry, Oregon State University, Corvallis, OR 97331.

Procedure[1]

■ **Caution.** *This reaction evolves CO, and should be conducted in a well-ventilated hood. The starting material and product have significant vapor pressures and contain CO; hence, they should be considered toxic.*

A 100-mL round-bottomed flask is equipped with a stirring bar and a reflux condenser. The condenser is fitted with a three-way stopcock that is attached to a vacuum line and an oil bubbler for pressure release. A Teflon sleeve or a liberal application of silicone grease is applied between the flask and the condenser. The flask is charged with molten dicyclopentadiene (16 mL, 0.13 mol; vacuum-distilled) and $Re_2(CO)_{10}$ (15.084 g, 23.117 mmol; used as received from commercial sources). The mixture is degassed with stirring under oil pump vacuum and saturated with nitrogen. This is repeated twice. The flask is placed in an oil bath, with the oil slightly below the level of the contents. The bath is heated to $\sim 200°C$ to reflux the reaction mixture. Stirring is commenced when the $Re_2(CO)_{10}$ dissolves.

The reaction is monitored by silica gel thin-layer chromatography (TLC) (with flourescent indicator) in 90 : 10 hexane : ethyl acetate (v/v). UV light visualization shows starting material $Re_2(CO)_{10}$ with $R_f = 0.72$, and product $Re(\eta^5\text{-}C_5H_5)(CO)_3$ with $R_f = 0.38$. After approximately 12 h, the reaction ceases to reflux, although gas evolution is still apparent. The oil bath temperature is raised to 230°C, and maintained for an additional 3 h. The reaction mixture is then allowed to cool to room temperature. It solidifies, and is extracted with acetone (2×40 mL). This gives a clear solution, and leaves behind a small amount of white polymer-like solid. The acetone extract is filtered through a coarse-frit sintered-glass funnel containing 1.5 cm of silica gel topped with 1.5 cm of Celite. Additional acetone (400 mL) is passed through the funnel to completely elute the product. Solvent is removed from the filtered extracts by rotary evaporation, giving a white solid. The solid is dissolved in CH_2Cl_2 (40 mL), and hexanes (100 mL) are added. The mixture is concentrated by rotary evaporation to ~ 50 mL, at which point colorless crystals form. These are collected by filtration and are air-dried overnight to give $Re(\eta^5\text{-}C_5H_5)(CO)_3$ (14.384 g, 42.89 mmol, 93%) as an off-white solid.

Properties

The product is air-stable and thermally stable for an indefinite period under ambient conditions. IR (cm^{-1}, CH_2Cl_2) ν_{CO} 2023 (s), 1937 (s), 1902 (s). ${}^1H$ NMR (δ, $CDCl_3$) 5.37 (s, C_5H_5). Additional purification can be effected by sublimation[42,43] (mp 110.5–111°C; lit.:[42] 111–112°C) or CH_2Cl_2/hexane recrystallization.[44]

Merits of Procedure

This preparation is based on the original synthesis of $Re(\eta^5\text{-}C_5H_5)(CO)_3$ reported by Green and Wilkinson.[42] The synthesis of $Re(\eta^5\text{-}C_5H_5)(CO)_3$ from $Re_2(CO)_{10}$ via $Re(CO)_5X$ (X = Cl, Br) has been reported by other investigators.[43-46] The two-step route is superior for the preparation of the deuterated complex $Re(\eta^5\text{-}C_5D_5)(CO)_3$.[8] Carbonyl complex $Re(\eta^5\text{-}C_5H_5)(CO)_3$ is also the entry point for elegant rhenium oxo chemistry.[47]

B. (η^5-CYCLOPENTADIENYL)DICARBONYL (NITROSYL)RHENIUM(I) TETRAFLUOROBORATE(1−)

Procedure[1]

■ **Caution.** *This reaction evolves carbon monoxide, and should be conducted in a well-ventilated hood;* $(NO)^+[BF_4]^-$ *is corrosive, and rubber gloves should be worn during all manipulations.* Etching of the flask may occur.

A 250-mL Schlenk flask is equipped with a stirring bar and a stopper and is attached to a vacuum line. The flask is charged with $Re(\eta^5\text{-}C_5H_5)(CO)_3$ (8.00 g, 23.9 mmol) and CH_2Cl_2 (50 mL). The solution is degassed under vacuum with stirring, and then saturated with nitrogen. This is repeated twice. The colorless solution is cooled to 0°C with an ice bath. The stopper is momentarily removed, and then crystalline $(NO)^+[BF_4]^-$ (4.00 g, 34.2 mmol), which has been freshly washed with CH_2Cl_2, is added with stirring. The solution immediately turns a deep clear yellow, and is stirred for 12 h while the ice bath is allowed to warm. During this period the reaction mixture darkens and a yellow precipitate forms.

Solvent is removed from the flask by rotary evaporation, leaving a brown solid. This is extracted with acetone (∼ 1 L) and the extracts are filtered through Whatman No. 1 filter paper. Solvent is removed from the filtrate by rotary evaporation to give a yellow-brown solid. This is washed with THF until the THF is colorless and dried under oil pump vacuum to give $[Re(\eta^5\text{-}C_5H_5)(NO)(CO)_2]^+ BF_4^-$ (9.12 g, 21.5 mmol, 90%) as a yellow solid.

Properties

The product is air-stable and thermally stable for an indefinite period under ambient conditions. IR (cm^{-1}, KBr) ν_{CO} 2102 (s), 2041 (s), ν_{NO} 1803 (s). ^{1}H NMR (δ, acetone-d_6) 6.67 (s, C_5H_5). Additional purification may be effected by recrystallization from acetone/diethyl ether, which gives yellow plates, mp 255–257°C dec.

Merits of Procedure

Essentially equivalent syntheses of $[Re(\eta^5\text{-}C_5H_5)(NO)(CO)_2]^+X^-$ complexes have been reported by Fischer,[48] Casey,[45] and Graham.[49] The latter two investigators have utilized this cation as an entry point for elegant C_1 chemistry.

C.   CARBONYL(η^5-CYCLOPENTADIENYL)NITROSYL-(TRIPHENYLPHOSPHINE)RHENIUM(I) TETRA-FLUOROBORATE(1−)

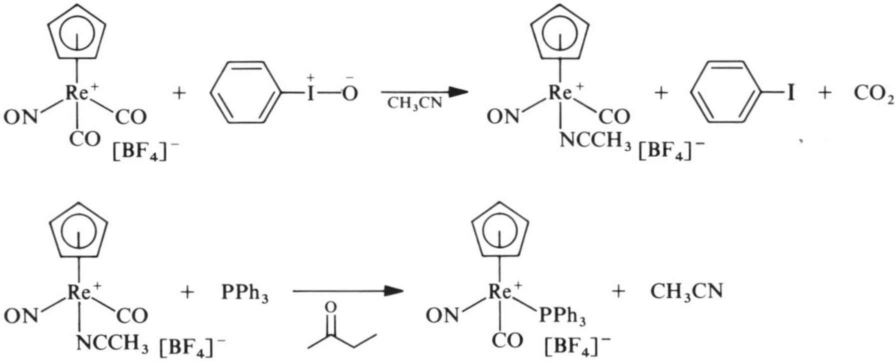

Procedure[1]

A 200-mL Schlenk flask is equipped with a stirring bar and a stopper. The flask is charged with $[Re(\eta^5\text{-}C_5H_5)(NO)(CO)_2]^+[BF_4]^-$ (5.00 g, 11.8 mmol) and CH_3CN (50 mL). Nitrogen is bubbled through the solution for 20 min. The solution is cooled to 0°C with an ice bath, and freshly prepared iodosylbenzene (2.60 g, 11.8 mmol)[50] is added as a solid.* The iodosylben-

---

* In larger-scale reactions (10–20 ×), uncontrolled foaming can occur. Thus, the iodosobenzene should be added slowly, and a proportionally larger reaction flask utilized.

zene is only slightly soluble, and the suspension is stirred for 3.5 h while gradually being brought to room temperature. The progress of the reaction can be directly monitored by ^{1}H NMR on a continuous-wave (CW) instrument (starting material η^5-C$_5$H$_5$, δ 6.21 s; [Re(η^5-C$_5$H$_5$)(NO)(CO) (NCCH$_3$)]$^+$ [BF$_4$]$^-$, δ 6.02 s).* The solvent is removed by rotary evaporation, leaving an oily brown residue. This is extracted with acetone and filtered, applying slight vacuum, through a short plug of silica gel on a medium-frit sintered-glass funnel. Solvent removal by rotary evaporation gives the acetonitrile complex [Re(η^5-C$_5$H$_5$)(NO)(CO)(NCCH$_3$)]$^+$ [BF$_4$]$^-$ as a brown. oil.† This is taken up in methyl ethyl ketone (50 mL) and transferred to a 200-mL round-bottomed flask that is equipped with a stirring bar and a reflux condenser. The condenser is attached to a nitrogen line and an oil bubbler for pressure release. Then PPh$_3$ (6.19 g, 23.6 mmol) is added, and a nitrogen atmosphere is established. The solution is refluxed (oil bath) with stirring for 24 h. A yellow solid begins to precipitate after a short period of time. The mixture is allowed to cool and is then filtered using a medium-frit sintered-glass funnel. The solid is collected, washed with diethyl ether, and dried under oil pump vacuum to give [Re(η^5-C$_5$H$_5$)(NO)(PPh$_3$)(CO)]$^+$ [BF$_4$]$^-$ (5.80 g, 8.81 mmol, 75%) as thin yellow needles.

Properties

The product is air-stable and thermally stable for an indefinite period under ambient conditions. Recrystallization from CH$_2$Cl$_2$/diethyl ether gives golden prisms, mp 277–278°C, dec.

Anal. Calcd. for C$_{24}$H$_{20}$BF$_4$NO$_2$PRe: C, 43.78; H, 3.06; N, 2.13; P, 4.70. Found: C, 43.28; H, 2.87; N, 2.49; P, 4.59. IR (cm^{-1}, CH$_2$Cl$_2$) v_{CO} 2001 (s), v_{NO} 1760 (s). ^{1}H NMR (δ, CD$_3$CN) 7.63 (m, PPh$_3$), 5.90 (s, C$_5$H$_5$). ^{13}C NMR (ppm, CD$_3$CN) 213.3 (s, CO); PPh$_3$ at 134.8 (d, J_{CP} = 11.3 Hz), 134.0 (s, para), 132.2 (d, J_{CP} = 61.5 Hz, ipso), 131.2 (d, J_{CP} = 11.3 Hz); 96.4 (s, C$_5$H$_5$).

---

* If required, more iodosylbenzene is added. This is seldom needed if the iodosylbenzene has been freshly prepared.

† The acetonitrile complex [Re(η^5-C$_5$H$_5$)(NO)(CO)(NCCH$_3$)]$^+$ [BF$_4$]$^-$ can be purified by recrystallization from acetone/diethyl ether. This gives air-stable, orange-yellow needles, mp 105–107 °C (dec. 175–179°C). *Anal.* Calcd. for C$_8$H$_8$BF$_4$N$_2$O$_2$Re: C, 21.97; H, 1.84; N, 6.40; Re, 42.57. Found: C, 21.88; H, 1.84; N, 6.35; Re, 42.30. IR (cm^{-1}, CH$_2$Cl$_2$) v_{CO} 2028 s, v_{NO} 1758 s.

D. (η^5-CYCLOPENTADIENYL) (METHOXYCARBONYL)-(NITROSYL)(TRIPHENYLPHOSPHINE) RHENIUM

Procedure[2,6]

A 25-mL round-bottomed flask is equipped with a stirring bar and charged with $[Re(\eta^5\text{-}C_5H_5)(NO)(PPh_3)(CO)]^+[BF_4]^-$ (3.00 g, 4.56 mmol) and methanol (5 mL). The suspension is treated with 4.37 M sodium methoxide in methanol (3.6 mL, 15.7 mmol; commercially available from Aldrich). The heterogeneous mixture warms slightly, and is stirred for 1.5 h. The greenish-yellow solid is collected by filtration, washed with methanol (3 × 2 mL), washed with water (2 × 5 mL), washed with methanol (2 × 3 mL), and dried under vacuum to give "methyl ester" $Re(\eta^5\text{-}C_5H_5)(NO)(PPh_3)(CO_2CH_3)$ (2.64 g, 4.38 mmol, 96%) as a light yellow solid.

Properties

The product, decomposition point (gradual darkening) 203–205°C, is air-stable and thermally stable for an indefinite period under ambient conditions.

Anal. Calcd. for $C_{25}H_{23}NO_3PRe$: C, 49.83; H, 3.85; N, 2.32; P, 5.14. Found: C, 49.23; H, 3.74; N, 2.50; P, 5.06. IR (cm^{-1}, CHCl$_3$) ν_{NO} 1670 s, ν_{CO} 1580 m. ^{1}H NMR (δ, CDCl$_3$) 7.97–7.37 (m, PPh$_3$), 5.25 (s, C$_5$H$_5$), 3.11 (s, CH$_3$). ^{13}C NMR (ppm, CDCl$_3$) 196.5 (d, J_{CP} = 4.8 Hz, CO), PPh$_3$ at 135.2 (d, J = 55 Hz, ipso), 133.2 (d, J = 11.0 Hz), 130.0 (s, para), 128.0 (d, J = 10.6 Hz); 91.6 (s, C$_5$H$_5$), 49.0 (s, CH$_3$).

This compound and Grignard reagents RMgX react to give acyl complexes $Re(\eta^5\text{-}C_5H_5)(NO)(PPh_3)(COR)$.[6] These may, in turn, be elaborated to vinylidene complexes $[Re(\eta^5\text{-}C_5H_5)(NO)(PPh_3)(=C=CHR)]^+X^-$, acetylide complexes $Re(\eta^5\text{-}C_5H_5)(NO)(PPh_3)(C\equiv CR)$,[18] and alkyl complexes $Re(\eta^5\text{-}C_5H_5)(NO)(PPh_3)(CH_2R)$.[6]

Optically active $Re(\eta^5\text{-}C_5H_5)(NO)(PPh_3)(CO_2CH_3)$ can be prepared by reaction of NaOCH$_3$ with optically active $[Re(\eta^5\text{-}C_5H_5)(NO)(PPh_3)(CO)]^+$ $[BF_4]^-$ (below), but for best optical purity the reaction should be conducted at -24°C or lower.

Merits of Procedure

This procedure has been scaled up by a factor of 27 without any diminution in yield.

E. (+)-(SR)-(η^5-CYCLOPENTADIENYL) [[[1-(1-NAPHTHALENYL)ETHYL]AMINO]- CARBONYL](NITROSYL)(TRIPHENYLPHOSPHINE) RHENIUM

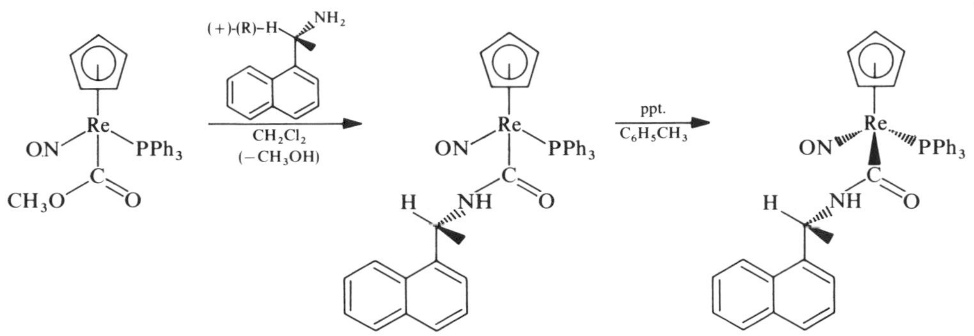

(+)-(SR)
(−)-(RR)

(+)-(SR)

Procedure[2]

A 25-mL Schlenk flask is charged with a stirring bar, Re(η^5-C_5H_5)(NO)(PPh$_3$)(CO$_2$CH$_3$) (1.600 g, 2.65 mmol),* a nitrogen atmosphere, and CH$_2$Cl$_2$ (4 mL, freshly distilled from P$_2$O$_5$).† Then commercial (+)-(R)-1-(1-naphthalenyl)ethylamine (0.52 mL, 3.20 mmol) is added by syringe, and the yellow solution is stirred for 4 h. Then toluene (10 mL, freshly distilled from sodium) is added, and the CH$_2$Cl$_2$ is removed by rotary evaporation at room temperature.‡ The solution is kept for 12 h at room temperature, during which time a yellow powder precipitates. The powder is collected by filtration, washed with toluene (3 × 1 mL), washed with hexanes (3 × 1 mL), and dried under vacuum to give a first-product crop (0.622 g, de > 98%). The

---

* It is important that the starting material be washed with water as described in the preceding preparation (Section D).

† If necessary, the solution can be filtered at this stage.

‡ Crystallization may commence during rotary evaporation, but this does not affect the de. The de is assayed by integration of the cyclopentadienyl ^{1}H NMR resonances of the two diastereomers.[2,26]

combined filtrates are concentrated to 2 mL by rotary evaporation at room temperature, and kept for 12 h at room temperature. This gives a second-product crop (0.182 g, ≥ 98% de), which is collected by filtration, washed with toluene (2 × 1 mL), washed with hexanes (1 mL), and dried under vacuum. The crops are combined to give "amide" (+)-(SR)-Re(η^5-C_5H_5)(NO)(PPh$_3$) [CONHCH(CH$_3$)C$_{10}$H$_7$] (0.804 g, 1.084 mmol, 82%).

Recrystallization is recommended when a product de of > 99% is sought. The amide (0.804 g, 1.084 mmol) is dissolved in CH$_2$Cl$_2$ (~ 10 mL), and toluene is added (~ 11 mL). Then the CH$_2$Cl$_2$ is removed as above.‡ The mixture is kept for 2 days at room temperature, after which time the amide is isolated at bright yellow needles analogously to the first crop obtained above (0.707 g, 0.953 mmol, 88%), ≥ 99.6% de.

Properties

The product, $[\alpha]_{589}^{24} = 113°$ ($c = 0.90$ mg/mL, CHCl$_3$), mp 241–243°C dec., is indefinitely stable under ambient conditions.

Anal. Calcd. for C$_{36}$H$_{32}$N$_2$O$_2$P Re: C, 58.29; H, 4.35; N, 3.78. Found: C, 58.40; H, 4.36; N, 3.74. IR (cm^{-1}, CHCl$_3$) ν_{NO} 1644 (s), ν_{CO} 1532 (m). ^{1}H NMR (δ, CDCl$_3$) naphthyl at 8.18 (d, J_{HH} = 7.8 Hz, 1H), 7.83 (dd, J_{HH} = 7.9, 2.2 Hz, 1H), 7.72 (dd, J_{HH} = 7.9, 2.2 Hz, 1H); naphthyl and phenyl at 7.56–7.40 (m, 19H); 5.74 (pseudoquintet, J_{HH} = 7 Hz, NCH), 5.58 (d, J_{HH} = 7.8 Hz, NH), 5.21 (s, C$_5$H$_5$) 0.99 (d, J_{HH} = 6.6 Hz, CH$_3$). ^{13}C{^{1}H} NMR (ppm, CDCl$_3$) 189.7 (d, J_{CP} = 12.0 Hz, CO), aryl carbons at 141.2 (Np), 136.2 (d, J_{CP} = 55.4 Hz, *ipso*-PPh$_3$), 133.8 (d, J_{CP} = 11.5 Hz), 131.4 (s), 130.2 (s), 128.5 (d, J_{CP} = 11.2 Hz), 128.2 (s), 127.2 (s), 126.9 (s), 125.6 (s), 125.4 (s), 125.3(s), 124.5 (s), 122.2 (s); 92.2 (s, C$_5$H$_5$), 44.2 (s, CH), 20.9 (s, CH$_3$).

Merits of Procedure

This procedure has been scaled up by a factor of 35 without any dimunation in yield or optical purity.

F. (+)-(S)-CARBONYL(η^5-CYCLOPENTADIENYL)-NITROSYL(TRIPHENYLPHOSPHINE)-RHENIUM(I) TETRAFLUOROBORATE(1−)

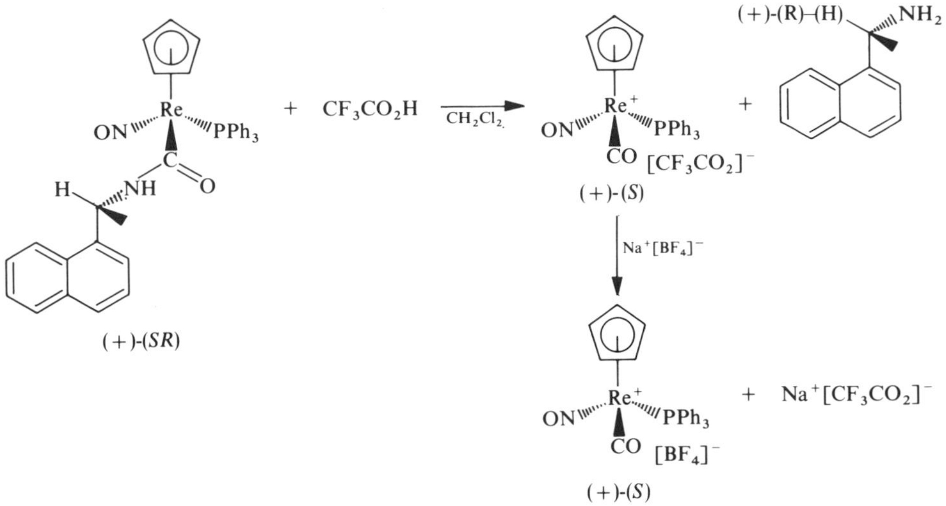

Procedure[2]

■ **Caution.** *Since* CF_3CO_2H *is corrosive and toxic, rubber gloves should be worn and all manipulations should be performed in a well-ventilated fume hood.*

A 20-mL round-bottomed flask is equipped with a stirring bar, charged with (+)-(SR)-Re(η^5-C$_5$H$_5$)(NO)(PPh$_3$)[CONHCH(CH$_3$)C$_{10}$H$_7$] (0.600 g, 0.809 mmol) and CH$_2$Cl$_2$ (3 mL), and cooled to 0°C in an ice bath. Then CF$_3$CO$_2$H (0.155 mL, 2.01 mmol) is added with stirring. The solvent is removed by rotary evaporation to give a bright yellow oil or foam. The residue is taken up in methanol (3 mL), and a solution of Na$^+$[BF$_4$]$^-$ (0.178 g, 1.62 mmol) in water (1.5 mL) is added with vigorous stirring. A bright yellow solid immediately precipitates. Additional water (5 mL) is added, and the mixture is stirred for 15 min. The yellow solid is collected by filtration, washed with water (4 × 5 mL), washed with THF (3 × 5 mL), and dried under vacuum to give carbonyl complex (+)-(S)-[Re(η^5-C$_5$H$_5$)(NO)(PPh$_3$)(CO)]$^+$[BF$_4$]$^-$ (0.513 g, 0.779 mmol, 96%).*

---

* If recovery of the amine is desired, the aqueous supernatant is neutralized with K$_2$CO$_3$ and the resulting milky-white suspension is extracted with CH$_2$Cl$_2$. The extract is washed with brine and dried over Na$_2$SO$_4$. The drying agent is removed by filtration, and solvent is removed under oil pump vacuum to give (+)-(R)-1-(1-naphthalenyl)ethylamine as a light yellow oil, $[\alpha]_{589}^{25} = 61°$ (c = 9.7 mg mL^{-1}, CH$_3$OH).

Properties

The product, $[\alpha]_{589}^{24} = 184°$ ($c = 0.43$ mg mL^{-1}, CH_2Cl_2), mp 278–279°C, shows less than a 1% optical activity loss after storage for 3 months under nitrogen in a refrigerator.

Anal. Calcd. for $C_{24}H_{20}BF_4NO_2PRe$: C, 43.78; H, 3.06; N, 2.13; P, 4.70. Found: C, 43.68; H, 3.16; N, 2.26; P, 4.69. Other properties are identical to those of the racemate, $(\pm)$-$[Re(\eta^5\text{-}C_5H_5)(NO)(PPh_3)(CO)]^+[BF_4]^-$, described above.

The enantiomer, $(-)$-(R)-$[Re(\eta^5\text{-}C_5H_5)(NO)(PPh_3)(CO)]^+[BF_4^-]$, is available by two routes. First, procedure E can be executed with commercial $(-)$-(S)-1(1-naphthalenyl)ethylamine and the enantiomeric product used in this procedure. Second, the supernatants that are enriched in $(-)$-(RR)-$Re(\eta^5\text{-}C_5H_5)(NO)(PPh_3)[CONHCH(CH_3)C_{10}H_7]$ from procedure E can be converted to enantiomerically enriched $(-)$-(R)-$Re[(\eta^5\text{-}C_5H_5)(NO)\text{-}(PPh_3)(CO)]^+[BF_4^-]$. The latter is then optically purified by repeating procedure E with $(-)$-(S)-1-(1-naphthalenyl)ethylamine.

Merits of Procedure

This procedure has been scaled up by a factor of 40 without any diminution of yield. A sequence of reactions corresponding to those in procedures D, E, and F can be utilized to resolve the enantiomers of the analogous pentamethylcyclopentadienyl carbonyl complex, $Re[\eta^5\text{-}C_5Me_5)(NO)(PPh_3)\text{-}(CO)]^+[BF_4^-]$.[26]

G.   (+)-(S)-(η^5-CYCLOPENTADIENYL)(METHYL)-NITROSYL(TRIPHENYLPHOSPHINE)RHENIUM

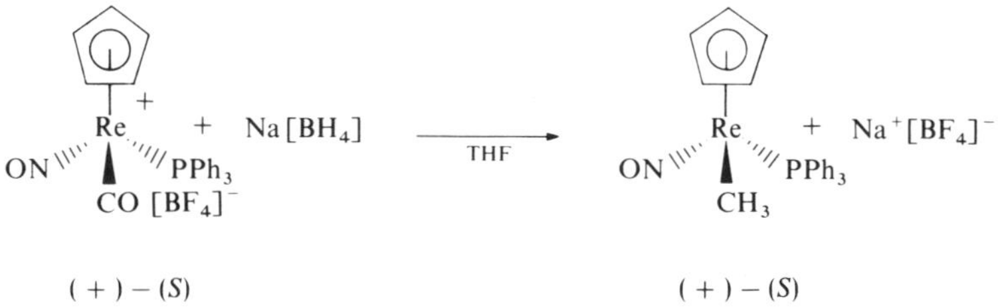

$(+) - (S)$ $(+) - (S)$

Procedure[1,2]

■ **Caution.** *Since benzene is a known carcinogen, all manipulations should be performed in a well-ventilated fume hood.*

A 100-mL Schlenk flask is equipped with a stirring bar and a stopper and is attached to a vacuum line. The flask is charged with (+)-(S)-[Re(η^5-C_5H_5)(NO)(PPh$_3$)(CO)]$^+$[BF$_4$]$^-$ (1.368 g, 2.08 mmol) and THF (20 mL). The mixture is degassed under vacuum with stirring, and then saturated with nitrogen. This is repeated two additional times. Then the stopper is momentarily removed and Na[BH$_4$] (0.237 g, 6.26 mmol) is added. The mixture is stirred for 2 h and then filtered through a plug of Celite on a coarse-frit sintered-glass funnel. Solvent is removed from the filtrate by rotary evaporation at room temperature (this precaution is unnecessary when preparing racemic compound) and the residue is taken up in benzene. This is filtered through a plug of silica gel on a coarse-frit sintered-glass funnel, yielding a bright orange solution. The benzene is removed by rotary evaporation under oil pump vacuum (room temperature). The residue is reprecipitated in a refrigerator from benzene–hexanes to give methyl complex (+)-(S)-Re(η^5-C_5H_5)(NO)(PPh$_3$)(CH$_3$) (0.923 g, 1.65 mmol, 79%) as a bright orange amorphous solid, which is collected by filtration and dried under oil pump vacuum.

Properties

The product, $[\alpha]^{24}_{579} = 178°$ ($c = 0.44$ mg mL^{-1}, CHCl$_3$), mp 192–194°C, is air-stable and thermally stable for a period of hours under ambient conditions. For longer periods, storage under nitrogen in a refrigerator is recommended. After 3 months, less than a 1% loss in optical activity occurs.

Anal. Calcd. for $C_{24}H_{23}$NOPRe: C, 51.60; H, 4.15; N, 2.51; P, 5.54. Found: C, 51.39; H, 4.23; N, 2.72; P, 5.45. IR (cm^{-1}, THF) ν_{NO} 1630s. ^{1}H NMR (δ, CD$_2$Cl$_2$) 7.8–6.8 (m, PPh$_3$), 4.96 (s, C_5H_5), 0.95 (d, $J_{CP} = 5$ Hz, CH$_3$). ^{13}C{^{1}H} NMR (ppm, CD$_2$Cl$_2$) PPh$_3$ at 136.3 (d, $J_{CP} = 53$ Hz, ipso), 133.8 (d, $J_{CP} = 11$ Hz), 130.4 (s, para), 128.7 (d, $J_{CP} = 10$ Hz); 90.2 (s, C_5H_5), -37.5 (d, $J_{CP} = 6$ Hz, CH$_3$). ^{31}P{^{1}H}NMR (ppm, C_6D_6) 25.5 (s).

The racemic methyl complex, ($\pm$)-Re(η^5-C_5H_5)(NO)(PPh$_3$)(CH$_3$), is prepared in an analogous fashion from ($\pm$)-[Re(η^5-C_5H_5)(NO)(PPh$_3$)-(CO)]$^+$[BF$_4$]$^-$ (procedure C).[1] It is obtained as an orange amorphous solid, mp 214–215°C, from CH$_2$Cl$_2$/hexanes or benzene–hexanes (refrigerator).

Anal. Calcd. for $C_{24}H_{23}$NOPRe: C, 51.60; H, 4.15; N, 2.51; P, 5.54. Found: C, 51.28; H, 3.96; N, 2.94; P, 5.54.

The optically active or racemic products react with (Ph$_3$C)$^+$[PF$_6$]$^-$ to give the optically active or racemic methylidene complexes [Re-(η^5-C_5H_5)(NO)(PPh$_3$)(=CH$_2$)]$^+$[PF$_6$]$^{-.2,5}$ These can, in turn, can be elaborated to a variety of alkyl complexes Re(η^5-C_5H_5)(NO)(PPh$_3$)(CH$_2$R) and alkylidene complexes $[Re(\eta^5\text{-}C_5H_5)(NO)(PPh_3)(=CHR)]^+X^-$.[2,3,5,40] The product readily reacts with acids HX to give complexes of the formula $Re(\eta^5\text{-}C_5H_5)(NO)(PPh_3)(X)$.[7] The product reacts with $H[BF_4]\cdot OEt_2$ and $H[PF_6]\cdot OEt_2$ in dichloromethane to give species that serve as the functional equivalent of the optically active chiral Lewis acid $[Re(\eta^5\text{-}C_5H_5)(NO)$-$(PPh_3)]^+$.[21,23,29,31,34-36,38]

H. (η^5-CYCLOPENTADIENYL)(FORMYL)-NITROSYL(TRIPHENYLPHOSPHINE)RHENIUM

Procedure[1]

■ **Caution.** *This reaction evolves hydrogen, and should be conducted in a well-ventilated hood away from any flames or ignition sources.*

A 50-mL Schlenk flask is equipped with a stirring bar and a stopper and is attached to a vaccum line. The flask is charged with $[Re(\eta^5\text{-}C_5H_5)$-$(NO)(PPh_3)(CO)]^+[BF_4]^-$ (1.021 g, 1.55 mmol) and THF/H_2O (25 mL, 1 : 1 v/v). The mixture is degassed under vacuum with stirring, and then saturated with nitrogen. This is repeated twice. The heterogeneous mixture is cooled to 0°C in an ice bath and $Na[BH_4]$ (0.155 g, 4.10 mmol) is added. The mixture is stirred for 1.5 h at 0°C, allowed to warm to room temperature over the course of 1 h, and then extracted with CH_2Cl_2 (3 × 10 mL). Each extract is transferred via cannula to a 100-mL Schlenk flask. Solvent is then removed via oil pump vacuum, and the flask is taken into a nitrogen atmosphere glove box. The yellow residue is extracted with THF (1 × 20 mL and 2 × 10 mL) and filtered through a medium-frit sintered-glass funnel. Hexane (~ 60 mL) is layered on top of the filtrate. Air-sensitive yellow microcrystals of formyl complex $Re(\eta^5\text{-}C_5H_5)(NO)(PPh_3)(CHO)$ form overnight and are collected by filtration and dried under vacuum (0.640 g, 1.12 mmol, 72%). The checker substituted a gastight syringe for the cannula in transferring the CH_2Cl_2 extracts, and reports a 57% yield.

Properties

The product is air-sensitive, especially in solution, and should be stored cold under a nitrogen atmosphere. Decomposition points vary somewhat between samples. Darkening (typically at 130°C) precedes gradual melting (155–158°C).

Anal. Calcd. for $C_{24}H_{21}NO_2PRe$: C, 50.34; H, 3.70. Found: C, 50.45; H, 3.68. The IR ν_{CO} is medium-dependent. The following data (cm^{-1}; 7 mg mL^{-1} for solutions) are for the same sample and are calibrated to polystyrene at 1602 cm^{-1}: ν_{CO} (s) 1549 (KBr), 1565 (THF), 1542 (CHCl$_3$); ν_{NO} (s) 1655 (KBr), 1662 (THF), 1661 (CHCl$_3$). ^{1}H NMR (δ, CD$_2$Cl$_2$) 16.48 (s, CHO), 7.50–7.36 (m, PPh$_3$), 5.25 (s, C$_5$H$_5$). ^{13}C{^{1}H}NMR (ppm, CD$_2$Cl$_2$, -30°C) 251.3 (d, J_{CP} = 11.0 Hz, CO); PPh$_3$ at 135.4 (d, J_{CP} = 55.0 Hz, ipso), 133.5 (d, J_{CP} = 11.0 Hz), 131.0 (s, para), 129.0 (d, J_{CP} = 11.0 Hz); 94.0 (s, C$_5$H$_5$).

The product is a useful hydride donor toward a variety of organometallic compounds.[1,31,35] The by-product $[Re(\eta^5\text{-}C_5H_5)(NO)(PPh_3)(CO)]^+X^-$ can be recovered in high yield. In contrast to borohydride reductants, no Lewis acidic by-products are generated.

Acknowledgment

We thank the Department of Energy for support of the research with racemic compounds, and the NIH for support of the research with optically active compounds. We also thank the numerous coworkers who have contributed to the development of these procedures. Most of them are listed as coauthors in refs. 1–41.

References

1. W. Tam, G.-Y. Lin, W.-K. Wong, W. A. Kiel, V. K. Wong, and J. A. Gladysz, *J. Am. Chem. Soc.*, **104**, 141 (1982).
2. J. H. Merrifield, C. E. Strouse, and J. A. Gladysz, *Organometallics*, **1**, 1204 (1982).
3. (a) W. A. Kiel, G.-Y. Lin, A. G. Constable, F. B. McCormick, C. E. Strouse, O. Eisenstein, and J. A. Gladysz, *J. Am. Chem. Soc.*, **104**, 4865 (1982); (b) W. A. Kiel, G.-Y. Lin, G. S. Bodner, and J. A. Gladysz, *J. Am. Chem. Soc.*, **105**, 4958 (1983); (c) W. A. Kiel, W. E. Buhro, and J. A. Gladysz, *Organometallics*, **3**, 879 (1984).
4. A. T. Patton, C. E. Strouse, C. B. Knobler, and J. A. Gladysz, *J. Am. Chem. Soc.*, **105**, 5804 (1983).
5. J. H. Merrifield, G.-Y. Lin, W. A. Kiel, and J. A. Gladysz, *J. Am. Chem. Soc.*, **105**, 5811 (1983).
6. W. E. Buhro, A. Wong, J. H. Merrifield, G.-Y. Lin, A. G. Constable, and J. A. Gladysz, *Organometallics*, **2**, 1852 (1983).
7. J. H. Merrifield, J. M. Fernández, W. E. Buhro, and J. A. Gladysz, *Inorg. Chem.*, **23**, 4022 (1984).
8. P. C. Heah, A. T. Patton, and J. A. Gladysz, *J. Am. Chem. Soc.*, **108**, 1185 (1986).
9. (a) W. E. Buhro, S. Georgiou, J. M. Fernández, A. T. Patton, C. E. Strouse, and J. A. Gladysz, *Organometallics*, **5**, 956 (1986); (b) W. E. Buhro, M. C. Etter, S. Georgiou, J. A. Gladysz, and F. B. McCormick, *Organometallics*, **6**, 1150 (1987).

10. F. B. McCormick, W. B. Gleason, X. Zhao, P. C. Heah, and J. A. Gladysz, *Organometallics*, **5**, 1778 (1986).

11. (a) G. S. Bodner, J. A. Gladysz, M. F. Nielsen, and V. D. Parker, *J. Am. Chem. Soc.*, **109**, 1757 (1987); (b) M. Tilset, G. S. Bodner, D. R. Senn, J. A. Gladysz, and V. D. Parker, *J. Am. Chem. Soc.*, **109**, 7551 (1987).

12. E. J. O'Connor, M. Kobayashi, H. G. Floss, and J. A. Gladysz, *J. Am. Chem. Soc.*, **109**, 4837 (1987).

13. D. R. Senn, K. Emerson, R. D. Larsen, and J. A. Gladysz, *Inorg. Chem.*, **26**, 2737 (1987).

14. B. D. Zwick, A. M. Arif, A. T. Patton, and J. A. Gladysz, *Angew. Chem. Int. Ed. Engl.*, **26**, 910 (1987).

15. G. S. Bodner, A. T. Patton, D. E. Smith, S. Georgiou, W. Tam, W.-K. Wong, C. E. Strouse, and J. A. Gladysz, *Organometallics*, **6**, 1954 (1987).

16. G. S. Bodner, D. E. Smith, W. G. Hatton, P. C. Heah, S. Georgiou, A. L. Rheingold, S. J. Geib, J. P. Hutchinson, and J. A. Gladysz, *J. Am. Chem. Soc.*, **109**, 7688 (1987).

17. (a) W. E. Buhro, B. D. Zwick, S. Georgiou, J. P. Hutchinson, and J. A. Gladysz, *J. Am. Chem. Soc.*, **110**, 2427 (1988); (b) W. E. Buhro, A. M. Arif, and J. A. Gladysz, *Inorg. Chem.*, **28**, 3837 (1989).

18. D. R. Senn, A. Wong, A. T. Patton, M. Marsi, C. E. Strouse, and J. A. Gladysz, *J. Am. Chem. Soc.*, **110**, 6096 (1988).

19. G. L. Crocco and J. A. Gladysz, *J. Am. Chem. Soc.*, **110**, 6110 (1988).

20. G. L. Crocco, C. S. Young, K. E. Lee, and J. A. Gladysz, *Organometallics*, **7**, 2158 (1988).

21. (a) J. M. Fernández and J. A. Gladysz, *Organometallics*, **8**, 207 (1989); (b) C. H. Winter and J. A. Gladysz, *J. Organomet. Chem.*, **354**, C33 (1988).

22. C. H. Winter, A. M. Arif, and J. A. Gladysz, *Organometallics*, **8**, 219 (1989).

23. (a) C. H. Winter, W. R. Veal, C. M. Garner, A. M. Arif, and J. A. Gladysz, *J. Am. Chem. Soc.*, **111**, 4766 (1989); (b) A. Igau and J. A. Gladysz, *Organometallics*, **10**, 2327 (1991).

24. C. M. Garner, J. M. Fernández, and J. A. Gladysz, *Tetrahedron Lett.*, **30**, 3931 (1989).

25. (a) G. S. Bodner, K. Emerson, R. D. Larsen, and J. A. Gladysz, *Organometallics*, **8**, 2399 (1989); (b) S. K. Agbossou, G. S. Bodner, A. T. Patton, and J. A. Gladysz, *Organometallics*, **9**, 1184 (1990).

26. Y.-H. Huang, F. Niedercorn, A. M. Arif, and J. A. Gladysz, *J. Organomet. Chem.*, **383**, 213 (1990).

27. (a) S. K. Agbossou, J. M. Fernández, and J. A. Gladysz, *Inorg. Chem.*, **29**, 476 (1990); (b) S. K. Agbossou, W. W. Smith, and J. A. Gladysz, *Chem. Ber.*, **123**, 1293 (1990).

28. (a) M. A. Dewey, A. M. Arif, and J. A. Gladysz, *J. Organomet. Chem.*, **384**, C29 (1990); (b) M. A. Dewey, J. M. Bakke, and J. A. Gladysz, *Organometallics*, **9**, 1349 (1990); (c) M. A. Dewey and J. A. Gladysz, *Organometallics*, **9**, 1351 (1990); (d) M. A. Dewey, A. M. Arif, and J. A. Gladysz, *J. Chem. Soc. Chem. Commun.*, **1991**, 712.

29. G. S. Bodner, T.-S. Peng, A. M. Arif, and J. A. Gladysz, *Organometallics*, **9**, 1191 (1990).

30. (a) T.-S. Peng and J. A. Gladysz, *J. Chem. Soc. Chem. Commun.*, **1990**, 902; (b) T.-S. Peng and J. A. Gladysz, *Organometallics*, **9**, 2884 (1990).

31. C. M. Garner, N. Quirós Méndez, J. J. Kowalczyk, J. M. Fernández, K. Emerson, R. D. Larsen, and J. A. Gladysz, *J. Am. Chem. Soc.*, **112**, 5146 (1990).

32. T.-S. Peng and J. A. Gladysz, *Tetrahedron Lett.*, **31**, 4417 (1990).

33. G. L. Crocco, K. E. Lee, and J. A. Gladysz, *Organometallics*, **9**, 2819 (1990).

34. J. J. Kowalcyzk, S. K. Agbossou, and J. A. Gladysz, *J. Organomet. Chem.*, **397**, 333 (1990).

35. D. M. Dalton, J. M. Fernández, K. Emerson, R. D. Larsen, A. M. Arif, and J. A. Gladysz, *J. Am. Chem. Soc.*, **112**, 9198 (1990).

36. (a) N. Quirós Méndez, A. M. Arif, and J. A. Gladysz, *Angew. Chem. Int. Ed. Engl.*, **29**, 1473 (1990); (b) N. Quirós Méndez, C. L. Mayne, and J. A. Gladysz, *Angew. Chem. Int. Ed. Engl.*, **29**, 1475 (1990).

37. (a) K. E. Lee, A. M. Arif, and J. A. Gladysz, *Organometallics*, **10**, 751 (1991); (b) K. E. Lee, A. M. Arif, and J. A. Gladysz, *Chem. Ber.*, **124**, 309 (1991).
38. (a) J. J. Kowalczyk, A. M. Arif, and J. A. Gladysz, *Organometallics*, **10**, 1079 (1991); (b) J. J. Kowalczyk, A. M. Arif, and J. A. Gladysz, *Chem. Ber.*, **124**, 729 (1991).
39. N. Quirós Méndez, A. M. Arif, and J. A. Gladysz, *Organometallics*, **10**, 2199 (1991).
40. C. Roger, G. S. Bodner, W. G. Hatton, and J. A. Gladysz, *Organometallics*, **10**, 3266 (1991).
41. I. Saura-Llamas, C. M. Garner, and J. A. Gladysz, *Organometallics*, **10**, 2533 (1991).
42. M. L. H. Green and G. Wilkinson, *J. Chem. Soc.*, **1958**, 4314.
43. E. O. Fischer and W. Fellman, *J. Organomet. Chem.*, **1**, 191 (1963).
44. R. B. King and R. H. Reimann, *Inorg. Chem.*, **15**, 179 (1976).
45. C. P. Casey, M. A. Andrews, D. R. McAlister, and J. E. Rinz, *J. Am. Chem. Soc.*, **102**, 1927 (1980).
46. R. L. Pruett and E. L. Morehouse, *Chem. Ind.*, **1958**, 980.
47. W. A. Herrmann, *Angew. Chem. Int. Ed. Engl.*, **27**, 1297 (1988).
48. E. O. Fischer and H. Strametz, *Z. Naturforsch. B*, **23**, 278 (1968).
49. J. R. Sweet and W. A. G. Graham, *J. Am. Chem. Soc.*, **104**, 2811 (1982).
50. H. Saltzman and J. G. Sharefkin, *Organic Syntheses*, Wiley, New York, Collect. Vol. V, p. 658.

52. DI-μ-CHLORO-BIS[(η⁵-PENTA-METHYLCYCLOPENTADIENYL) CHLORORUTHENIUM(III)], [Cp*RuCl₂]₂ AND DI-μ-METHOXO-BIS(η⁵-PENTAMETHYLCYCLO-PENTADIENYL)DIRUTHENIUM(II), [Cp*RuOMe]₂

Submitted by U. KOELLE* and J. KOSSAKOWSKI*
Checked by D. GRUMBINE† and T. DON TILLEY†

Di-μ-chloro-bis[(η⁵-pentamethylcyclopentadienyl)chlororuthenium(III)],[1,2] [Cp*RuCl₂]₂[3] (1) is the general entry into Cp*(η⁵-C₅Me₅) Ru sandwich and half-sandwich chemistry and has been used in numerous transformations since its introduction in 1984. It is easily prepared according to Eq. 1. Analogous bromo and iodo complexes are obtained by halide exchange with NaX in MeOH.[4]

[Cp*RuOMe]₂[4,5] (2) is a versatile reagent for the preparation of all kinds of complexes Cp*RuL₂X, where the dimer is complexed or cleaved by ligands L and the OMe group is exchanged for X concomitantly or separately.[4] It is also easily transformed into the chloride [Cp*RuCl]₄ by treatment with either Me₃SiCl[5] or LiCl[6] in ether. Acid cleavage gives a cationic Cp*Ru⁺ fragment that can be complexed by addition of suitable ligands.[4,7]

* Institute for Inorganic Chemistry, Technical University of Aachen, 5100 Aachen, Germany.
† Department of Chemistry, University of California, San Diego, La Jolla, CA 92093.

Complex **2** is prepared from **1** according to Eq. 2 with yields up to quantitative depending only on the rigorous exclusion of oxygen in the work-up procedure.

Alternative routes to $[Cp^*RuOMe]_2$ are the reaction of $[Cp^*RuCl]_4$ with $Na(Li)OMe^8$ or the treatment of $Cp^*Ru(1,5$-cyclooctadiene$)Cl$ with LiOMe in MeOH.[6] The procedure given below gives the highest yield starting directly from $[Cp^*RuCl_2]_2$, the ultimate precursor for the other procedures as well.

$$RuCl_3 \cdot 3H_2O + C_5Me_5H \xrightarrow[-\text{ HCl}]{\text{ROH, 6-h reflux}} \tfrac{1}{2} [Cp^*RuCl_2]_2 + RuCp_2^* \qquad (1)$$

$$[Cp^*RuCl_2]_2 + 2K_2CO_3 \xrightarrow{\text{MeOH}} [Cp^*RuOMe]_2 + CH_2O + 4KCl$$
$$+ 2H_2O + 2CO_2 \qquad (2)$$

A. $[Cp^*RuCl_2]_2$ (1)[9]

All operations are carried out in a 500-mL round-bottomed flask under an inert atmosphere.

To a filtered solution of 8.0 g (31.6 mmol) of $RuCl_3 \cdot 3H_2O^{10}$ ($\sim 42\%$ Ru) in 200 mL of MeOH is added 9.6 g (72 mmol)[11] of C_5Me_5H, and the mixture is refluxed for 4 h.[12,13] The brown solution, which contains part of the product as a microcrystalline precipitate, is chilled to $-80°C$ for 12 h and filtered cold through a porous frit. The solid is dried *in vacuo* and washed twice with 30 mL of pentane in order to remove decamethylruthenocene, which can be collected in 10–20% yield from the pentane washings. The yield of **1** is 80% of the theoretical based on Ru.

In case the methanolic mother liquid is green, the reaction has not gone to completion. It can be refluxed for a second time, and more of the product can be recovered using the above work-up procedure.

Properties

$[Cp^*RuCl_2]_2$ is soluble in dichloromethane and chloroform and slightly soluble in acetone and methanol, where it tends to react. As a solid it is stable toward air for shorter periods. Storage in a refrigerator for longer periods is recommended, since samples having been at ambient temperature for weeks showed diminished solubility.

The ^{1}H-NMR spectrum shows a broad singlet with a chemical shift depending on the solvent[3]: δ 6.8 (CD_2Cl_2), 5.1 ($CDCl_3$).[4] IR (KBr): 2958 (w), 1436 (br, s), 1376 (s), 1151 (w), 1075 (w), 1024 (s), 436 (m) cm^{-1}.

Anal. Calcd. (found): C39.1 (38.9), H4.9 (4.9), C123.1 (22.4).

B.   [Cp*RuOMe]$_2$ (2)[14]

Into a 100-mL Schlenk flask is placed 1 g of [Cp*RuCl$_2$]$_2$ and 3 g of dry K$_2$CO$_3$, which has been heated to 200°C for several hours and cooled under nitrogen atmosphere. The flask is evacuated, and filled with nitrogen, and 30 mL of pure, dry, nitrogen-saturated methanol is added. The mixture is stirred at ambient temperature for about 6 h or, alternatively, refluxed for 1 h. During that time the brown Ru(III)–halide complex dissolves and the mixture turns intensely red.

For many transformations of [Cp*RuOMe]$_2$, this methanolic solution can be used directly after filtration from excess K$_2$CO$_3$. It contains the product along with CH$_2$O and KCl. It can be assayed photometrically by measuring the absorbance at 490 nm in an optical Schlenk cuvette taking $\varepsilon = 365\ M^{-1}\,cm^{-1}$.

To isolate the methoxo complex in pure form, methanol is evaporated *in vacuo* and the residue is extracted with several portions of either pentane or diethyl ether. The higher solubility in the latter solvent facilitates extraction, although the ensuing red solution contains nearly all of the KCl. Extraction with pentane requires more solvent (~ 70–100 mL) but gives the complex free from any salt impurities. After extraction a white residue, devoid of Ru complexes or salts, remains. Crystalline [Cp*RuOMe]$_2$ is obtained in 75% yield by slowly cooling the saturated pentane extract to − 30°C.

Properties

[Cp*RuOMe]$_2$ is air-sensitive, especially in solution, and solutions are preferentially transferred by canule or syringe techniques. Air oxidation is indicated by the formation of a green color and later a brown precipitate. This latter can be reconverted into [Cp*RuOMe]$_2$ by stirring it again with K$_2$CO$_3$ in MeOH.

The compound is soluble without decomposition in hydrocarbon and ether solvents as well as in acetone, benzene, and toluene. It slowly reacts with halocarbon and coordinating solvents. The solid state molecular structure is that of a folded dimer,[7a,8] and it is dimeric also in solution. ¹H NMR (C$_6$D$_{12}$): δ 1.62 (Cp*), 4.82 (OMe).

Anal. Calcd. (found): C49.4 (49.3), H6.8 (6.7).

The extinction coefficient of [Cp*RuOMe]$_2$ in pentane has been determined as 290 $M^{-1}\,cm^{-1}$ at 490 nm.

References and Notes

1. T. D. Tilley, R. H. Grubbs, and J. E. Bercaw, *Organometallics*, **3**, 274 (1984).
2. N. Oshima, H. Suzuki, and Y. Moro-oka, *Chem. Lett.*, **1984**, 1161.
3. [Cp*RuCl₂]₂ was shown by an X-ray molecular structure determination to be dimeric; U. Koelle, J. Kossakowski, N. Klaff, L. Wesemann, U. Englert, and G. E. Herberich, *Angew. Chem. Int. Ed. Engl.*, **30**, 690 (1991).
4. U. Koelle and J. Kossakowski, *J. Organomet. Chem.*, **362**, 383 (1989).
5. U. Koelle and J. Kossakowski, *J. Chem. Soc. Chem. Commun.*, **1988**, 549.
6. U. Koelle, B.-S. Kang, G. Raabe, and C. Krüger, *J. Organomet. Chem.*, **386**, 261 (1990).
7. (a) U. Koelle, J. Kossakowski, and R. Boese, *J. Organomet. Chem.*, **378**, 449 (1989). (b) U. Koelle and M. H. Wang, *Organometallics*, **9**, 195 (1990).
8. S. D. Loren, B. K. Campion, R. H. Heyn, T. Don Tilley, B. E. Bursten, and K. W. Luth, *J. Am. Chem. Soc.*, **111**, 4712 (1989).
9. Starting from ethyltetramethylcyclopentadiene, the ethyltetramethyl derivative (**1a**) is obtained following the same procedure. ¹H-NMR (CDCl₃): δ 5.1, 4.75, 3.48, v ≈ 20 cps.
10. Johnson Matthey Ltd., Reading, United Kingdom.
11. We found that this excess is necessary in order to completely convert the RuCl₃.
12. Alternatively, EtOH can be used as the solvent (see ref. 2). Reflux time in this case is 3 h. The yield of [Cp*RuCl₂]₂ is 60%, and more decamethylruthenocene is formed (20%). Further details are given in ref. 4.
13. Pentamethylcyclopentadiene is prepared from acetaldehyde and 3-pentanone F. X. Kohl and P. Jutzi, *J. Organomet. Chem.*, **243**, 119 (1983)] or from 2-bromobutene and ethyl acetate [(a) J. M. Manriquez, P. J. Fagan, L. D. Schertz, and T. J. Marks, *Inorg. Synth.*, **21**, 181 (1982); (b) R. S. Threlkel, J. E. Bercaw, P. F. Seidler, J. M. Stryker, and R. G. Bergman, *Org. Synth.*, **65**, 42 (1987)].
14. Starting from **1a** the ethyltetramethyl derivative is obtained following the same procedure; ¹H NMR (C₆D₆): δ 1.22 (t, 3 H), 1.74 (br, s, 12 H), 2.23 (q, 2 H), 4.80 (s, OMe).

53.   (η⁵-PENTAMETHYLCYCLOPENTADIENYL)RHODIUM AND -IRIDIUM COMPOUNDS

Submitted by C. WHITE,* A. YATES,* and P. M. MAITLIS*
Checked by D. M. HEINEKEY†

A significant development of organometallic chemistry in recent years has been the increasing use of pentamethylcyclopentadienyl compounds. Not only are such compounds usually more soluble and more readily crystallized than their unsubstituted cyclopentadienyl analogs, but they are generally more stable as a result of the steric and electron-donating effects of the five methyl groups. This is particularly so for the (η⁵-pentamethylcyclopentadienyl)rhodium and -iridium complexes, where the η⁵-C₅Me₅ acts as an

* Department of Chemistry, The University, Sheffield S3 7HF, United Kingdom.
† Department of Chemistry, Yale University, New Haven, CT 06511.

excellent stabilizing ligand toward Rh or Ir since it is displaced only with considerable difficulty. The chlorides $[M(\eta^5\text{-}C_5Me_5)Cl_2]_2$ (M = Rh or Ir) are the precursors to a wide range of (η^5-pentamethylcyclopentadienyl)rhodium and -iridium complexes. They have previously been prepared by the reaction of hexamethylbicyclo[2.2.0]hexa-2,5-diene (hexamethyl Dewar benzene) with the corresponding metal trichlorides;[1] however, the procedure reported here utilizes the more readily available pentamethylcyclopentadiene.[2] It is also convenient for large-scale preparations and consistently gives high yields and pure products.

A useful property of the complexes $[M(\eta^5\text{-}C_5Me_5)Cl_2]_2$ (M = Rh or Ir) is their ability to react with AgY in the presence of coordinating solvents to form the tris-solvent complexes $[M(\eta^5\text{-}C_5Me_5)(\text{solvent})_3](Y)_2$ (Y = $[BF_4]^-$ or $[PF_6]^-$). With weakly coordinating solvents (e.g., CH_2Cl_2, Me_2CO, or MeOH), these may be generated *in situ* and the coordinated solvent molecules can be displaced by a wide range of donor ligands under extremely mild conditions. Alternatively, with more strongly coordinating solvents, e.g., MeCN, dimethylsulfoxide (DMSO), (Py) the tris-solvent complexes may be isolated, stored if necessary, and subsequently used as precursors to other (η^5-pentamethylcyclopentadienyl)rhodium or -iridium compounds.

The preparations and utility of these solvent complexes are illustrated by the syntheses of $[M(\eta^5\text{-}C_5Me_5)(NCMe)_3]^{2+}$ (M = Rh or Ir) and the synthesis of the fluorene complex $[Ir(\eta^5\text{-}C_5Me_5)(\eta^6\text{-}C_{13}H_{10})][PF_6]_2$ from the *in situ* reaction of fluorene with the acetone complex $[Ir(\eta^5\text{-}C_5Me_5)(Me_2CO)_3]$-$[PF_6]_2$.

The rhodium and iridium trichloride hydrates are nonstoichiometric materials that vary in composition and that can give variable results. We find that the materials supplied by Johnson Matthey give good and reproducible yields.

A.   DI-μ-CHLORO-DICHLOROBIS(η^5-PENTAMETHYLCYCLOPENTADIENYL)DIRHODIUM(III)

$$2RhCl_3\cdot3H_2O + 2C_5Me_5H \rightarrow [Rh(\eta^5\text{-}C_5Me_5)Cl_2]_2 + 2HCl$$

Rhodium trichloride trihydrate (10 g, 0.042 mol, Johnson Matthey) pentamethylcyclopentadiene (6 g, 0.044 mol),[2] reagent-grade methanol (300 mL) and a magnetic stirring bar are placed in a 500-mL round-bottomed flask fitted with a reflux condenser. A nitrogen bubbler is attached to the top of the condenser, the apparatus purged with nitrogen for 5 min, and the mixture then refluxed gently under nitrogen for 48 h with stirring. The reaction mixture is allowed to cool to room temperature and the dark red precipitate is filtered off in air through a glass sinter. The red filtrate is reduced in volume to ~ 50 mL using a rotary evaporator to give more red crystals that were combined with the first crop and washed with diethyl ether (3×50 cm^3). Air drying gives 11.25 g (95% yield) of [Rh(η^5-C$_5$Me$_5$)Cl$_2$]$_2$, which is pure enough for most purposes. If required, the product may be recrystallized by dissolving in a minimum volume of chloroform, filtering if necessary, and slowly adding twice that volume of hexane.

Anal. Calcd. for Rh$_2$C$_{20}$H$_{30}$Cl$_4$: C, 38.9; H, 4.9; Cl, 22.9. Found: C, 38.8; H, 4.8; Cl. 22.3.

B. DI-μ-CHLORO-DICHLOROBIS(η^5-PENTAMETHYL-CYCLOPENTADIENYL)DIIRIDIUM(III)

$$2\text{IrCl}_3 \; x\text{H}_2\text{O} + 2\text{C}_5\text{Me}_5\text{H} \rightarrow [\text{Ir}(\eta^5\text{-C}_5\text{Me}_5)\text{Cl}_2]_2 + 2\text{HCl}$$

This may be prepared by the procedure described in Section A using iridium trichloride hydrate (10 g, 0.026 mol, Johnson Matthey) and penta-methylcyclopentadiene (5.0 g, 0.036 mol). Recrystallization of the product from chloroform–hexane gives 10.7 g (85% yield) of [Ir(C$_5$Me$_5$)Cl$_2$]$_2$ as an orange microcrystalline solid.

Anal. Calcd. for Ir$_2$C$_{20}$H$_{30}$Cl$_4$: C, 30.2; H, 3.8; Cl, 17.8. Found: C, 30.5; H, 3.7; Cl, 16.5.

Properties

Both the chloro complexes [M(η^5-C$_5$Me$_5$)Cl$_2$]$_2$ (M = Rh or Ir) are thermally stable (mp $> 230°$C) and have been stored in air at room temperature for several years without any apparent decomposition. They are readily soluble in chlorinated solvents but only sparingly soluble in acetone or alcohols. The rhodium complex is also somewhat soluble in water. The simplicity of their PMR spectra [i.e., CDCl$_3$, δ 1.60 s (M = Rh); 1.73 s (M = Ir)] allows their reactions to be conveniently followed by monitoring the shift in this C$_5$Me$_5$ resonance.

The corresponding iodo complexes are readily prepared by metathesis reactions with sodium iodide.[1] Other reactions have been reviewed[3] and recently [M(η^5-C$_5$Me$_5$)Cl$_2$]$_2$ have proved to be valuable precursors to novel bridged carbene complexes of the type [{M(η^5-C$_5$Me$_5$)Y}$_2$(μ-CH$_2$)$_2$].[4] In the presence of base, [Rh(η^5-C$_5$Me$_5$)Cl$_2$]$_2$ is an effective catalyst for the hydrogenation of olefins[5] and for the hydrosilylation of olefins[6] and the disproportionation of aldehydes.[7]

C. TRIS(ACETONITRILE)(η^5-PENTAMETHYL-
CYCLOPENTADIENYL)RHODIUM(2+)
HEXAFLUOROPHOSPHATE

$$[Rh(\eta^5\text{-}C_5Me_5)Cl_2]_2 + 4Ag[PF_6] + 6MeCN \longrightarrow$$
$$2[Rh(\eta^5\text{-}C_5Me_5)(NCMe)_3][PF_6]_2 + 4AgCl$$

Procedure

■ **Note.** *Silver hexafluorophosphate is moisture sensitive and therefore for convenience it is best to accurately weigh out an approximate desired amount of silver salt and then to adjust the weight of* [$Rh(\eta^5$-$C_5Me_5)Cl_2]_2$ *used according to the stoichiometry of the reaction.*

Silver hexafluorophosphate (1.30 g, 5.2 mmol, Aldrich) is weighed out into a stoppered dry sample tube. The $[Rh(\eta^5\text{-}C_5Me_5)Cl_2]_2$ (0.8 g, 1.3 mmol) is weighed into a dry 50-mL conical flask; a magnetic stirring bar together with acetonitrile (10 mL, dried over $MgSO_4$) is added and a 'T piece' connected to a standard-type joint fitted into the neck of the flask. The mixture is stirred gently while a slow stream of dry nitrogen is passed over the top of the flask via the 'T piece'. After 5 min the stirring rate is increased to a brisk pace and the silver hexafluorophosphate is tipped into the flask; any mixture remaining in the sample tube is washed into the reaction flask with additional dry acetonitrile (∼ 1 mL) using a dropping pipette. The reaction mixture becomes warm, and fast stirring is continued for 10 min. A precipitate of silver chloride forms and is removed by centrifugation in air; the yellow supernatant liquid is decanted, and the solid is washed with acetone (5 mL). The combined acetonitrile solution and acetone washings are taken to dryness using a rotary evaporator to leave an orange oil. Any remaining silver salts are removed by dissolving the oil in acetone (10 mL) and allowing this acetone solution to flow under gravity through a short column, ∼ 15 cm long and 1.5 cm diameter, packed with Hyflo-Supercel filter aid (BDH Chemicals Ltd., Broom Road, Poole, BH12 4NN, U.K.). The checker substituted Celatom FW-50 filter agent (Aldrich) and obtained the same or slightly better yields of product cited below.

The column is washed with additional portions of acetone (2×4 mL), and the solvent is then removed from the combined eluate using a rotary evaporator. An orange oil remains, and this solidifies on pumping (∼ 0.1 mm Hg) for several hours. This orange solid is pure enough for most purposes, but if desired the product may be recrystallized with difficulty by dissolving in a minimum volume of acetone and adding 1 drop of acetonitrile followed by the slow dropwise addition of diethyl ether until precipitation of the yellow-orange solid is complete. This is filtered off and dried *in vacuo* to give 1.10 g (65% yield) of yellow-orange $[Rh(C_5Me_5)(NCMe)_3](PF_6)_2$.

Anal. Calcd. for $RhC_{16}H_{24}F_{12}N_3P_2$: C, 29.5; H, 3.7; N, 6.5. Found: C, 29.4; H, 3.7; N, 6.6.

D. TRIS(ACETONITRILE)(η^5-PENTAMETHYL-CYCLOPENTADIENYL)IRIDIUM(2+) HEXAFLUOROPHOSPHATE

$$[Ir(\eta^5\text{-}C_5Me_5)Cl_2]_2 + 4Ag[PF_6] + 6MeCN \longrightarrow$$
$$2[Ir((\eta^5\text{-}C_5Me_5)(NCMe)_3](PF_6)_2 + 4AgCl$$

This may be prepared as white crystals by the procedure described in Section C using $[Ir(\eta^5\text{-}C_5Me_5)Cl_2]_2$ (0.8 g, 1 mmol) and silver hexafluorophosphate (1.0 g, 4 mmol). Yield: 1.11 g (84%).

Anal. Calcd. for $IrC_{16}H_{24}F_{12}N_3P_2$: C, 25.9; H, 3.3; N, 5.7. Found: C, 25.9; H, 3.3; N, 4.7.

Properties

The complexes $[M(\eta^5\text{-}C_5Me_5)(NCMe)_3][PF_6]_2$ (M = Rh or Ir) are air-stable and may be stored in a dry atmosphere for weeks without any significant decomposition. Their infrared spectra (Nujol) exhibit $\nu(CN)$ at 2298 and 2320 cm^{-1} (Rh) and 2300 and 2330 cm^{-1} (Ir); this compares with $\nu(CN)$ at 2254 and 2295 cm^{-1} observed for uncoordinated acetonitrile. The ^1H NMR spectrum of each complex consists of two singlets in the ratio of 5 : 3 [in acetone-d_6 M = Rh, δ 1.85 (C_5Me_5) and 2.53; M = Ir 1.89 (C_5Me_5) and 2.76].

Partial or complete displacement of the coordinated acetonitrile ligands may be achieved with monodentate ligands (e.g., phosphines[8]), bidentate ligands (e.g., dienes),[9] or tridentate ligands, providing a convenient entry to a range of (η^5-pentamethylcyclopentadienyl)rhodium and -iridium complexes.

E. PREPARATION OF TRIS(ACETONE)(η^5-PENTA-METHYLCYCLOPENTADIENYL)IRIDIUM(2+) AND ITS REACTION WITH FLUORENE TO GIVE (η^6-FLUORENE)(η^5-PENTAMETHYLCYCLOPENTA-DIENYL)IRIDIUM(2+)

$$[Ir(\eta^5\text{-}C_5Me_5)Cl_2]_2 + 4Ag[PF_6] + 6Me_2CO \longrightarrow$$
$$2[Ir(\eta^5\text{-}C_5Me_5)(OCMe_2)_3][PF_6]_2$$

$$[Ir(\eta^5\text{-}C_5Me_5)(OCMe_2)_3][PF_6]_2 + C_{13}H_{10} \longrightarrow$$
$$[Ir(\eta^5\text{-}C_5Me_5)(C_{13}H_{10})][PF_6]_2$$

Procedure

Tris(acetone)(η^5-pentamethylcyclopentadienyl)iridium(2 +) is prepared from [Ir(η^5-C_5Me_5)Cl_2]$_2$ (0.200 g, 0.25 mmol) and silver hexafluorophosphate (0.254 g, 1.0 mmol) by the procedure described in Section C using acetone (dried over $MgSO_4$) in place of acetonitrile. After centrifuging off the precipitated silver chloride, the last traces of silver salts are removed by allowing the pale yellow solution to flow under gravity through a short column (~ 10 cm × 1.5 cm) packed with Hyflo supercel (or Celatom FW-50) filter aid (see procedure C above). The eluate is collected in a 50-mL conical flask containing fluorene (0.30 g, 1.80 mmol) and a magnetic stirring bar; slow stirring is commenced as the eluate begins to drip into the flask. The silver chloride precipitate is washed with acetone (5 mL) and the washings passed down the column. To remove any remaining iridium complex from the column, it is washed with a further portion of acetone (5 mL). Stirring is continued for a further 5 min after the last drop of the acetone washings enters the flask, and then the solvent is removed using a rotary evaporator. The oil remaining is washed with diethyl ether (2 × 5 mL) and dichloromethane (5 mL) to remove excess fluorene and then dissolved in acetone (~ 3 mL). Dropwise addition of chloroform to the acetone solution precipitates out [Ir(η^5-C_5Me_5)(η^6-$C_{13}H_{10}$)][PF_6]$_2$ as a white solid that is filtered off and dried *in vacuo* (0.358 g, 91% yield).

Anal. Calcd. for $IrC_{23}H_{25}F_{12}P_2$: C, 35.3; H, 3.2. Found: C, 35.7; H, 3.2.

Properties

Although [Ir(η^5-C_5Me_5)($OCMe_2$)$_3$[PF_6]$_2$ cannot be isolated, it can be readily identified by the characteristic η^5-C_5Me_5 resonance in its ^1H NMR spectrum, specifically, δ 1.75 in acetone-d_6; the ^1H NMR spectra of the rhodium analog and also of the corresponding methanol and dichloromethane solvent complexes have been reported.[10] If left in solution for several days at room temperature, [Ir(η^5-C_5Me_5)($OCMe_2$)$_3$][PF_6]$_2$ and the corresponding rhodium complex undergo a series of complex rearrangements.[11] However, if used immediately, such complications do not arise and the extreme lability of these solvent complexes, as illustrated by the mild conditions under which fluorene reacts, makes them valuable synthons for a wide range of pentamethylcyclopentadienyl rhodium and iridium complexes.[10,12]

The chemistry and NMR data of the fluorene complex [Ir(η^5-C_5Me_5)-(η^6-$C_{13}H_{10}$)][PF_6]$_2$ have been reported.[10] It is pertinent to note that this complex is remarkably stable, considering that it contains an arene ligand bonded to a metal in a 3 + oxidation state, and this exemplifies the stabilizing influence of the bulky, strongly electron-donating pentamethyl-cyclopentadienyl ligand.

References

1. J. W. Kang, K. Moseley, and P. M. Maitlis, *J. Am. Chem. Soc.*, **91**, 5970 (1969).
2. J. M. Manriquez, P. J. Fagan, L. D. Schertz, and T. J. Marks, *Inorg. Synth.*, **21** 181 (1982); R. S. Threlkel, J. E. Bercaw, P. F. Seidler, J. M. Stryker, and R. G. Bergman, *Org. Synth.*, **65**, 42 (1987).
3. P. M. Maitlis, *Chem. Soc. Rev.*, **10**, 1 (1981).
4. K. Isobe, A. Vázquez de Miguel, P. M. Bailey, S. Okeya, and P. M. Maitlis, *J. Chem. Soc. Dalton Trans.*, **1983**, 1441; A. Vázquez de Miguel, M. Gómez, K. Isobe, B. F. Taylor, B. E. Mann, and P. M. Maitlis, *Organometallics*, **2**, 1724 (1983).
5. D. S. Gill, C. White, and P. M. Maitlis, *J. Chem. Soc. Dalton Trans.*, **1978**, 617.
6. A. Millan, E. Towns, and P. M. Maitlis, *J. Chem. Soc. Chem. Commun.*, **1981**, 673; A. Millan, M.-J. Fernandez, P. O. Bentz, and P. M. Maitlis, *J. Mol. Catal.*, **26**, 89 (1984).
7. J. Cook, J. E. Hamlin, A. Nutton, and P. M. Maitlis, *J. Chem. Soc. Dalton Trans.*, **1981**, 2342.
8. S. J. Thompson, C. White, and P. M. Maitlis, *J. Organomet. Chem.*, **136**, 87 (1977).
9. C. White, S. J. Thompson, and P. M. Maitlis, *J. Chem. Soc. Dalton Trans.*, **1978**, 1305.
10. C. White, S. J. Thompson, and P. M. Maitlis, *J. Chem. Soc. Dalton Trans.*, **1977**, 1654.
11. S. J. Thompson, C. White, and P. M. Maitlis, *J. Organomet. Chem.*, **134**, 319 (1977).
12. S. L. Grundy, A. J. Smith, H. Adams, and P. M. Maitlis, *J. Chem. Soc. Dalton Trans.*, **1984**, 1747.

54. BIS(η^5-CYCLOPENTADIENYL)BIS(DIETHYLAMIDO)-URANIUM(IV) AND TRIS(η^5-CYCLOPENTADIENYL)(DIETHYLAMIDO)URANIUM(IV)

**Submitted by F. OSSOLA,* G. ROSSETTO,* P. ZANELLA,* A. ARUDINI,†
J. D. JAMERSON,† and J. TAKATS†
Checked by A. DORMOND‡**

Organouranium amides are useful precursors for the synthesis of a large variety of organouranium derivatives since they are known to undergo very easy cleavage of the uranium nitrogen bond by acidic hydrogen or to give insertion of molecules with polar multiple bonds.[1] The compounds $UCp_2(NEt_2)_2$ and UCp_3NEt_2 ($Cp = \eta^5\text{-}C_5H_5$, $Et = C_2H_5$) are the most

* Consiglio Nazionale della Ricerca, Istituto di Chimica e Tecnologia dei Radioelementi, 35100 Padua, Italy.
† Department of Chemistry, University of Alberta, Edmonton, Alberta, Canada, T6G 2G2.
‡ Laboratoire de synthese et d'electrosynthese associe au C.N.R.S. (U.A. 33), Faculte des Sciences, 6 Bd. Gabriel, 21100 Dijon, France.

convenient organouranium amides for these purposes. These molecules were first synthesized in 1974 by the stoichiometric reaction of CpH with $U(NEt_2)_4$.[2] More recently an alternative synthetic route, which utilizes the displacement of chloride ions from UCp_nCl_{4-n} by $LiNEt_2$, has been proposed.[3] Both methods are useful for the preparation of UCp_3NEt_2, but $UCp_2(NEt_2)_2$ is often contaminated by unwanted UCp_3NEt_2, and its purification is rendered difficult by the similar solubilities of the two compounds in common organic solvents. We describe here improvements over the procedures reported previously, which provide convenient, high-purity, and high-yield syntheses of $UCp_2(NEt_2)_2$ and UCp_3NEt_2.

■ **Caution.** *Organouranium compounds are very air-sensitive. They must be handled in a glove box or in Schlenk vessels under rigorously purified nitrogen or argon. All procedures described here were carried out in a glove box. The solvents used must be dried by heating at reflux, under purified nitrogen, with potassium benzophenone and distilled just prior to use.*

A. BIS(η^5-CYCLOPENTADIENYL)BIS(DIETHYLAMIDO)-URANIUM(IV)

$$UCl_4 + 2TlC_5H_5 \xrightarrow[\text{room temp., 24 h}]{\text{THF}} \text{``}U(\eta^5\text{-}C_5H_5)_2Cl_2\text{''}* + 2TlCl$$

$$\text{``}U(\eta^5\text{-}C_5H_5)_2Cl_2\text{''} + 2LiNEt_2 \xrightarrow[\text{room temp., 24 h}]{\text{THF}} U(\eta^5\text{-}C_5H_5)_2(NEt_2)_2 + 2LiCl$$

Procedure

■ **Caution.** *Thallium compounds are extremely toxic. The one used here is volatile and should be handled with extreme care.*

Cyclopentadienylthallium,[4,†] 5.39 g (20 mmol), is added to a vigorously stirred suspension of uranium tetrachloride,[5,‡] 3.80 g (10 mmol) in 100 mL of freshly distilled tetrahydrofuran (THF) contained in a 200-mL, single-necked, round-bottomed flask. The suspension is stirred at room temperature for

---

* Although "UCp_2Cl_2" as a pure solid cannot be isolated, it can be prepared *in situ* in equilibrium with UCp_3Cl and $UCpCl_3 \cdot 2THF$:

$$\text{``}2UCp_2Cl_2\text{''} \underset{}{\overset{\text{THF}}{\rightleftharpoons}} UCp_3Cl + UCpCl_3$$

The formation of only $UCp_2(NEt_2)_2$ strongly indicates that "UCp_2Cl_2" reacts with $LiNEt_2$ much faster than do $UCpCl_3 \cdot 2THF$ and UCp_3Cl.

† Cyclopentadienylthallium is available from Strem Chemicals, Inc., P.O. Box 108, Newbury Port, MA 01950.

‡ Uranium tetrachloride is available from Cerac Inc., Box 1178 Milwaukee, WI 53201.

24 h. After this time, the mixture is filtered to remove the thallium(I) chloride formed. The residue is then washed with THF until the washings are colorless. Freshly prepared (diethylamido)lithium,[6] 1.58 g (20 mmol) dissolved in 50 mL of THF, is added dropwise over 2 h with constant stirring to the combined filtrates of "$U(\eta^5-C_5H_5)_2Cl_2$". After 24 h, the volatile components are removed from the brown solution under vacuum, and the residue is treated with 200 mL of pentane and then filtered. The solvent is removed again, and the yellow solid is dried for 5 h under vacuum at room temperature. Yield 4.35 g (85%).

Anal. Calcd. for $C_{18}H_{30}N_2U$: C, 42.19; H, 5.90. Found: C, 42.31; H, 6.04.

Properties

Bis(η^5-cyclopentadienyl)bis(diethylamido)uranium(IV) is a golden yellow, air-sensitive solid that is very soluble in aromatic and aliphatic solvents, as well as in THF and diethyl ether. Its ^{1}H NMR spectrum, which can serve as a satisfactory check for the purity of the compound, exhibits a singlet at δ -13.70 ppm due to the cyclopentadienyl moiety and a quartet and a triplet at δ 8.78 and 1.54 ppm due to the CH_2 and CH_3 moieties of the diethylamido group, respectively (C_6D_6, 27°C, internal standard C_6D_5H, shifts in ppm calculated with respect to $SiMe_4$).

B.   TRIS(η^5-CYCLOPENTADIENYL)(DIETHYLAMIDO)-URANIUM(IV)

$$UCl_4 + LiNEt_2 \xrightarrow[\text{room temp., 30 h}]{Et_2O} U(NEt_2)_4 + 4LiCl$$

$$U(NEt_2)_4 + 3C_5H_6 \xrightarrow[\text{room temp., 24 h}]{\text{pentane}} U(\eta^5-C_5H_5)_3NEt_2 + 3Et_2NH$$

$$U(\eta^5-C_5H_5)_3Cl + LiNEt_2 \xrightarrow[\text{room temp., 24 h}]{THF} U(\eta^5-C_5H_5)_3NEt_2 + LiCl$$

Procedure

Method I. A 250-mL, single-necked, round-bottomed flask containing a magnetic stirring bar and wrapped in foil to exclude light, is charged with 3.80 g (10 mmol) of uranium tetrachloride, 3.32 g (42 mmol) of (diethylamido)lithium, and 100 mL of anhydrous diethyl ether. The reaction mixture is stirred at room temperature for 30 h, then filtered, and the solvent is removed under vacuum. The resulting dark green oil is dissolved in 30 mL of pentane. To this solution, 2.5 mL (30 mmol) of freshly distilled cyclopentadiene, dissolved in 10 mL of pentane, is added. The solution is stirred for 24 h, then filtered, and the red solid is washed with pentane until the washings are colorless. The solid residue is dried for 2 h under vacuum at room temperature. Yield 4.14 g (82%).

Method II. A 200-mL, single-necked, round-bottomed flask containing a magnetic stirring bar is charged with 4.68 g (10 mmol) of chlorotris (η^5-cyclopentadienyl)uranium(IV),[7] 100 mL of freshly distilled THF, and 0.79 g (10 mmol) of (diethylamido)lithium.[6] The red solution is stirred for 24 h at room temperature. The solvent is then removed under vacuum. The residue obtained is dissolved in 100 mL of toluene, and the solution is filtered to separate the lithium chloride, which is washed with toluene until the washings are colorless. The solvent is then removed under vacuum, giving a red solid that is dried for 2 h under vacuum at room temperature. Yield 4.29 g (85%).

Anal. Calcd. for $C_{19}H_{25}NU$: C, 44.27, H, 4.89. Found: C, 45.23; H 5.06.

Properties

Tris(η^5-cyclopentadienyl)(diethylamido)uranium(IV) is a red, air-sensitive solid that is very soluble in toluene, THF, and diethyl ether and slightly soluble in aliphatic hydrocarbons. It sublimes at 150°C at 10^{-4} mm Hg. The purity of the compound can be checked by its ^{1}H NMR spectrum, which shows a singlet at δ − 11.62 ppm due to the cyclopentadienyl hydrogens, a quartet at δ 0.84 ppm, and a triplet at δ − 1.16 ppm due to the CH_2 and CH_3 moieties of the diethylamido group, respectively (C_6D_6, 27°C, internal standard C_6D_5H, shifts in ppm calculated with respect to $SiMe_4$).

References

1. (a) T. J. Marks and R. D. Ernst, in *Comprehensive Organometallic Chemistry*, G. Wilkinson, F. G. A. Stone and E. W. Abel (eds.), Pergamon Press, Oxford, 1982, Vol. 3, Chap. 21, p. 173; (b) J. Takats, in *Fundamental and Technological Aspects of Organo-f-Element Chemistry*, T. J. Marks and I. Fragala (eds.), Reidel, Dordrecht, 1984, p. 159; (c) G. Paolucci, G. Rossetto, P. Zanella, K. Yünlü, R. D. Fischer, *J. Organomet. Chem.*, **272**, 363 (1984); (d) A. Dormond, A. Aaliti, and C. Moise, *J. Chem. Soc. Chem. Commun.*, **1985**, 1231.
2. (a) A. L. Arduini, N. M. Edelstein, J. D. Jamerson, J. G. Reynolds, K. Schmid, and J. Takats, *Inorg. Chem.*, **20**, 2470 (1981); (b) J. D. Jamerson, J. Takats, *J. Organomet. Chem.*, **78**, C23 (1974).

3. F. Ossola, G. Rossetto, P. Zanella, G. Paolucci, and R. D. Fischer, *J. Organomet. Chem.*, **309**, 55 (1986).
4. A. J. Nielson, C. E. F. Rickard, and J. M. Smith, *Inorg. Synth.*, **24**, 97 (1985).
5. J. A. Hermann and J. F. Suttle, *Inorg. Synth.*, **5**, 143 (1957).
6. R. G. Jones, G. Karmas, G. A. Martin Jr., and H. Gilman, *J. Am. Chem. Soc.*, **78**, 4285 (1956).
7. T. J. Marks, A. M. Seyam, and W. A. Vachter, *Inorg. Synth.*, **16**, 147 (1976).

Chapter Five

CLUSTER AND CAGE COMPOUNDS CONTAINING TRANSITION METALS

55. DISODIUM TRIS(TETRABUTYLAMMONIUM) [β-HEXATRICONTAOXO (μ_{12}-TETRAOXOSILICATO) (10,11,12-TRINIOBIUMNONATUNGSTEN)ATO (7−)] (η^5-PENTAMETHYLCYCLOPENTADIENYL) RHODATE (5−)*, $Na_2(Bu_4N)_3[Rh[\beta$-Nb_3 $SiW_9O_{40}]\{\eta^5$-$C_5(CH_3)_5\}]$

Submitted by R. G. FINKE,† K. NOMIYA,† C. A. GREEN,†
and M. W. DROEGE†
Checked by A. R. SIEDLE‡

The interest in polyoxoanions, such as $[Nb_3SiW_9O_{40}]^{7-}$,[1,2] stems from the fundamental interest in this large class of compounds as well as their broad range of potential applications; these points have been discussed in a recent volume of *Inorganic Syntheses*.[3] In the present case, $[Rh[\beta$-$Nb_3SiW_9O_{40}]$ $\{\eta^5$-$C_5(CH_3)_5\}]^{5-}$ is a prototypical example of a transition metal supported on the surface oxygen atoms of a *Keggin-type* polyoxoanion. Other related complexes, notably $[Ir(Nb_3SiW_9O_{40})(COD)]^{6-}$ (where COD = 1,5-cyclo-octadiene), have proved to be interesting catalyst precursors.[4]

In the present synthesis, the initial polyoxoanion product results from oligomerization, under acidic conditions, to form three bridging Nb—O—Nb

* For numbering of Keggin structures and nomenclature, see "Nomenclature of Polyanions," *Pure Appl. Chem.*, **59** (11), 1529–1548 (1987).

† Department of Chemistry, University of Oregon, Eugene, OR 97403.

‡ 3M Central Research Laboratories, St. Paul, MN 55144.

bonds (detectable by a strong IR band at 690 cm^{-1}): $2[Nb_3SiW_9O_{40}]^{7-}$ $+ 6H^+ \rightarrow [Nb_6Si_2W_{18}O_{77}]^{8-} + 3H_2O$. The product, $[(C_4H_9)_4N]_6H_2$ $[Nb_6Si_2W_{18}O_{77}]$, has been characterized by elemental analysis, fast-atom bombardment mass spectroscopy (FAB–MS), solution ultracentrifugation molecular-weight measurements, IR, and ^{183}W NMR spectroscopy.[1,2] Reversal of the oligomerization reaction (cleavage of the three Nb—O—Nb bonds) is readily effected by tetrabutylammonium hydroxide, $[(C_4H_9)_4N]OH$. The stoichiometry of this reaction, $[H_2(Nb_6Si_2W_{18}O_{77})]^{6-} + 8OH^- \rightarrow 2[Nb_3SiW_9O_{40}]^{7-} + 5H_2O$, has been quantitatively established by monitoring the loss of the 690-cm^{-1} IR band. The resultant $[(C_4H_9)N]_7[Nb_3SiW_9O_{40}]$ has been characterized by FAB–MS, solution molecular-weight measurements, IR, and its two-line ^{183}W NMR spectrum.[1,2] The title complex, A-β-$[(C_4H_9)_4N]_3Na_2[Rh[\beta$-$Nb_3SiW_9O_{40}]\{\eta^5$-$C_5(CH_3)_5\}]$, is then formed by the addition of $[Rh\{\eta^5$-$C_5(CH_3)_5\}(CH_3CN)_3]^{2+}$ and characterized by a full elemental analysis and IR, ^{1}H NMR, and ^{183}W NMR spectroscopy.[2b] Earlier, we characterized the $[(C_4H_9)_4N]_5$ salt by each of these techniques as well as by solution molecular-weight measurements and studies employing ion exchange resins.[1,2a] The data show that $[Rh\{\eta^5$-$C_5(CH_3)_5\}]^{2+}$ is firmly and covalently attached to the surface oxygen atoms of $(Nb_3SiW_9O_{40})^{7-}$ in a fashion that yields a C_S symmetry complex.

A.   HEXAKIS(TETRABUTYLAMMONIUM)DIHYDROGEN 10d:10d′,10f:10f′, 11f:11f′-TRI-μ_4-OXOBIS- [β-TRITRICONTAOXO(μ_{12}-TETRAOXOSILICATO) (10,11,12-TRINIOBIUMNONATUNGSTEN)]ATE (8−), [(C$_4$H$_9$)$_4$N]$_6$H$_2$[(μ_4-O)$_3$[β-Nb$_3$SiW$_9$O$_{37}$]$_2$]

$\{H[Nb_6O_{19}]\}^{7-} + 2\{H[SiW_9O_{34}]\}^{9-} + (17 + x)H^+ + (8 - x)[(C_4H_9)_4N]^+$

$$\xrightarrow[\text{(2) NaHSO}_3]{\text{(1) H}_2\text{O}_2} [(C_4H_9)_4N]_{8-x}H_x[Nb_6Si_2W_{18}O_{77}] + 10H_2O$$

$[(C_4H_9)_4N]_{8-x}H_x[Nb_6Si_2W_{18}O_{77}] + (2 - x)H^+ \rightarrow$

$$[(C_4H_9)_4N]_6H_2[Nb_6Si_2W_{18}O_{77}] + (2 - x)[(C_4H_9)_4N]^+$$

$\{H[Nb_6O_{19}]\}^{7-} + 2\{H[SiW_9O_{34}]\}^{9-} + 19H^+ + 6[(C_4H_9)_4N]^+ \rightarrow$

$$[(C_4H_9)_4N]_6H_2[Nb_6Si_2W_{18}O_{77}] + 10H_2O$$

The preparation of the title complex is based on the reaction of the lacunary $\{H[SiW_9O_{34}]\}^{9-}$ with Nb(V).[1,2] Excess H_2O_2 is present with the

$[Nb_6O_{19}]^{8-}$ to inhibit the formation of Nb_2O_5 that would otherwise occur on acidification in aqueous solution.[5] Work-up consists of slow $NaHSO_3$ addition to destroy the peroxide,[2] precipitation of crude $[(C_4H_9)_4N]_{8-x}$ $H_x[Nb_6Si_2W_{18}O_{77}]$ ($x \le 2$) with tetrabutylammonium bromide, precipitation of crude $[(C_4H_9)_4N]_6H_2[Nb_6Si_2W_{18}O_{77}]$ from CH_3CN by the addition of aqueous HCl, and then recrystallization of the $[(C_4H_9)_4N]_6$ $H_2[Nb_6Si_2W_{18}O_{77}]$ from CH_3CN. The more soluble monoprotonated complex, $[(C_4H_9)_4N]_7H[Nb_6Si_2W_{18}O_{77}]$, can also be isolated in $\le 50\%$ yield by recrystallization of the initial precipitate from CH_3CN.[2]

Preparation

In a hood, $K_7H[Nb_6O_{19}] \cdot 13H_2O$* (12.0 g, 8.76 mmol) is added to a vigorously stirred solution of 0.6 M H_2O_2 (1.5 L, 0.90 mol). As soon as the $K_7H[Nb_6O_{19}]$ dissolves (a colorless solution is initially formed), 20.4 mL of 6 M HCL (0.122 mol) is immediately added, and the solution turns yellow. Crude $Na_9H[SiW_9O_{34}] \cdot 23H_2O$[3] (50.0 g, 17.5 mmol) is quickly added,[†] and the solution turns to yellow-orange and effervesces as the lacunary tungstosilicate dissolves. The resulting solution is stirred for 15 min, and then excess sodium hydrogen sulfite (125 g, 1.2 mol) is slowly added (over 10–15 min)

---

* The preparation and characterization of this complex have previously been reported.[6,7] The preparation described here is similar to that described by Flynn and Stucky[7] and provides a simple means of obtaining the complex in good yield as a powder. A mixture of Nb_2O_5 (10 g, 37.6 mmol) and 85% KOH (35 g, 530 mmol) is combined in a nickel crucible. In a hood, the crucible is heated cautiously with a low flame (to prevent boilover and minimize splattering), until the contents occasionally boil. During the heating, the reaction mixture changes from white to gray through the formation of a transparent melt. The crucible is then heated using a high flame for 10 min and then allowed to cool to room temperature ($\sim$ 30 min). The crucible and melt (now gray in color) are placed in a 250-mL beaker with 200 mL of H_2O. After the melt has dissolved (this may be hastened by stirring with a glass rod), the crucible is removed and the warm solution is filtered through Celite to remove any insoluble material. The solution is transferred to a beaker and, while rapidly stirring with a glass rod, 200 mL of 95% ethanol is added to the solution. The product separates first as an oil that is readily transformed into a white powder with continued stirring ($\sim$ 15 min). After standing for 30 min, the product is collected on a filter (medium glass frit), washed twice with 25 mL of ethanol/water (60/40), twice with 25 mL of 95% ethanol, and with 25 mL of diethyl ether. The product is then air-dried for 4–6 h; yield 14.2–17.0 g (83–99%).

† Sodium metasilicate (60 g, 0.21 mol) is dissolved in 500 mL of water in a 1-L beaker with vigorous stirring. Sodium tungstate (362 g, 1.1 mol) is dissolved in the clear solution, and then 200 mL of 6M HCl (1.2 mol) is added *in 1–2 min* with vigorous stirring (during which a white gelatinous silica precipitate forms). The mixture is rapidly stirred for 10 min, and then filtered through glass wool on a coarse glass frit. After the filtrate has been refrigerated at 4° overnight, the white crystalline precipitate is collected on a filter and dried at 80°. Yield: 66 g (19%) of crude A-β-$Na_9H[SiW_9O_{34}] \cdot 23H_2O$.

to destroy the peroxides; a color change from yellow-orange to colorless accompanies this step.[2]

■ **Caution.** *This reaction is very exothermic, and care should be taken during this addition.*

The hot, colorless solution is vigorously stirred for 2 h, as it cools, and then filtered through Celite, if necessary, to remove undissolved material (occasionally, a small amount of white material, presumably hydrous niobium oxide, is present). The solution is cooled to ice temperature and tetrabutylammonium bromide, $[(C_4H_9)_4N]Br$ (25.1 g, 78 mmol), is added over 10 min to the well-stirred solution. The product precipitates as a white powder. After stirring for 1 h, the mixture is allowed to stand overnight in a refrigerator at 4°. [By cooling the mixture before addition of the $[(C_4H_9)_4N]Br$, and by allowing the mixture to stand overnight, the material is more easily collected and washed.] The white precipitate is collected by vacuum filtration on a 600-mL 90M glass-fritted Büchner funnel. The material is extensively washed: three times (i.e., 450 mL total) with 150 mL of water, three times with 150 mL of 95% ethanol, and three times with 150 mL of diethyl ether. The white powder is dried overnight at 50°.

■ **Caution.** *An explosion-proof oven should be used because of ether fumes. Alternatively, air drying for ≥ 1 day can be used.*

This typically yields 60 g of crude material, $[(C_4H_9)_4N]_{8-x}H_x$ $[Nb_6Si_2W_{18}O_{77}]$ ($x \leq 2$). This material is dissolved in 360 mL of hot (80°) acetonitrile and immediately filtered through Celite to remove any undissolved material (on a 350-mL coarse glass frit to avoid long filtration times). The slightly yellow filtrate is transferred to a 600-mL beaker and rapidly stirred as 150 mL of 8 M HCl (excess, 1.2 mol) is added as follows: 30 mL of acid is added quickly to the solution (which turns colorless), 20 mL of acid is added in 5-mL portions (over 10 min) to cause the solution to cloud, the mixture is stirred for 20–30 min (during which a white microcrystalline powder precipitates), and then the remaining 100 mL of acid is added (over 1 h in 10–15-mL portions). The slurry is stirred for 1 h and then refrigerated at 4° overnight. The crude $[(C_4H_9)_4N]_6H_2[Nb_6 Si_2W_{18}O_{77}]$ is collected on a 600-mL 90M glass-fritted Büchner funnel and extensively washed, twice (i.e., 200 mL total) with 100 mL of H_2O, twice with 100 mL of 95% ethanol, and twice with 100 mL of diethyl ether. It is then dried at 50° overnight (yield: 51.7 g, 89%).*

---

* In the hands of one of us, the (lowest observed) yield was 32.7 g. In this case, a second crop can be isolated as follows. The combined filtrate and washings are evaporated to 400-mL total volume on a steam bath. This produces an oily residue, and 100 mL of H_2O is added to aid separation of the material from the solution. After stirring the mixture for 30 min, the residue is collected on a filter. The residue is dissolved in 150 mL of hot (80°) acetonitrile, 40 mL of 8 M HCL is added, and the mixture is rapidly stirred for 1 h. The product is collected and washed as before. [Additional yield: 8.7 g for a combined yield: 41.4 g (71%).]

■ **Caution.** *An explosion-proof oven should be used. Alternatively, air drying for* ≥ 1 *day can be used.*

This material is recrystallized by dissolving it in 360 mL of hot (80°) acetonitrile, rotary-evaporating the solution to 150 mL, transferring the solution to a 250-mL beaker, tightly covering the beaker, and allowing the mixture to stand overnight in a refrigerator at 4°. The crystalline precipitate is filtered (medium glass frit), washed with 15 mL of ice-cold acetonitrile, and dried overnight at 50° (yield: 24.8 g). After rotary evaporating the filtrate and wash to 30-mL total volume, a second crop is similarly isolated (yield: 5.2 g). Combined yield: 30.0 g (52%).[8]

Anal. Calcd. for $[(C_4H_9)_4N]_6H_2[Nb_6Si_2W_{18}O_{77}]$: C, 17.4; H, 3.32; N, 1.27; Si, 0.85; W, 50.0; Nb, 8.43. Found: C, 17.4; H, 3.41; N, 1.33; Si, 0.85; W, 50.5; Nb, 8.15.

Properties

The product is soluble in polar aprotic solvents such as acetonitrile and dimethylformamide. It is very slightly soluble to insoluble in ethanol, methanol, and water. The three Nb—O—Nb bridges in $[Nb_6Si_2W_{18}O_{77}]^{8-}$ are readily cleaved (as monitored by the loss of the 690 cm^{-1} IR band) by OH$^-$, H$_2$O$_2$, 2-aminoethanol, and other nontertiary amines[1,2] in reactions that closely parallel those of hydrous niobium oxide, $[Nb_2O_5 \cdot xH_2O]$. Spectroscopic data (FAB–MS, ^{183}W NMR, IR) are also available.[1,2]

B. DISODIUM TRIS(TETRABUTYLAMMONIUM) [β-HEXATRICONTAOXO(μ_{12}-TETRAOXOSILICATO) (10,11,12-TRINIOBIUMNONATUNGSTEN)ATO(7−)] [η^5-PENTAMETHYLCYCLOPENTADIENYL)RHODATE(5−), $Na_2[(C_4H_9)_4N]_3[Rh[β-Nb_3SiW_9O_{40}\{η^5-C_5(CH_3)_5\}]$

$$\{H_2[Nb_6Si_2W_{18}O_{77}]\}^{6-} + 8OH^- \rightarrow 2[Nb_3SiW_9O_{40}]^{7-} + 5H_2O$$

$$[Rh(C_5(CH_3)_5)Cl_2]_2 + 4Ag^+ + 6CH_3CN \rightarrow$$
$$2[Rh(C_5(CH_3)_5)(CH_3CN)_3]^{2+} + 4AgCl$$

$$2[Nb_3SiW_9O_{40}]^{7-} + 2[Rh(C_5(CH_3)_5)(CH_3CN)_3]^{2+} + 6[(C_4H_9)_4N]^+ + 4Na^+ \rightarrow$$
$$2Na_2[(C_4H_9)_4N]_3[Rh[β-Nb_3SiW_9O_{40}\{η^5-C_5(CH_3)_5\}] + 6CH_3CN$$

$$\{H_2[Nb_6Si_2W_{18}O_{77}]\}^{6-} + 8OH^- + [Rh(C_5(CH_3)_5)Cl_2]_2 + 4Ag^+$$
$$+ 6[(C_4H_9)_4N]^+ + 4Na^+ \rightarrow$$
$$2[(C_4H_9)_4N]_3Na_2[Rh[β-Nb_3SiW_9O_{40}]\{η^5-C_5(CH_3)_5\}] + 5H_2O + 4AgCl$$

The deprotonated, organic solvent-soluble $[(C_4H_9)_4N]_7[Nb_3SiW_9O_{40}]$ is formed by cleavage of the $[(C_4H_9)_4N]_6H_2[Nb_6Si_2W_{18}O_{77}]$ complex with 8 equiv of tetrabutylammonium hydroxide, $[(C_4H_9)_4N]OH$. The resultant triniobium-substituted Keggin-type anion is then heated in refluxing CH_3CN with $[Rh(C_5(CH_3)_5)(CH_3CN)_3]^{2+9}$ to form a single isomer of the polyoxoanion-supported organometallic complex. The product is isolated in the best yield, and in analytically pure form and free of solvate, with the counterion composition $\{[(C_4H_9)_4N]_3Na_2\}^{5+}$. The product is also readily isolated, but in somewhat lower yield, as its $\{[(C_4H_9)_4N]_5\}^{5+}$ salt; most of our earlier spectroscopic, characterization, and other studies were done with this salt.[1,2] In the present case, contamination of the desired product[1,2a] by $[(C_4H_9)_4N]^+[BF_4]^-$ (from the $Ag[BF_4]$) is avoided, and purification accomplished, by reprecipitating at least twice from acetonitrile using ethyl acetate, since $[(C_4H_9)_4N]^+[BF_4]^-$ is very soluble in this solvent.

Preparation

A solution containing $[Nb_3SiW_9O_{40}]^{7-}$ is prepared by dissolving $[(C_4H_9)_4N]_6H_2[Nb_6Si_2W_{18}O_{77}]$ (5.00 g, 0.756 mmol) in 50 mL of CH_3CN and adding dropwise an aqueous solution of tetrabutylammonium hydroxide, $[(C_4H_9)_4N]OH$ (34% solution, 4.65 g, 6.05 mmol, Aldrich). The solvent is then removed by rotary evaporation, and the resulting residue, $[(C_4H_9)_4N]_7[Nb_3SiW_9O_{40}]$, is redissolved in 50 mL of CH_3CN. Next, $[Rh(C_5(CH_3)_5)Cl_2]_2$* (0.467 g, 0.756 mmol) is suspended in 10 mL of CH_3CN and $Ag[BF_4]$ (0.589 g, 3.03 mmol) is added to the well-stirred solution. The solution turns almost immediately from red to yellow-orange, and the AgCl precipitate forms. After stirring for 10 min, the yellow-orange solution is filtered (using a medium glass-fritted funnel) directly into the well-stirred $[Nb_3SiW_9O_{40}]^{7-}$ solution. The AgCl collected is washed with 5 mL of CH_3CN, and the wash is also added to the solution. The orange solution is transferred to a 200-mL round-bottomed flask and then refluxed (under dry N_2) for 1 h (without this step, several isomers are observed in the [183]W NMR

---

* The preparation of $[Rh(C_5(CH_3)_5)Cl_2]_2$ is based on the direct reaction of pentamethylcyclopentadiene with $Rh(III)Cl_3·3H_2O$.[10] First, $Rh(III)Cl_3·3H_2O$ (2.75 g, 10.4 mmol) is added to 95% $C_5(CH_3)_5H$ (2.75 g, 19 mmol, Aldrich) dissolved in 90 mL of methanol. The mixture is refluxed, under dry N_2, for 4 h and then refrigerated at 4° overnight. The brick-red crystals are collected by filtering the cold mixture through a medium glass-fritted funnel and washing twice with 3 mL of ice-cold methanol. The product is air-dried and then dried overnight under high vacuum at room temperature; yield 2.81 g (87%). Additional product can be isolated by rotary evaporation of the filtrate to 15–20 mL and refrigeration of the mixture overnight. The product is collected as before; yield 0.13 g (combined yield of 91%). The product is readily characterized by ^{1}H NMR ($\delta = 1.60$ in dimethyl-d_3 sulfoxide (DMSO-d_6).

spectrum).[1, 2] While the solution is hot, solid $NaBF_4$ (0.332 g, 3.024 mmol) is added,* and the system is well stirred. [Alternatively, a homogeneous solution of $Na[BF_4]$ (0.332 g) in roughly 100 mL of CH_3CN can be prepared, then added.] After all the $Na[BF_4]$ dissolves, the solution is cooled to room temperature. The resulting orange-red solution is filtered once through paper (Whatman No. 5) and then rotary-evaporated to form a thick red-orange oil. After the remaining solvent is removed under high vacuum for 5 h, an orange-red solid is obtained [which, at this point, contains 4 equiv of $[(C_4H_9)_4N][BF_4]$ as a contaminant].

About 5 mL of CH_3CN is added to the orange-red solid in a round-bottomed flask, and the solution is then transferred to a 600-mL beaker. The round-bottomed flask is washed with small amounts of CH_3CN, and the washings are added to the beaker. Ethyl acetate (320 mL total) is added in 40-mL portions with efficient stirring. With slow addition of the second or the third portion of ethyl acetate, a yellow-orange solid begins to deposit. (The smaller the amount of total CH_3CN used the better. The total amount of

---

* When desired, the work-up can proceed without the addition of $Na[BF_4]$ and $[(C_4H_9)_4N]_5[Rh[\beta-Nb_3SiW_9O_{40}]\{\eta^5-C_5(CH_3)_5\}]$ can be obtained in $Bu_4N^+[BF_4]^-$-free form. In this case, adding excess ethyl acetate to the CH_3CN solution produces a reddish paste or oil, but not a precipitate. After the mother liquor is decanted and the residue is washed several times with diethyl ether, the product is dried overnight under vacuum at room temperature to give a yellow-orange powder [yields of the first and final EtOAc reprecipitations are 3.15 g (50%) and 1.75 g (38.5%), respectively]. Numerous attempts at crystallizations of this salt failed. The [183]W NMR of this material at 20°C in DMSO-d_6 (0.12 M), obtained as previously described[1] and relative to external, saturated Na_2WO_4 in D_2O, is [δ,(intensity)]: $-50.67(2)$, $-90.76(2)$, $-102.81(2)$, 148.27(2), 152.75(1).

Anal. Calcd. for $[(C_4H_9)_4N]_5[Rh[\beta-Nb_3SiW_9O_{40}]\{\eta^5-C_5(CH_3)_5\}]$: C,26.68; H, 4.85; N, 1.73; Rh, 2.54; Si, 0.69; W, 40.8; Nb, 0.88; O, 15.8. Found: C, 26.10; H, 4.19; N, 1.71; Rh, 2.34; Si, 0.72; W, 41.6; Nb, 0.85; O, 16.1; total, 100.4%.

Even if the amounts of $Na[BF_4]$ used are changed, no salts other than the $[(C_4H_9)_4N]_3Na_2$ and sodium-free $[(C_4H_9)_4N]_5$ could be obtained in significant yields. For example, when 1 equiv of $Na[BF_4]$ (0.166 g, 1.512 mmol) is used, the $[(C_4H_9)_4N]_4Na_1$ salt is not obtained, but rather, the $[(C_4H_9)_4N]_3Na_2$ salt is obtained in lower yields [for the first and final reprecipitations: 2.83 g (52%) and 1.98 g (36%), respectively]. The reddish filtrate of the first reprecipitation contains the $[(C_4H_9)_4N]_5$ salt as its main product in this case. The $> 50\%$ yield suggests that the product is slightly contaminated with $[(C_4H_9)_4N]_{5-x}Na_x$ ($x = 0$ or 1).

As part of our efforts to obtain salts that might afford crystallographic-quality single crystals, the tetrapropylammonium salt $[(C_3H_7)_4N]_5$ has been prepared by making the appropriate substitutions of $[(C_3H_7)_4N]^+$ for $[(C_4H_9)_4N]^+$ and then twice reprecipitating from CH_3CN using EtOAc and drying overnight at 60°C at 1 atm. The product is not quite analytically pure.

Anal. Calcd. for $[(C_3H_7)_4N]_5Rh[\beta-Nb_3SiW_9O_{40}]\{\eta^5-C_5(CH_3)_5\}$: C, 22.29; H, 4.14; N, 1.86; Rh, 2.73; Si, 0.74; W, 43.9; Nb, 7.39; O, 17.0. Found: C, 21.19, H, 4.19; N, 1.73; Rh, 2.48; Si, 0.76; W, 44.6; Nb, 7.85; O, 17.8; total, 100.6%.

CH_3CN should be less than 10 mL, otherwise the yield is *markedly* lowered, even if a large excess of ethyl acetate is used.) After the addition of a final 40-mL portion of ethyl acetate, the suspension is stirred for 30 min. The precipitate is filtered with a medium glass-fritted funnel, washed three times with 40 mL of diethyl ether, and dried under high vacuum overnight at *room temperature* (since some CH_3CN insoluble material is formed if 50°C is used in this step) [yield: 4.32 g (79%)]. This material, still contaminated by small amounts of $[(C_4H_9)_4N][BF_4]$ (as determined by its 1084-cm^{-1} KBr IR band), is redissolved in *less than* 10 mL of CH_3CN. The solution is filtered through paper (Whatman No. 2) and the desired compound reprecipitated with eight 40-mL portions of ethyl acetate. The now BF_4^--free yellow-orange solid is isolated as before [final yield: 3.49 g (64%)].

Anal. Calcd. for $[(C_4H_9)_4N]_3Na_2[Rh[\beta-Nb_3SiW_9O_{40}]\{\eta^5-C_5(CH_3)_5\}]$: C, 19.28; H, 3.43; N, 1.16; Na, 1.27; Rh, 2.85; Si, 0.78; W, 45.8; Nb, 7.71; O, 17.7. Found: C, 19.59; H, 3.78; N, 1.18; Na, 1.29; Rh, 2.62; Si, 0.81; W, 45.6; Nb, 7.30; O, 17.9; total, 100.1%. This material is slightly hygroscopic.

Properties

The $[(C_4H_9)_4N]_3Na_2$ salt of the product is very soluble in CH_3CN, *N,N*-dimethylformamide and DMSO, slightly soluble in acetone, and insoluble in ethyl acetate, diethyl ether, and all nonpolar organic solvents. In water, alcohol, and chloroform, it seems to react partly or to decompose, although the nature of these reactions was not investigated except to confirm (by IR) that $[Nb_6Si_2W_{15}O_{77}]^{8-}$ is produced in H_2O.

A solution molecular-weight measurement is consistent with the title (monomeric) formulation.[1,2] The heteropolyanion-supported organometallic complex moiety $[Rh[\beta-Nb_3SiW_9O_{40}]\{\eta^5-C_5(CH_3)_5\}]^{5-}$ is further characterized by its ^1H NMR in CD_3CN [1.82 ppm, $C_5(CH_3)_5$], its IR spectrum [measured as a KBr disk, characteristic bands at 996, 957, $\sim$ 923 (sh), 888, $\sim$ 839 (sh), $\sim$ 801 (sh), and 781 cm^{-1}], and especially its five-line ^{183}W NMR [signals of 2 : 2 : 2 : 2 : 1 intensity, which requires C_S symmetry in the product, δ (DMSO-d_6) of a 0.11 M solution, 21°C, relative to external, saturated Na_2WO_4 in D_2O at 21°C: $-$ 52.28, $-$ 90.95, " $-$ " 103.33, $-$ 148.8, $-$ 153.1]. Coupling constants ($^2J_{W-O-W}$), the non-ion exchangeability of the $[Rh(C_5(CH_3)_5)]^{2+}$ moiety, and other data are also available.[1,2]

References and Notes

1. R. G. Finke and M. W. Droege, *J. Am. Chem. Soc.*, **106**, 7274 (1984).
2. (a) M. W. Droege, "Initial Studies on the Use of Trisubstituted Heteropolytungstates as

Soluble Metal Oxide Analogues in Catalysis," Ph.D. dissertation, University of Oregon, 1984, Chapter 3; (b) R. G. Finke, M. W. Droege, and K. Nomiya, manuscript in preparation.

3. R. G. Finke, C. A. Green, and B. Rapko, *Inorg. Synth.*, **27**, 128 (1989).

4. R. G. Finke, D. Lyon, *Inorg. Chem.*, **29**, 1787 (1990).

5. F. Fairbrother, *The Chemistry of Niobium and Tantalum*, Elsevier, Amsterdam, 1967, p. 38. The use of $[Nb_6O_{19}]^{8-}$ and H_2O_2 as a reagent for the synthesis of heteropolytungstates was first reported by M. Dabbabi and M. Boyer, *J. Inorg. Nucl. Chem.*, **38**, 1011 (1976).

6. M. Filowitz, R. K. C. Ho, W. G. Klemperer, and W. Shum, *Inorg. Chem.*, **18**, 93 (1979).

7. C. M. Flynn and G. D. Stucky, *Inorg. Chem.*, **8**, 178 (1969).

8. Previous synthesis of this complex at twice the present scale gave a yield of 70% (see refs. 1 and 2a), the present synthesis gives a 52% yield, and syntheses done at one-third of the present scale are less satisfactory, giving overall yields of approximately 30%. Clearly, the observed yields are significantly concentration-dependent (as perhaps would be expected given the polyanion–polycation nature of the complex), and it appears that yields > 70% are possible from syntheses done at larger scale.

9. The use of $[Rh(C_5(CH_3)_5)(solvate)_x]^{2+}$ as a reactive intermediate is well documented; see C. J. Besecker, V. W. Day, and W. G. Klemperer, *Organometallics*, **4**, 564 (1985); L. A. Oro, M. Valderama, P. Cifuentes, C. Foces-Foces, and F. H. Cano, *J. Organomet. Chem.*, **276**, 67 (1984); and C. White, S. J. Thompson, and P. M. Maitlis, *J. Chem. Soc. Dalton Trans.*, **1977**, 1654.

10. Previous reports have based the synthesis on the reaction of hexamethylbicyclo[2.2.0] hexadiene with $Rh(III)Cl_3 \cdot 3H_2O$. It has been suggested that pentamethylcyclopentadiene, which was not then commercially available, was the actual reactant in the hexamethylbicyclo[2.2.0]hexadiene synthesis; see B. L. Booth, R. N. Haszeldine, and M. Hill, *J. Chem. Soc. (A)*, **1969**, 1299 and P. M. Maitlis, *Chem. Soc. Rev.*, **10**, 1 (1981).

56. η^5-CYCLOPENTADIENYLCHROMIUM COMPLEXES OF PHOSPHORUS, TETRACARBONYL-μ-DIPHOSPHIDO-BIS(η^5-CYCLOPENTADIENYLCHROMIUM), $[CpCr(CO)_2]_2(\mu\text{-}\eta^2\text{-}P_2)$ AND DICARBONYL-CYCLOTRIPHOSPHIDO-(η^5-CYCLOPENTADIENYLCHROMIUM), $CpCr(CO)_2(\eta^3\text{-}P_3)$

Submitted by LAI YOONG GOH* and RICHARD C. S. WONG*
Checked by O. J. SCHERER† and A. SCHNEIDER†

There is much current interest in the synthesis and reactivity of transition metal complexes containing P atoms or aggregates.[1] In particular, $(\mu\text{-}\eta^2\text{-}P_2)$ dinuclear and cyclo $(\eta^3\text{-}P_3)$ mononuclear complexes have received special attention, on account of their ability to act as complex ligands in the formation of polynuclear complexes.[2-8] Whereas reported syntheses of these

* Department of Chemistry, University of Malaya, 59100 Kuala Lumpur, Malaysia.
† Fachbereich Chemie der Universität Kaiserslautern, Erwin-Schrödinger-Strasse, D-6750 Kaiserslautern, Germany.

P_2 and P_3 complexes of Group 6 metals from the reaction of the triply bonded dinuclear complexes $[MCp(CO)_2]_2$ (M = Cr,[9] Mo,[10] and W;[11] Cp = η^5-cyclopentadienyl) with elemental phosphorus have generally produced low product yields, we recently obtained a high-yield process for the synthesis of $[CrCp(CO)_2]_2(\mu\text{-}\eta^2\text{-}P_2)$ (1) and $CrCp(CO)_2(\eta^3\text{-}P_3)$ (2) from the following reaction:

(1)   (2)

General Procedures

The syntheses are performed either in a Vacuum Atmospheres Drilab, or using conventional Schlenk-tube techniques[12] under an atmosphere of dry oxygen–free argon. Toluene and *n*-hexane are distilled under nitrogen from sodium benzophenone. The compound $[CrCp(CO)_3]_2$ is synthesised as described previously.[13] Yellow phosphorus is obtained from BDH Chemicals Ltd., United Kingdom. Chromatographic materials are silica gel (Merck Kiesel gel 60, 35/70-mesh ASTM), alumina standardized (Merck) and florisil (Sigma, 100/200 mesh), dried at 140°C overnight before use.

■ **Caution.** *Yellow phosphorus* (P_4) *is extremely toxic and causes severe burns. Owing to the known toxicity of metal carbonyls and gaseous CO (which is evolved in this synthesis), all preparative procedures must be carried out in an efficient fume hood. No other hazardous steps are involved, although care must be exercised when handling metallic sodium.*

Procedure

The time required for the complete work-up is 2–3 days. A deep green suspension of $[CrCp(CO)_3]_2$ (1.0 g, 2.49 mmol) and yellow phosphorus* (0.31 g, 9.94 mmol) in toluene (∼ 25 mL) is stirred in a 50-mL Schlenk flask fitted with a condenser and immersed in an oil bath maintained at 85–90°C for 3.5 h.† A gradual color change to reddish brown is observed. The resultant

---

* Washed with ether followed by toluene. The checkers dried P_4 in an oil pump vacuum.
† The resulting product composition, and hence the yields of (1) and (2), are very sensitive to the Cr–P mole ratio and the temperature and duration of the reaction, owing to the formation of $[CrCp(CO)_2]_2$ and $Cr_2Cp_2P_5$ as thermolytic by-products.[14]

product solution is allowed to cool and filtered through a 2-cm-thick disk of silica gel, which is then eluted with two 5-mL portions of toluene. The filtrate cum washings is concentrated to 2–3 mL under reduced pressure. Addition of ~ 0.5 mL hexane, followed by overnight cooling at $-30°$C, gives fine lustrous deep magenta crystals of $[CrCp(CO)_2]_2P_2$ (1) (400 mg, 0.98 mmol, 39% yield).

The mother liquor is concentrated to ~ 1 mL and loaded onto a column (2 cm × 11 cm) of florisil or alumina prepared in *n*-hexane. Elution with *n*-hexane (80 mL) gives a yellow fraction that on concentration to ~ 2 mL and cooling at $-30°$C yields yellowish brown flakes of $CrCp(CO)_2P_3$ (2) (525 mg, 1.97 mmol, 40% yield). Further elution with 1 : 1 *n*-hexane–toluene (30 mL) followed by toluene (30 mL) gives a magenta-colored fraction, which on concentration to 1–2 mL, addition of a little hexane, and cooling to $-30°$C leads to isolation of a further lot of (1) (120 mg, 0.29 mmol, 12% yield).

An alternative work-up involving chromatography of the total product solution on an alumina column (2.5 cm × 14 cm) may be employed.*

Anal. Calcd. for $C_{14}H_{10}Cr_2O_4P_2$ (1): C, 41.19; H, 2.47; Cr, 25.47; P, 15.19. Found: C, 40.74; H, 2.56; Cr, 24.95; P, 15.14.

Anal. Calcd. for $C_7H_5CrO_2P_3$ (2): C, 31.59; H, 1.90; Cr, 19.53; P, 34.95. Found: C, 31.96; H, 1.80; Cr, 18.80; P, 35.30.

Properties

The complex $[CrCp(CO)_2]_2P_2$ (1) crystallizes as fine deep magenta crystals, stable in the solid state in air and for prolonged periods under an inert atmosphere at ambient temperature. It is soluble in most organic solvents, giving magenta solutions that are stable at ambient temperature under an inert atmosphere. The infrared spectrum shows v (CO) (± 5 cm^{-1}) at 1970 (ssh), 1950 (vs), 1925 (ssh), and 1890 (vs) cm^{-1} (toluene). The NMR spectrum in benzene-d_6 shows singlet resonances for the Cp rings (^{1}H, δ 4.15; ^{13}C, δ 86.4), the carbonyl ligands (δ 238.6) and the μ-η^2-P_2 ligand (^{31}P, δ 110.5). The mass spectrum exhibits the parent ion m/z 408 and a fragmentation pattern indicative of successive losses of CO ligands, as well as the fragmentation

---

* The checkers found that in both methods of work-up, they could get yields of only 213–264 mg (21–26%) of (1) and 185–238 mg (14–18%) of (2), and that higher volumes of eluent were required. Using Kieselgel 60 (0.063–0.2 mm, MacLerey-Nagel) heated before use to 250°C for 5 h and reactivated with 3% H_2O, their yields could be improved to 380 mg (37%) and 350 mg (26%), respectively, via chromatography of the total product solution.

patterns of $CrCp(CO)_2P_3$ and $Cr_2Cp_2P_5$. The structure has been determined and reveals the presence of a $Cr_2(\mu\text{-}\eta^2\text{-}P_2)$ group.[15]

The complex $CrCp(CO)_2P_3$ (2) exists as moderately air-stable yellowish brown crystalline flakes. These are very soluble in organic solvents, giving a yellow solution that is fairly stable at room temperature under an inert atmosphere. The infrared spectrum shows v (CO) ($\pm$ 5 cm^{-1}) at 1975 (vs) and 1920 (vs) cm^{-1} (toluene). The NMR spectrum in benzene-d_6 possesses singlet resonances for the Cp ring (^{1}H, δ 3.92; ^{13}C, δ 84.9), the carbonyl ligands (δ 233.7), and the *cyclo*-P_3 ligand (^{31}P, δ $-$ 285.7). The mass spectrum exhibits the parent ion m/z 266 and a fragmentation pattern showing successive losses of the CO ligands, as well as the presence of the mass ion $Cr_2Cp_2P_5$ and a trace of $Cr_2Cp_2P_6$. The structure has been determined and shown to possess a *cyclo*-η^3-P_3 ligand.[15]

Acknowledgment

Financial support from the University of Malaya and IRPA R&D Grant No. 04-07-04-127 is gratefully acknowledged.

References

1. M. DiVaira, P. Stoppioni, and M. Peruzzini, *Polyhedron*, **6**, 351 (1987) and references cited therein.
2. S. Midollini, A. Orlandini, and L. Sacconi, *Angew. Chem. Int. Ed. Engl.*, **18**, 81 (1979).
3. C. A. Ghilardi, S. Midollini, A. Orlandini, and L. Sacconi, *Inorg. Chem.*, **19**, 301 (1980).
4. A. Vizi-Orosz, V. Galamb, G. Palyi, and L. Markó, *J. Organomet. Chem.*, **216**, 105 (1981).
5. F. Cecconi, C. A. Ghilardi, S. Midollini, and A. Orlandini, *J. Chem. Soc. Chem. Commun.*, **1982**, 229; *Angew. Chem. Int. Ed. Engl.*, **22**, 554 (1983).
6. O. J. Scherer, H. Sitzmann, and G. Wolmershäuser, *Angew. Chem. Int. Ed. Engl.*, **23**, 968 (1984).
7. A. Vizi-Orosz, G. Palyi, L. Markó, R. Boese, and G. Schmid, *J. Organomet. Chem.*, **288**, 179 (1985).
8. L. Y. Goh, R. C. S. Wong, and T. C. W. Mak, *J. Organomet. Chem.*, **364**, 363 (1989); **373**, 71 (1989).
9. O. J. Scherer, J. Schwalb, G. Wolmershäuser, W. Kaim, and R. Gross, *Angew. Chem. Int. Ed. Engl.*, **25**, 363 (1986).
10. O. J. Scherer, H. Sitzmann, and G. Wolmershäuser, *J. Organomet. Chem.*, **268**, C9 (1984); *Angew. Chem. Int. Ed. Engl.*, **24**, 351 (1985).
11. O. J. Scherer, J. Schwalb, H. Swarowsky, G. Wolmershäuser, W. Kaim, and R. Gross, *Chem. Ber.*, **121**, 443 (1988).
12. D. F. Shiver and M. A. Drezdzon, *The Manipulation of Air-Senstive Compounds*, 2nd ed., Wiley-Interscience, New York, 1986.
13. R. B. King and F. G. A. Stone, *Inorg. Synth.*, **7**, 104 (1963).
14. L. Y. Goh, R. S. C. Wong, C. K. Chu, and T. W. Hambley, *J. Chem. Soc. Dalton Trans.*, **1990**, 977.
15. L. Y. Goh, C. K. Chu, R. C. S. Wong, and T. W. Hambley, *J. Chem. Soc. Dalton Trans.*, **1989**, 1951.

57. η^5-CYCLOPENTADIENYLCHROMIUM COMPLEXES OF SULFUR

Submitted by LAI YOONG GOH*
Checked by MARCETTA Y. DARENSBOURG† and YUI-MAY HSIAO†

There is much current interest on transition metal chalcogen complexes. Recent investigations focused on the reactivity the μ—S_2 ligand,[1-4, 7-9, 12] and the linear M≡E≡M (M = S, Se) multiple bond[5-11] show their potential in the generation of heterometallic clusters[13,14] and in organic synthesis.[4] We describe here the facile high-yield synthesis of $[CpCr(CO)_2]_2S$ (1) and the related complex $Cp_2Cr_2(CO)_5S_2$ (2) from the reaction of $[CpCr(CO)_3]_2$ with elemental sulfur. These complexes are of interest as precursors to the homometallic cubanes $Cp_4Cr_4(CO)_2S_2$ and $Cp_4Cr_4S_4$[8] and potential precursors to heteronuclear clusters.[13-15]

General Procedures

The syntheses are performed either in a Vacuum Atmospheres Drilab or using conventional Schlenk-tube techniques[16] under an atmosphere of dry oxygen free argon. Tetrahydrofuran (THF), toluene, and *n*-hexane were distilled under nitrogen from sodium–benzophenone. $[CpCr(CO)_3]_2$ is synthesized as described previously.[17] Silica gel used is Merck Kieselgel 60 (35/70 mesh ASTM). Convenient scales of preparation are described below, but the reaction can easily be scaled up several times if necessary.

■ **Caution.** *Owing to the known toxicity of metal carbonyls and gaseous CO (which is evolved in these syntheses), all preparative procedures must be carried out in an efficient fume hood. No other hazardous steps are involved, although care must be exercised when handling metallic sodium.*

A. TETRACARBONYL-μ-SULFIDO-BIS(η^5-CYCLO-PENTADIENYLDICHROMIUM)

$$[CpCr(CO)_3]_2 + \tfrac{1}{8}S_8 \rightarrow [CpCr(CO)_2]_2S$$

(1)

*Department of Chemistry, University of Malaya, 59100 Kuala Lumpur, Malaysia.
†Department of Chemistry, Texas A&M University, College Station, TX 77843.

Procedure

A deep green suspension of [CpCr(CO)$_3$]$_2$ (400 mg, 1.0 mmol) and S$_8$ (32 mg, 1.0 mmol S) in THF or toluene (20 mL) is stirred in a 50-mL Schlenk flask at ambient temperature for 30 min. A rapid color change to a deep greenish brown occurs, accompanied by some effervescence. The resultant homogeneous solution is filtered through a disk of silica gel (1.5 cm × 2.0 cm), which is then washed with THF (~10 mL), leaving a deep green rim at the top. The filtrate cum washings is concentrated under vacuum to 1–2 mL when crystallization begins. Addition of ether (2 mL) followed by cooling several hours, preferably overnight, at −30°C gives deep green crystals of [CpCr(CO)$_2$]$_2$S (140 mg, 0.37 mmol, 37%) which are washed with cold ether–hexane. The mother liquor and washings is concentrated under vacuum to almost dryness, and redissolved in ether-*n*-hexane. Cooling overnight at −30°C deposited more crystals (190 mg, 0.50 mmol, 50%).*

Anal. Calcd. for C$_{14}$H$_{10}$Cr$_2$O$_4$S: C, 44.45; H, 2.66; Cr, 27.49; S, 8.48. Found: C, 44.56; H, 2.77; Cr, 26.9; S, 8.27.

Properties

The complex, [CpCr(CO)$_2$]$_2$S, crystallizes as dark green needles, stable in the solid state in air and for prolonged periods under an inert atmosphere at ambient temperature. The crystals melt at 111–112°C, accompanied by decomposition with effervescence to a dark brown liquid. It is readily soluble in most organic solvents, giving dark greenish-brown solutions, which decompose in air. The infrared spectrum shows ν(CO) at 2005 (m), 1964 (vs), 1935 (ms), 1928 (sh) cm^{-1} (*n*-hexane), and at 1994 (vs), 1955 (vs), 1938 (vs), 1916 (ssh), 1906 (vs), 1885 (vs) cm^{-1} (KBr). The NMR spectra in benzene-d_6 or toluene-d_8 show singlet resonances for the Cp rings [^{1}H, δ 4.36; ^{13}C, δ 89.20] and the carbonyl ligands (δ 246.7). The mass spectrum exhibits the parent ion m/z 377.9110 and a fragmentation pattern indicative of successive losses of CO ligands. The structure has been determined and reveals a linear Cr≡S≡Cr group.[5]

B.   PENTACARBONYL-μ-(η^1:η^2- DISULFIDO)BIS(η^5-CYCLO-PENTADIENYLCHROMIUM)

$$[CpCr(CO)_3]_2 + \tfrac{2}{8}S_8 \rightarrow Cp_2Cr_2(CO)_5S_2$$

(2)

---

* The analogous Se complex, [CpCr(CO)$_2$]$_2$Se, can be similarly prepared as dark brown crystals in ~ 85% yield.

Procedure

A deep green suspension of [CpCr(CO)₃]₂ (400 mg, 1.0 mmol) and S₈ (64 mg, 2.0 mmol) in THF (20 mL) is stirred in a 100-mL Schlenk flask at ambient temperature, whereupon the solution instantaneously turns blood-red. (Owing to the high lability of the complex in solution⁷ a rapid work-up is essential for isolation of the pure complex.) The homogeneous product solution is immediately filtered through Celite, which is then washed with THF (10 mL). The filtrate from washing is rapidly concentrated under reduced pressure to ~ 2 mL. Addition of *n*-hexane (~ 3 mL), followed by cooling at − 30°C for 15 min, gives a fine reddish-brown air-stable crystalline product (260 mg, 0.60 mmol, 60% yield) which is washed with cold hexane. A second crop (120 mg, 0.28 mmol, 28%) is obtained with addition of more *n*-hexane followed by cooling.

Anal. Calcd. for $C_{15}H_{10}Cr_2O_5S_2$: C, 41.10; H, 2.30; Cr, 23.72; S, 14.63. Found: C, 41.05; H, 2.46; Cr, 22.8; S, 14.0.

Properties

The dark brown air-stable crystals of $Cp_2Cr_2(CO)_5S_2$ readily dissolve in most organic solvents to give blood-red solutions which rapidly undergo extrusion of S and CO, giving a greenish brown solution.⁷ The infrared spectrum shows $v(CO)$ at 2031 (m), 1979 (vs), 1953 (m), 1940 (m), 1906 (m), 1865 (w) (THF) and at 2014 (s), 1969 (vs), 1939 (vs), 1924 (vs), 1919 (vs), 1852 (s), 1821 (wsh) cm⁻¹ (KBr). The NMR spectra in benzene-d_6 exhibit two singlets for the two Cp rings (¹H, δ 4.17 and 4.51), (¹³C, δ 91.33 and 93.28). The mass spectrum does not show the parent ion, but rather the fragmentation pattern of [CpCr(CO)₂]₂S. The crystal structure shows the presence of a $Cr_2(\eta^1{:}\eta^2\text{-}S_2)$ group.⁷

References

1. A. Müller, W. Jaegermann, and J. H. Enemark, *Coord. Chem. Rev.*, **46**, 245 (1982) and references cited therein.
2. D. Seyferth, G. B. Womack, M. K. Gallagher, M. Cowie, B. W. Hames, J. P. Fackler, Jr., and A. M. Mazany, *Organometallics*, **6**, 283 (1987) and references cited therein.
3. A. M. Mazany, J. P. Fackler, Jr., M. K. Gallagher, and D. Seyferth, *Inorg. Chem.*, **22**, 2593 (1983).
4. C. J. Casewit, D. E. Coons, L. L. Wright, W. K. Miller, and M. R. DuBois, *Organometallics*, **5**, 951 (1986) and references cited therein.
5. T. J. Greenhough, B. W. S. Kolthammer, P. Legzdins, and J. Trotter, *Inorg. Chem.*, **18**, 3543 (1979).

6. W. A. Herrmann, *Angew. Chem. Int. Ed. Engl.*, **25**, 56 (1986) (a review) and ref. 97 cited therein.

7. L. Y. Goh, T. W. Hambley, and G. B. Robertson, *Organometallics*, **6**, 1051 (1987) and references cited therein.

8. Chen Wei, L. Y. Goh, and T. C. W. Mak, *Organometallics*, **5**, 1997 (1986).

9. Chen Wei, L. Y. Goh, and E. Sinn, *Organometallics*, **7**, 2020 (1988).

10. L. Y. Goh, *J. Chem. Soc. Dalton Trans.*, **1989**, 431.

11. W. A. Herrmann, J. Rohrmann, E. Herdtweck, H. Bock, and A. Veltmann, *J. Am. Chem. Soc.*, **108**, 3134 (1986); J. Rohrmann, W.A. Herrmann, E. Herdtweck, J. Riede, M. Ziegler, and G. Sergeson, *Chem. Ber.*, **119**, 3544 (1986).

12. W. A. Herrmann and J. Rohrmann, *Chem. Ber.*, **119**, 1437 (1986).

13. M. D. Curtis, P. D. Williams, and W. M. Butler, *Inorg. Chem.*, **27**, 2853 (1988).

14. I. L. Eremenko, A. A. Pasynskii, B. Orazsakhatov, A. F. Shestakov, G. Sh. Gasanov, A. S. Katugin, Yu. T. Struchkov, and V. E. Shklover, *J. Organomet. Chem.*, **338**, 369 (1988).

15. L. Y. Goh, work in progress.

16. D. F. Shriver, *The Manipulation of Air-Sensitive Compounds*, McGraw-Hill, New York, 1969.

17. R. B. King and F. G. A. Stone, *Inorg. Synth.*, **7**, 104 (1963); R. Birdwhistell, P. Hackett, and A. R. Manning, *J. Organomet. Chem.*, **157**, 239 (1978).

58. SULFUR-BRIDGED DINUCLEAR MOLYBDENUM(V) COMPLEXES WITH Mo_2O_3S AND $Mo_2O_2S_2$ CORES

Submitted by TAKASHI SHIBAHARA* and HARUO AKASHI*
Checked by ANDREAS TOUPADAKIS† and DIMITRI COUCOUVANIS†

Methods of preparing molybdenum(V) complexes with Mo_2O_4 cores have been published.[1] However, those of sulfur-bridged molybdenum(V) complexes with Mo_2O_3S and $Mo_2O_2S_2$ cores remain difficult. Molybdenum–sulfur cluster compounds attract much attention since almost all molybdenum enzymes contain both molybdenum and sulfur, being always accompanied by iron.

Facile syntheses of two types of molybdenum(V) complexes with Mo_2O_3S and $Mo_2O_2S_2$ cores—especially $Na_2[Mo_2O_3S(cys)_2]\cdot4H_2O$ [cys = cysteinato(2 −)] and $Na_2[Mo_2O_2S_2(cys)_2]\cdot4H_2O$—are described here. The cores, Mo_2O_3S and $Mo_2O_2S_2$, of the compounds described here have the geometry of syn (Fig. 1).

The purity of the cysteinato and analogous edta [edta = (1,2-ethane-diyldinitrilo)tetraacetato(4 −) = ethylenediaminetetraacetato(4 −)] com-

* Department of Chemistry, Okayama University of Science, 1-1 Ridai-cho, Okayama 700, Japan.
† Department of Chemistry, University of Michigan, Ann Arbor, MI 48109.

Fig. 1.   The syn geometry of the Mo_2O_3S (**A**) and $Mo_2O_2S_2$ (**B**) cores.

plexes has been checked against the complexes prepared from the pure aqua ion and the corresponding ligands by comparison of their electronic spectra. Once the cysteinato complexes are prepared, many other complexes with Mo_2O_3S and $Mo_2O_2S_2$ cores can be prepared through substitution reactions. The methods described below can be scaled up, and a relatively large amount of compound can be obtainable.

The sulfur bridged complexes described here are stable to aerobic oxidation, in contrast to di-μ-oxo complexes.[1]

Starting Materials

Sodium molybdate dihydrate, hydrazine dihydrochloride, and L-cysteine hydrochloride monohydrate were purchased from Nacalai Tesque.

A.   SODIUM μ-OXO-μ-SULFIDO-BIS[(L-CYSTEINATO)-OXOMOLYBDATE(V)] TETRAHYDRATE, $Na_2[Mo_2O_3S(cys)_2]\cdot 4H_2O$ (A)

$$4MoO_4^{2-} + N_2H_4\cdot 2HCl + 10H^+ \longrightarrow 2[Mo_2O_4(aq)]^{2+} + N_2$$
$$+ 8H_2O + 2Cl^-$$
$$[Mo_2O_4(aq)]^{2+} + Na_2S + 2H^+ \longrightarrow [Mo_2O_3S(aq)]^{2+} + 2Na^+ + H_2O$$
$$[Mo_2O_3S(aq)]^{2+} + 2H_2cys \longrightarrow [Mo_2O_3S(cys)]_2^{2-} + 4H^+$$

Water-soluble dimeric molybdenum(V) complexes containing μ-oxo μ-sulfido cores (Mo_2O_3S) were first prepared by Schultz and coworkers.[2] Studies on the complexes with Mo_2O_3S cores in organic solvents have also been made.[3] We use here the easily obtainable sodium molybdate and hydrazine hydrochloride instead of molybdenum pentachloride, and sodium sulfide instead of H_2S gas.[2,4]

Procedure

Concentrated HCl (50 mL) is added to sodium molybdate dihydrate (9.6 g, 0.04 mol) dissolved in water (150 mL).* Hydrazine dihydrochloride (10 g, 0.1 mol) is added to the solution, which is then heated to $> 90°C$ for 1.5 h. The hot solution is then poured into a polyethylene bottle containing sodium sulfide nonahydrate (10 g, 0.04 mol) and a magnetic stirrer, and the bottle is tightly stoppered. The solution is allowed to stand with magnetic stirring until the solution comes to room temperature (~ 2 h). (The bottle may partially collapse as a result of the low inside pressure). L-Cysteine hydrochloride monohydrate (7 g, 0.04 mol) is dissolved in the solution, which is then filtered. After the solution is cooled in an ice bath, the pH of the solution is raised to ~ 12[†] by addition of concentrated sodium hydroxide solution (~ 13 M) and the solution is kept in a refrigerator overnight. The precipitate is filtered and then washed with methanol and diethyl ether; yield ~ 7.5 g (58%). Recrystallization is as follows: The crude solid is dissolved in hot water (60 mL), the solution is filtered, and the volume of the solution is reduced to about one-third by heating in a water bath. The solution is kept in a refrigerator, and the crystals that deposit are collected and washed as above. Yield: 5.7 g.[‡]

Anal. Calcd. for $C_6H_{18}N_2O_{11}S_3Na_2Mo_2$: C, 11.47; H, 2.88; N, 4.45%. Found: C, 11.60; H, 2.97; N, 4.39%. Electronic spectral data are given in Table I.

B. MAGNESIUM [μ-(1,2-ETHANEDIYLDINITRILO)-TETRAACETATO)]-μ-OXO-μ- SULFIDO-BIS [OXOMOLYBDATE(V)] HEXAHYDRATE, Mg[Mo₂O₃S(edta)]·6H₂O (B)

$$[Mo_2O_3S(cys)_2]^{2-} + edta^{4-} \longrightarrow [Mo_2O_3S(edta)]^{2-} + 2cys^{2-}$$

---

* Relatively high acidity of 3 *M* HCl as solvent is employed to minimize the formation of the di-μ-sulfido species. In the final stage of the preparation, the pH of the solution is raised to ~ 12, which gives the cysteinato complex in a fairly pure form, since the di-μ-oxo-Mo(V) cysteinato complex, which is the main by-product in the preparation, is decomposed in high-pH solution. Before recrystallization the cysteinato complex is of sufficiently high purity ($> 95\%$) to serve as the starting material for the preparation of other complexes with Mo_2O_3S cores (e.g., the edta complex described in Section B).

† During the addition of the NaOH solution, a yellow precipitate appears at $\sim$ pH 3 but disappears at pH ~ 6.

‡ The corresponding calcium, barium, and magnesium salts can be easily prepared by the addition of $CaCl_2 \cdot 2H_2O$, $BaCl_2 \cdot 2H_2O$, and $MgCl_2 \cdot 6H_2O$ to the solution of the sodium salt, respectively.[5]

TABLE I. Electronic Spectral Data for Mo(V) Dimers[a,b]

Compound	λ_{max} (ε M^{-1} cm^{-1})
1. $[Mo_2O_3S(aq)]^{2+}$	220 (sh)(5600), 278(3950), 314(3250), 450 (sh)(57)
	$\begin{bmatrix} 220 \text{ (sh)}(7000),\ 277(4290), \\ 310 \text{ (sh)}(3440),\ 450 \text{ (sh)}(53) \end{bmatrix}$
2. $Na_2[Mo_2O_3S(cys)_2]\cdot4H_2O$	214(26,000), 245 (sh)(11,230), 284(12,610), 320 (sh)(8010), 470 (sh)(192)
3. $Mg[Mo_2O_3S(edta)]\cdot6H_2O$	235(9500), 281(7950), 311(5920), 480 (sh)(110)
4. $[Mo_2O_2S_2(aq)]^{2+}$	225(5800), 245(4120), 280(5510), 300 (sh)(4360), 370(1940)
	$\begin{bmatrix} 225 \text{ (sh)}(6000),\ 245 \text{ (sh)}(4320), \\ 282(5550),\ 300 \text{ (sh)}(4640),\ \ 372(1830) \end{bmatrix}$
5. $Na_2[Mo_2O_2S_2(cys)_2]\cdot4H_2O$	229(32,160), 280 (sh)(10,100), 310 (sh)(8790), 370 (sh)(4060)
6. $Na_2[Mo_2O_2S_2(edta)]\cdot H_2O$	255 (sh)(7160), 278(11,800), 305 (sh)(6710), 350 (sh)(2430), 460 (sh)(54)

[a] Data for compounds 1 and 4 are for compounds in 1 M $HClO_4$; for compounds 2, 3, 5, and 6, in water. The values in square brackets are those for 1 and 4 in 1 M HCl, respectively.
[b] ε values per dimer.

The cysteinato complex A (1.2 g, 0.0019 mol) and disodium ethylenediamine-tetraacetate dihydrate $Na_2H_2edta\cdot2H_2O$ (0.80 g, 0.0021 mol) are dissolved in hot water (18 mL), to which concentrated HCl (2.5 mL) is added. The temperature of the solution is kept above 90°C for ~ 5 min. After the solution is ice-cooled, the pH of the solution is adjusted to ~ 6 by use of concentrated NaOH. Then, $MgCl_2$ (0.8 g)* dissolved in water (2 mL) is added and the solution is kept in a refrigerator overnight. The crystals that deposited are filtered and washed with methanol and diethyl ether successively and air-dried. Yield: 1.0 g (75%).

---

* The corresponding barium salt, $Ba[Mo_2O_3S(edta)]\cdot3H_2O$, can be prepared by the addition of $BaCl_2\cdot2H_2O$ instead of $MgCl_2\cdot6H_2O$, using the following amounts of materials: the cysteinato complex, 1.2 g; $Na_2H_2edta\cdot2H_2O$, 0.8 g; water, 50 mL; concentrated NaOH (to adjust the pH of the solution to ~ 6); $BaCl_2$, 0.9 g in 20 mL of water; yield 1.2 g (83%).

Anal. Calcd. for $C_{10}H_{18}N_2O_{14}SBaMo_2$: C, 17.34; H, 3.49; N, 4.04%. Found: C, 17.28; H, 3.48; N, 4.05%.

Anal. Calcd. for $C_{10}H_{24}N_2O_{17}SMgMo_2$: C, 17.34; H, 3.49; N, 4.04%. Found: C, 17.28; H, 3.48; N, 4.05%. Electronic spectral data are given in Table I.

C. μ-OXO-μ-SULFIDO-BIS[OXOMOLYBDENUM(V)](2+) AQUA DIMER, $[Mo_2O_3S(aq)]^{2+}$ (C)

$$[Mo_2O_3S(cys)_2]^{2-} + nH_2O \longrightarrow [Mo_2O_3S(aq)]^{2+} + 2cys^{2-} \quad (n \approx 6)$$

The cysteinato compound A (7 g, 0.011 mol) is dissolved in 1 M HCl (70 mL), the solution is stirred for $\sim$ 30 min and filtered, and Sephadex G-10 column chromatography (column diameter 3.5 cm, length 60 cm) is used for separation (eluent: 1 M HCl; $\sim$ 5 h). The second yellow band contains the aqua ion [80%: typically 300 mL of a 0.03 M (per dimer) solution].* The first (yellow) and third (yellow) bands contain small amounts of $[Mo_2O_4(aq)]^{2+}$ and $[Mo_2O_2S_2(aq)]^{2+}$, respectively. Electronic spectral data are given in Table I.†

D. SODIUM DI-μ-SULFIDO-BIS[(L-CYSTEINATO)-OXOMOLYBDATE(V)] TETRAHYDRATE, $Na_2[Mo_2O_2S_2(cys)_2]\cdot 4H_2O$ (D)

$$2MoO_4^{2-} + 3Na_2S + 12H^+ \longrightarrow [Mo_2O_2S_2(aq)]^{2+} + 6Na^+ + S + 6H_2O$$

$$[Mo_2O_2S_2(aq)]^{2+} + 2H_2cys \longrightarrow [Mo_2O_2S_2(cys)_2]^{2-} + 4H^+$$

So far, two synthetic methods for the preparation of the di-μ-sulfido cysteinato complex have been reported: (1) bubbling H_2S gas into the di-μ-oxo complex, $[Mo_2O_4(cys)_2]^{2-}$, in water;[7] (2) reduction of a one-to-one (1 : 1) mixture of thiomolybdate and molybdate by sodium dithionite, followed by the addition of the ligand.[8]

In the method described here, both reduction of molybdate and introduction of sulfide bridges are achieved by the addition of sodium sulfide to a solution of sodium molybdate. The addition of L-cysteine hydrochloride monohydrate to the resultant solution then gives the desired product.

Procedure

Concentrated hydrochloric acid (3.5 mL, 0.042 mmol) is added to 100 mL of water containing sodium molybdate dihydrate (4.8 g, 0.020 mol). Sodium

---

* Perchloric acid (1 M) can also be used instead of 1 M HCl here.[6]
† Isolation of the aqua ion as a solid has not been successful.

sulfide nonahydrate (10 g, 0.042 mol) is added to the solution,* which is then heated above 90°C until the solution becomes red-orange and transparent (~ 20 min). To the solution, L-cysteine hydrochloride monohydrate (6.0 g, 0.0342 mol) is added.† The solution in a beaker is heated above 90°C with stirring until its volume reduces to 40–50 mL (~ 3 h).† During the concentration, orange crystals appear. The solution, together with the crystals, is kept in a refrigerator overnight; the amount of crystals increases. The crystals are filtered by suction and washed successively with ethanol and diethyl ether. Yield: 4.8 g (75%).‡ Recrystallization is accomplished as follows. The sample (4.8 g) is dissolved in hot water (35 mL), the solution is filtered, and the volume of the solution is reduced to ~ 15 mL by heating in a water bath. The solution is kept in a refrigerator overnight, and the crystals that deposit are collected and washed as above. Yield: 4.5 g (70%).

Anal. Calcd. for $C_6H_{18}N_2O_{10}S_4Na_2Mo_2$: C, 11.18; H, 2.81; N, 4.34%. Found: C, 11.10; H, 2.77; N, 4.31%. Electronic spectral data are given in Table I.

E. SODIUM μ-(1,2-ETHANEDIYLDINITRILO)TETRA-ACETATO)BIS-μ-SULFIDO-BIS[OXOMOLYBDATE(V)]-MONOHYDRATE, $Na_2[Mo_2O_2S_2(edta)] \cdot H_2O$ (E)

$$[Mo_2O_2S_2(cys)_2]^2 + edta^{4-} \longrightarrow [Mo_2O_2S_2(edta)]^{2-} + 2cys^{2-}$$

The cysteinato complex D (1.2 g, 0.00186 mol) and $Na_2H_2edta \cdot 2H_2O$ (0.70 g, 0.00188 mol) are dissolved in hot water (75 mL), to which concentrated HCl (7 mL) is added. The solution is maintained at above 90°C for ~ 30 min. After the solution is cooled to ~ 40°C, the pH of the solution is adjusted to ~ 6 by addition of concentrated NaOH. Then the solution is kept in a refrigerator overnight. The crystals are filtered and washed with methanol and diethyl ether successively and stored *in vacuo* over P_2O_5. Yield: 1.0 g (90%).§

---

* The pH of the solution, after the addition of sodium sulfide, is ~ 13.5 and, after heating for 3 h, is ~ 11. No pH measurement or adjustment, however, is required in this method.

† A smaller amount of L-cysteine hydrochloride monohydrate (e.g., 3.51 g, 0.02 mol) gives a smaller amount of the desired compound (2.3 g).

‡ The cysteinato complex before recrystallization is of sufficiently high purity (> 95%) to serve as a starting material for the preparation of other complexes with $Mo_2O_2S_2$ cores.

§ Before dehydration in vacuum, the compound has about six water molecules of crystallization.

Anal. Calcd. for $C_{10}H_{14}N_2O_{11}S_2Na_2Mo_2$: C, 18.76; H, 2.20; N, 4.37%. Found: C, 18.57; H, 2.21; N, 4.48%. Electronic spectral data are given in Table I.

F.   DI-μ-SULFIDO-BIS[OXOMOLYBDENUM(V)](2+) AQUA DIMER, $[Mo_2O_2S_2(aq)]^{2+}$ (F)

The pure aqua ion, $[Mo_2O_2S_2(aq)]^{2+}$, is obtained by the aquation of the corresponding cysteinato complex **D** followed by Sephade G-10 separation. as described in the Section C. Electronic spectral data are given in Table I.

References

1. E. I. Stiefel, *Progr. Inorg. Chem.*, **22**, 1 (1977).
2. V. R. Ott, D. S. Swieter, and F. A. Schultz, *Inorg. Chem.*, **16**, 2538 (1977).
3. (a) J. Dirand-Colin, L. Ricard, and R. Weiss, *Inorg. Chim Acta*, **18**, L212 (1976); (b) W. E. Newton and J. W. McDonald, *Abstracts of Papers*, Proceedings of the Second International Conference on the Chemistry and Uses of Molybdenum, Oxford, U.K., P. C. H. Mitchell (ed.), 1976, p. 25; *J. Less-Common Met.*, **54**, 51 (1977); (c) D. M. L. Goodgame, R. W. Rollins, and A. C. Skapski, *Inorg. Chim. Acta*, **96**, L61 (1985).
4. T. Shibahara, H. Akashi, S. Nagaharta, H. Hattori, and H. Kuroya, *Inorg. Chem.*, **28**, 362 (1989).
5. T. Shibahara, H. Kuroya, K. Matsumoto, and S. Ooi, *Bull. Chem. Soc. Jpn.*, **60**, 2277 (1987).
6. T. Shibahara, S. Ooi, and H. Kuroya, *Bull. Chem. Soc. Jpn.*, **55**, 3742 (1982).
7. (a) B. Spivack and Z. Dori, *J. Chem. Soc., Chem. Commun.*, **1970**, 1716; (b) A. Kay and P. C. A. Mitchell, *J. Chem. Soc. (A)*, **1970**, 2421; (c) F. A. Armstrong, T. Shibahara, and A. G. Sykes, *Inorg. Chem.*, **17**, 189 (1978).
8. S. F. Gheller, T. W. Hambley, R. T. C. Brownlee, M. J. O'Connor, M. R. Snow, and A. G. Wedd, *J. Am. Chem. Soc.*, **105**, 1527 (1983).

59.   SULFUR-BRIDGED CUBANE-TYPE (Mo_4S_4) AND INCOMPLETE CUBANE-TYPE ($Mo_3O_{4-n}S_n$; $n = 1-4$) MOLYBDENUM AQUA IONS

Submitted by TAKASHI SHIBAHARA* and HARUO AKASHI*
Checked by ANDREAS TOUPADAKIS† and DIMITRI COUCOUVANIS†

Molybdenum clusters have attracted much attention, and a number of compounds with cubane-type or incomplete cubane-type cores have been

* Department of Chemistry, Okayama University of Science, 1-1 Ridai-cho, Okayama 700, Japan.
† Department of Chemistry, University of Michigan, Ann Arbor, MI 48109.

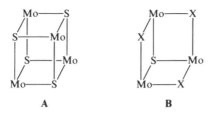

Fig. 1. Cubane-type (**A**) and incomplete cubane-type (**B**) cores. [X = O or S in (**B**)].

prepared and characterized.[1] Many reports have appeared on cubane-type and incomplete cubane-type aqua ions with sulfur bridge(s),[2-18] and the core structures of the aqua ions or their derivatives have been confirmed by X-ray analyses: cubane type, $[Mo_4S_4(aq)]^{n+}$ ($n = 5,6$)[2,3,6] and $[Mo_4OS_3(aq)]^{5+}$;[3] incomplete cubane type, $[Mo_3S_4(aq)]^{4+}$,[10-12] $[Mo_3OS_3(aq)]^{4+}$,[15] $[Mo_3O_2S_2(aq)]^{4+}$,[16] and $[Mo_3O_3S(aq)]^{4+}$.[17,18]

The sulfur-bridged molybdenum aqua ions have been prepared by the following methods: (1) reduction of $[Mo_2O_2S_2(cys)_2]^{2-}$ with $Na[BH_4]$,[10,11,15,16] (2) reduction of $[Mo_2O_3S(aq)]^{2+}$ with $Na[BH_4]$,[3,4,18] (3) degradation–disproportionation of $[Mo_2O_3S(edta)]^{2-}$ in a weak alkaline solution,[17] (4) reaction of $Mo(CO)_6$ with Na_2S,[6,12] (5) electrolysis of $[Mo_2O_3S(cys)_2]^{2-}$ or $[Mo_2O_2S_2(cys)_2]^{2-}$,[4,5,7,13] (6) reaction of $[Mo_2O_3S(aq)]^{2+}$ or $[Mo_2O_2S_2(aq)]^{2+}$ with $[MoCl_6]^{3-}$,[4] (7) aquation of $[Mo_4S_4(edta)_2]^{3-}$,[2] (8) reaction of $[Mo_3OS_3(aq)]^{4+}$ with iron metal,[3] and (9) electrolysis of $[Mo_4S_4(aq)]^{5+}$.[2,9]

Detailed preparative methods of sulfur-bridged cubane-type and incomplete cubane-type aqua ions except $[Mo_4OS_3(aq)]^{5+}$ are described here (see also Fig. 1).

No air-free or special techniques and apparatus are required for these syntheses, and the procedures can be scaled up or down easily. Once the aqua ions are prepared, the derivatives are easily prepared by substitution reactions.[6,10-12,15-18]

The facile preparative methods of Mo(V) dimers, $Na_2[Mo_2O_3S(cys)_2]\cdot4H_2O$,[18,19] $Na_2[Mo_2O_2S_2(cys)_2]\cdot4H_2O$,[19] $[Mo_2O_3S(aq)]^{2+}$,[18,19] and $[Mo_2O_2S_2(aq)]^{2+}$,[19] which are the starting material for the cubane-type and incomplete cubane-type aqua ions, have been reported recently. The synthesis and characterization of the oxo-bridged aqua ion, $[Mo_3O_4(aq)]^{4+}$, and its derivatives have appeared lately.[20]

A. INCOMPLETE CUBANE-TYPE AQUA ION, [Mo$_3$(μ-O)$_3$ (μ_3-S)(aq)]$^{4+}$

Procedure

To Mo(V) aqua dimer [Mo$^V_2$O$_3$S(aq)]$^{2+}$ (300 mL, 0.03 M per dimer in 1 M HCl)[18, 19] is added Na[BH$_4$] (7 g, Na[BH$_4$]/Mo $\approx$ 10) gradually with stirring. Concentrated HCl (20 mL) is then added slowly, and the solution turns from orange-red to brown.

■ **Caution.** *Rapid addition of Na[BH$_4$] or concentrated HCl will cause violent H$_2$ evolution.*

The resultant brown solution is heated in a boiling-water bath for $\sim$ 2 h with introduction of an air stream to give a red-brown solution with a tint of purple. The volume of the solution is decreased to $\sim$ 100 mL by use of a rotary evaporator and the solution is cooled to room temperature. After the precipitates of boric acid are filtered off by means of suction, the filtrate is subjected to Sephadex G-15 (or G-10) column chromatography (eluent, 1 M HCl; diameter 4 cm, length 80 cm). The following species are eluted from the column in order: [Mo$_3$O$_3$S(aq)]$^{4+}$ (red-purple, 70%: the first part of red solution that may contain [Mo$_3$O$_4$(aq)]$^{4+}$ is discarded ($\sim$ 10 mL), and typically 210 mL of a 0.02 M (per trimer) solution is obtained), [Mo$_3$O$_2$S$_2$(aq)]$^{4+}$ (grayish green, 5%), [Mo$_2$O$_4$(aq)]$^{2+}$ (orange, 1%), [Mo$_3$OS$_3$(aq)]$^{4+}$ (green, 1%), [Mo$_2$O$_3$S(aq)]$^{2+}$ (orange, 15%), [Mo$_3$S$_4$(aq)]$^{4+}$ (green, 0.1%), and unidentified green species.

The Sephadex resin used can be easily refreshed by the following method. About 100 mL of hydrogen peroxide ($\sim$ 3%) was passed though the column, and the resin was washed well with water (> 6 L).

In order to obtain a solution in 2 M HPTS (p-toluenesulfonic acid), the [Mo$_3$O$_3$S(aq)]$^{4+}$ solution in 1 M HCl is diluted to five times its original volume with water, and was absorbed to Dowex 50W-X2 cation exchanger (column diameter 2 cm, length 10 cm). The column is then washed with 0.1 M HPTS to remove Cl$^-$ ion ($\sim$ 200 mL; the column can be washed rapidly, and full removal of Cl$^-$ ion is checked by AgNO$_3$ solution) and eluted slowly with 2 M HPTS. Typically 120 mL of a 0.04 M (per trimer) red solution is obtained.

Properties

The aqua ion, [Mo$_3$O$_3$S(aq)]$^{4+}$, is air-stable, contrary to the case of [Mo$_3$O$_4$(aq)]$^{4+}$. Electronic spectral data are given in Table I. On reduction

TABLE I. **Electronic Spectral Data for $[Mo_3O_{4-n}S_n(aq)]^{4+}$ ($n=0-4$) and $[Mo_4S_4(aq)]^{5+}$ Aqua Ions[a,b]**

Compound	$\lambda_{max}(\varepsilon M^{-1} cm^{-1})$	Ref.
1. $[Mo_3O_4(aq)]^{4+}$	303 (795), 505 (189)	20f
2. $[Mo_3O_3S(aq)]^{4+}$	333 (932), 512 (153)	17
	225 (8480), 339 (930), 522 (153)	This work
3. $[Mo_3O_2S_2(aq)]^{4+}$	338 (2760), 450 (sh) (177), 572 (202)	16
	260 (sh) (5760) , 336 (2770), 580 (186)	This work
4. $[Mo_3OS_3(aq)]^{4+}$	307 (3080), 370 (sh) (2200), 410 (sh) (1760)	
	588 (263)	15
	252 (8930), 310 (sh) (2960) , 361 (2180),	
	414 (sh) (1720) , 605 (243)	This work
5. $[Mo_3S_4(aq)]^{4+}$	367 (5190), 500 (sh) (290), 602 (351)	11
	256 (10,350) , 372 (5510), 620 (315)	This work
6. $[Mo_4S_4(aq)]^{5+}$	370 (sh) (1008), 643 (445), 1100 (128)	2
	300 (sh) (4380) , 380 (sh) (1140) , 647 (444)	This work
	1116 (120)	

[a] λ_{max} in nanometers (ε values: per trimer for incomplete cubane-type aqua ions and per tetramer for cubane-type aqua ion).
[b] In 2 M HPTS: underlined values are in 1 M HCl for $[Mo_3O_3S(aq)]^{4+}$, $[Mo_3O_2S_2(aq)]^{4+}$, and $[Mo_3OS_3(aq)]^{4+}$ and in 2 M HCl for $[Mo_3S_4(aq)]^{4+}$ and $[Mo_4S_4(aq)]^{5+}$.

of the aqua ion, $[Mo^{IV}_3O_3S(aq)]^{4+}$, two different reduced species, Mo^{III}_3 and $Mo^{IV}Mo^{III}_2$ trimers, are identified.

B.   CUBANE-TYPE $[Mo_4(\mu_3\text{-}S)_4(aq)]^{5+}$ AND INCOMPLETE CUBANE-TYPE $[Mo_3(\mu\text{-}O)_2(\mu\text{-}S)(\mu_3\text{-}S)(aq)]^{4+}$ AQUA IONS

1.   Cubane-Type $[Mo_4S_4(aq)]^{5+}$ and Incomplete Cubane-Type $[Mo_3O_2S_2(aq)]^{2+}$ Aqua Ions in Solution

Procedure

To the ice-cooled Mo(V) aqua dimer $[Mo_2O_2S_2(aq)]^{2+}$ (0.035 M per dimer, 200 mL in 1 M HCl)[19] in a beaker, is added $Na[BH_4]$ (5.5 g, $Na[BH_4]/Mo \approx 10$) gradually with stirring. Concentrated HCl (10 mL) is added slowly and the solution turns from orange-red to brown.

■ **Caution.** *Rapid addition of Na[BH$_4$] or concentrated HCl will cause violent H$_2$ evolution.*

The resultant brown solution is heated in a boiling-water bath for ~ 30 min (see *Properties*) with introduction of an air stream to give dark green solution. Then, the solution is cooled to room temperature. After the precipitates of boric acid are filtered off by suction, the filtrate is subjected to Sephadex G-15 (or G-10)* column chromatography (eluent, 1 *M* HCl; diameter 4 cm, length 80 cm; Fig. 2).

The order of elution is as follows: [Mo$_4$S$_4$(aq)]$^{5+}$ [green, 20%: typically 150 mL of a 0.005 *M* (per tetramer) solution], [Mo$_3$O$_2$S$_2$(aq)]$^{4+}$ [greyish green,*^{2} 17%: 70 mL of a 0.01 *M* (per trimer) solution], [Mo$_3$OS$_3$(aq)]$^{4+}$ [green, 22%: 350 mL of a 0.003 *M* (per trimer) solution], [Mo$_3$S$_4$(aq)]$^{4+}$ (green, 4%), and [Mo$_2$O$_2$S$_2$(aq)]$^{2+}$ (orange, 1%). The first and second bands, containing [Mo$_4$S$_4$(aq)]$^{5+}$ and [Mo$_3$O$_2$S$_2$(aq)]$^{4+}$, respectively, overlap each other slightly. To purify the solution of [Mo$_4$S$_4$(aq)]$^{5+}$ aqua ion, the solution obtained from the Sephadex separation is diluted to five times its original volume with water, and Dowex 50W-X2 column chromatography is performed (diameter 2 cm, length 15 cm). A small amount of [Mo$_3$O$_2$S$_2$(aq)]$^{4+}$ elutes with 1 *M* HCl at first, then [Mo$_4$S$_4$(aq)]$^{5+}$ is eluted with 2*M* HCl. Typically 130 mL of a 0.005*M* (per tetramer) solution is obtained.

To purify the solution of [Mo$_3$O$_2$S$_2$(aq)]$^{4+}$ aqua ion, the solution obtained from the Sephadex separation is diluted to five times its original

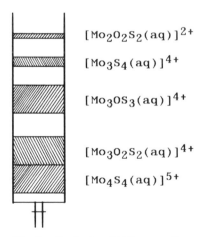

$$[Mo_2O_2S_2(aq)]^{2+}$$

$$[Mo_3S_4(aq)]^{4+}$$

$$[Mo_3OS_3(aq)]^{4+}$$

$$[Mo_3O_2S_2(aq)]^{4+}$$

$$[Mo_4S_4(aq)]^{5+}$$

Fig. 2. Sephadex G-10 separation.

* Sephadex resin can be refreshed as described in Section A.

volume with water, and Dowex 50W-X2 column chromatography is performed (diameter 2 cm, length 15 cm). Typically 500 mL of a 0.007 M (per trimer) solution in 1 M HCl is obtained. A more concentrated solution is obtainable by loading the solution in 1 M HCl onto a short 50W-X2 cation exchanger and eluting with 2 M HCl. The gray-violet* aqua ion in 2 M HPTS is obtained by loading the aqua ion in 1 or 2 M HCl onto a short 50W-X2 cation exchanger followed by elution with 2 M HPTS. Typically a 0.05 M (per trimer) solution is obtained.

Yields of the aqua ions, especially that of $[Mo_4S_4(aq)]^{5+}$, depend heavily on the period of heating as shown in Fig. 3. When longer heating (> 1 h) was employed, the yield of $[Mo_4S_4(aq)]^{5+}$ and $[Mo_2O_2S_2(aq)]^{4+}$ decreased, and the yield of $[Mo_3S_4(aq)]^{4+}$ increased. The aqua ions, $[Mo_3OS_3(aq)]^{4+}$ and $[Mo_3S_4(aq)]^{4+}$, can be prepared more easily from $Na_2[Mo_2O_2S_2(cys)_2] \cdot 4H_2O$ (see Section C).

Properties

Since the aqua ion $[Mo_4S_4(aq)]^{5+}$ is slightly air-sensitive, it should be kept in a refrigerator under a nitrogen atmosphere when it is stored more than a day, although air-free techniques are not required during the synthesis. The

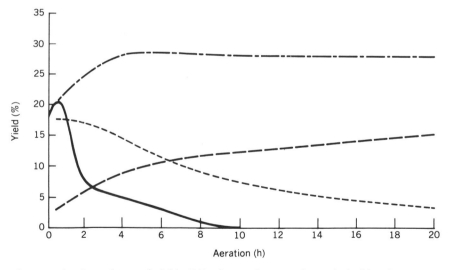

Fig. 3. The dependence of yields (%) of aqua ions on the period of heating of the mixture of $[Mo_2O_2S_2(aq)]^{2+}$ and $NaBH_4$ in diluted HCl (see Section B-1).

---

* The color of the aqua ion varies considerably by the change of the solvent from HCl to HPTS.

$[Mo_3O_2S_2(aq)]^{4+}$ is much more stable than $[Mo_4S_4(aq)]^{5+}$ and can be stored in the air for several days. A mean $Mo^{3.25+}$ state or a formal oxidation state of $Mo^{IV}Mo^{III}_3$ is assigned for the cubane-type aqua ion. The reduced species, $[Mo_4S_4(aq)]^{4+}$, can be obtained from the aqua ion.[2,9] Spectral data are given in Table I.

2. $[Mo_4S_4(H_2O)_{12}](CH_3C_6H_4SO_3)_5 \cdot 14H_2O$

Procedure

A solution of $[Mo_4S_4(aq)]^{5+}$ in 2 M HCl (0.005 M per tetramer, 100 mL)* is diluted to five times its original volume with water, and is absorbed to a short Dowex 50W-X2 cation exchanger (column diameter 1 cm, length 8 cm).† The column is then washed with 0.1 M HPTS (*p*-toluenesulfonic acid) to remove Cl⁻ ion (~ 50 mL), and elution with 4 M HPTS very slowly (0.3 mL/min) gives green solution [4 mL of a 0.1 M (per tetramer) solution]. The eluted solution is kept in a freezer (~ − 10°C) for a few days. The green plate crystals deposited are filtered quickly at room temperature and then washed with ethyl acetate. Yield: 0.9 g (~ 75%) based on $[Mo_4S_4(aq)]^{5+}$ in 2 M HCl.

Anal. Found (calcd.): C, 23.08 (22.89); H, 4.60 (4.77)%. X-Ray structure has been determined.[3]

Properties

The solid aqua compound is much more stable than that dissolved in HCl or HPTS; however, it is recommended to keep the crystals in a refrigerator under a nitrogen atmosphere. The crystals dissolve in several organic solvents, including acetonitrile, acetone, methanol, and ethanol. Spectral data are given in Table I.

C.   INCOMPLETE CUBANE-TYPE $[Mo_3(\mu\text{-}O)(\mu\text{-}S)_2(\mu_3\text{-}S)(aq)]^{4+}$ AND $[Mo_3(\mu\text{-}S)_3(\mu_3\text{-}S)(aq)]^{4+}$ AQUA IONS

1.   Incomplete Cubane-Type $[Mo_3OS_3(aq)]^{4+}$ and $[Mo_3S_4(aq)]^{4+}$ Aqua Ions in Solution

The method described here (method C) uses $Na_2[Mo_2O_2S_2(cys)_2] \cdot 4H_2O$ as it is and is more convenient than that described in Section B (method B),

---

* Much more diluted solution of $[Mo_3S_4(aq)]^{4+}$ in 1 or 2 M HCl can also be used.

because no aquation–separation step using Sephadex G-15 is required. Furthermore, method C gives higher yield ($\sim 25\%$) of $[Mo_3S_4(aq)]^{4+}$ than method B using $[Mo_2O_2S_2(aq)]^{2+}$ as a starting material [yield $\sim 15\%$ even by longer (e.g., 20 h) aeration with heating]. By method C, the yields of $[Mo_4S_4(aq)]^{5+}$ and $[Mo_3O_2S_2(aq)]^{4+}$ are low, even if the period of heating is short and method B is suitable for the preparation of $[Mo_4S_4(aq)]^{5+}$ and $[Mo_3O_2S_2(aq)]^{4+}$.

Procedure

The compound $Na_2[Mo_2O_2S_2(cys)_2] \cdot 4H_2O$ (10 g, 0.0155 mol) is dissolved in water (1 L) with stirring and 1 M HCl (30 mL) is added, which is followed by the slow addition of $Na[BH_4]$ (5 g, $Na[BH_4]/Mo \approx 10$). After the addition of $Na_2S \cdot 9H_2O$ (5 g)* to the solution, concentrated HCl (150 mL) is added very slowly.

■ **Caution.** *Rapid addition of $Na[BH_4]$ or concentrated HCl will cause violent H_2 evolution.*

After 10 min, the resulting dark brown solution is heated in a boiling-water bath for ~ 20 h (which need not be continuous) with an air stream bubbling through the solution to give a dark green solution. During heating in a water bath, a small amount of 1 M HCl is added occasionally if the volume of the solution drops below 200 mL. After the solution is cooled to room temperature, the precipitates of boric acid are filtered off by means of suction and the filtrate is subjected to Sephadex G-15 (or G-10)* column chromatography (eluent, 1 M HCl; diameter 4 cm, length 80 cm). The order of elution is as follows: $[Mo_4S_4(aq)]^{5+}$ (green, trace), $[Mo_3O_2S_2(aq)]^{4+}$ (grayish green, 1.5%), $[Mo_3OS_3(aq)]^{4+}$ [green, 27%: typically 900 mL of a 0.003 M (per trimer) solution], $[Mo_3S_4(aq)]^{4+}$ [green, 25%: 750 mL of a 0.0035 M (per trimer) solution], and $[Mo_2O_2S_2(aq)]^{2+}$ (orange, 17%). To obtain a more concentrated solution, the aqua ion $[Mo_3S_4(aq)]^{4+}$ is diluted to five times its original volume with water, and applied to Dowex 50W-X2 column chromatography (eluent, 2 M HCl; diameter 2 cm, length 10 cm). Typically 25 mL of a 0.1 M (per trimer) solution is obtained.

For the preparation of the aqua ion $[Mo_3OS_3(aq)]^{4+}$, heating for 6 h gives nearly the same yield as does 20 h heating. The ion is also concentrated similarly.

Properties

The aqua ions $[Mo_3OS_3(aq)]^{4+}$ and $[Mo_3S_4(aq)]^{4+}$ are very stable toward air oxidation and do not require a nitrogen atmosphere for storage. Spectral

---

*The resin (H^+ form) should be nearly replaced by the aqua ion to obtain a concentrated solution.

data are given in Table I. The incomplete cubane-type aqua ion $[Mo_3S_4(aq)]^{4+}$ reacts with metals to give cubane-type aqua ions with Mo_3MS_4 (M = Fe, Co, Ni, Cu, and Sn, etc.) cores.[21]

2. $[Mo_3S_4(H_2O)_9](CH_3C_6H_4SO_3)_4 \cdot 9H_2O$

A procedure similar to that for the preparation of $[Mo_4S_4(H_2O)_{12}]$-$(CH_3C_6H_4SO_3)_5 \cdot 14H_2O$ (see Section B.2) is employed. A solution of $[Mo_3S_4(aq)]^{4+}$ in 2 M HCl (0.1 M per trimer, 15 mL)* is diluted to five times its original volume with water (resulting in ~ 0.4 acid concentration), and is absorbed on a Dowex 50W-X2 cation exchanger (column diameter 1 cm, length 8 cm).† The column is washed with 0.1 M HPTS (*p*-toluenesulfonic acid) to remove Cl$^-$ ion (~ 100 mL), and very slow elution (~ 0.3 mL min^{-1}) is effected with 4 M HPTS to give a green solution [~ 12 mL of a 0.1 M (per trimer) solution]. The eluted solution is cooled in a freezer ($\sim -10°C$) for a few days to give green plate-like crystals. The crystals are filtered and washed with ethyl acetate. Yield: 1.5 g ($\sim 70\%$) based on $[Mo_3S_4(aq)]^{4+}$ in 2 M HCl.

Anal. Found (calcd.): C, 23.94 (23.59); H, 4.02 (4.52)%.

Properties

The aqua compound dissolves in several organic solvents, including aceto-nitrile, acetone, methanol, and ethanol. Spectral data are given in Table I.

References

1. (a) A. Muller, R. Jostes, and F. A. Cotton, *Angew. Chem. Int. Ed. Engl.*, **19**, 875 (1980);
   (b) P. Zanello, *Coord. Chem. Rev.*, **83**, 199 (1988).
2. T. Shibahara, H. Kuroya, K. Matsumoto, and S. Ooi, *Inorg. Chim. Acta*, **116**, L25 (1986).
3. H. Akashi, T. Shibahara, T. Narahara, H. Tsuru, and H. Kuroya, *Chem. Lett.*, **1989**, 129.
4. M. Martinez, B.-L. Ooi, and A. G. Sykes, *J. Am. Chem. Soc.*, **109**, 4615 (1987).
5. P. Kathirgamanathan, M. Martinez, and A. G. Sykes, *Polyhedron*, **5**, 505 (1986).
6. F. A. Cotton, M. P. Diebold, Z. Dori, R. Llusar, and W. Schwotzer, *J. Am. Chem. Soc.*, **107**, 6735 (1985).
7. P. Kathirgamanathan, M. Martinez, and A. G. Sykes, *J. Chem. Soc. Chem. Commun.*, **1985**, 953.
8. B.-L. Ooi, C. Sharp, and A. G. Sykes, *J. Am. Chem. Soc.*, **111**, 125 (1989).
9. C. Sharp and A. G. Sykes, *J. Chem. Soc. Dalton Trans.*, **1988**, 2579.

---

* Addition of $Na_2S \cdot 9H_2O$ gives a slightly higher yield of $[Mo_3S_4(aq)]^{4+}$.
† Sephadex resin can be refreshed as described in Section A.

10. T. Shibahara and H. Kuroya, *Abstracts of Papers*, 5th International Conference on the Chemistry and Uses of Molybdenum, Newcastle upon Tyne, U.K., July 1985, p. 59.

11. T. Shibahara and H. Kuroya, *Polyhedron*, **5**, 357 (1986).

12. F. A. Cotton, Z. Dori, R. Llusar, and W. Schwotzer, *J. Am. Chem. Soc.*, **107**, 6734 (1985).

13. P. Kathirgamanathan, M. Martinez, and A. G. Sykes, *J. Chem. Soc. Chem. Commun.*, **1985**, 1437.

14. B.-L. Ooi, M. Martinez, and A. G. Sykes, *J. Chem. Soc. Chem. Commun.*, **1988**, 1324.

15. T. Shibahara, H. Miyake, K. Kobayashi, and H. Kuroya, *Chim. Lett.*, **1986**, 139.

16. T. Shibahara, T. Yamada, H. Kuroya, E. F. Hills, P. Kathirgamanathan, and A. G. Sykes, *Inorg. Chim. Acta*, **113**, L19 (1986).

17. T. Shibahara, H. Hattori, and H. Kuroya, *J. Am. Chem. Soc.*, **106**, 2710 (1984).

18. T. Shibahara, H. Akashi, S. Nagahata, H. Hattori, and H. Kuroya, *Inorg. Chem.*, **28**, 362 (1989).

19. T. Shibahara and H. Akashi, *Inorg. Synth.*, **29**, 254 (1992).

20. (a) F. A. Benory, A. Bino, D. Gibson, F. A. Cotton, and Z. Dori, *Inorg. Chim. Acta*, **99**, 137 (1985); (b) S. F. Gheller, T. W. Hambley, R. T. C. Brownlee, M. R. O'Conner, and A. J. Wedd, *J. Am. Chem. Soc.*, **105**, 1527 (1983); (c) M. T. Paffett and F. C. Anson, *Inorg. Chem.*, **22**, 1347 (1983); (d) K. R. Rodgers, R. K. Murmann, E. O. Schlemper, and M. E. Schelton, *Inorg. Chem.*, **24**, 1313 (1985); (e) D. T. Richens and A. G. Sykes, *Inorg. Synth.*, **23**, 130 (1985); (f) M. A. Harmer, D. T. Richens, A. B. Soares, A. T. Thornton, and A. G. Sykes, *Inorg. Chem.*, **20**, 4155 (1981).

21. T. Shibahara, M. Yamasaki, H. Akashi, and T. Katayama, *Inorg. Chem.*, **30**, 2693 (1991).

60. PREPARATION OF DINUCLEAR AND TRINUCLEAR FERRABORANES, $Fe_2(CO)_6B_2H_6$ AND $Fe_3H(CO)_{10}BH_2$

Submitted by XIANGSHENG MENG* and THOMAS P. FEHLNER*
Checked by KENTON H. WHITMIRE†

The dinuclear ferraborane, $Fe_2(CO)_6B_2H_6$, was first isolated in 1978.[1] Since then further investigation of its properties have been carried out[2] as the compound serves as a good reagent for the preparation of other new compounds. That is, it serves as a "pattern maker" for the synthesis of other ferraboranes. Although the preparation of $Fe_2(CO)_6B_2H_6$ was later modified, the yield of the compound was still low ($\approx 1\%$) and the procedure complicated. Hence, further investigations of the ferraborane were limited by its inaccessability. Herein we report a new procedure that is simple (requires two reagents without acidification rather than four reagents plus acidification), provides a higher yield (15% rather than < 1%), and involves a short reaction and separation time (10 min and 2 h rather than 4 h and 24 h), respectively. This method not only yields $Fe_2(CO)_6B_2H_6$ (Section A), but

*Department of Chemistry, University of Notre Dame, Notre Dame, IN 46556.

†Department of Chemistry, Rice University, Houston TX 77251.

also produces the trinuclear ferraborane $Fe_3H(CO)_{10}BH_2$ (30% yield; Section B). The latter compound was first isolated in low yield from the reaction of $Na[Fe(CO)_4C(O)CH_3]$ in tetrahydrofuran (THF) with $BH_3 \cdot THF$ and $Fe(CO)_5$ followed by acidification.[3]

Procedure

■   **Caution.**   *The reaction must be run in a hood because of the toxicity of the borane complex and iron carbonyls.*

All the following manipulations are performed under nitrogen using Schlenk techniques or a nitrogen-atmosphere dry bag. All solvents are degassed and distilled under nitrogen before use unless stated otherwise. $Fe_2(CO)_9$, $BH_3 \cdot SMe_2$ (10.0 M) and CF_3CO_2H (99%) were purchased from Aldrich Chemical Company and used as received.

A. $Fe_2(CO)_6B_2H_6$

$$Fe_2(CO)_9 + BH_3 \cdot SMe_2 \rightarrow Fe_2(CO)_6B_2H_6$$

A 1.5-g (4.1-mmol) sample of $Fe_2(CO)_9$ is transferred into a double-necked, round-bottomed, 250-mL flask containing a magnetic stirring bar. The flask is fitted with a water cooled reflux condenser that is connected to a vacuum line and nitrogen source. The flask is evacuated for 2 min [slowly at first prevent flakes of $Fe_2(CO)_9$ from being pulled into the vacuum line] and is refilled with nitrogen, and 45 mL of hexane and 0.82 mL of $BH_3 \cdot SMe_2$ (8.2 mmol) are then added. The flask is immediately immersed in a hot oil bath (75°C) of sufficient volume such that a substantial decrease in bath temperature is avoided, and the reaction is allowed to proceed for about 10 min. After cooling to room temperature and then 0°C, the solution is concentrated to about 20 mL volume under reduced pressure using the vacuum line. The solution remaining in the flask is placed on a short (about 3 cm diameter, 8 cm long) silica gel chromatographic column and eluted with degassed hexane. The first band (yellow) is collected. For the column preparation silica gel 60, EM Science 70/230 mesh, is dried in an oven overnight before use. Air contained in the silica gel in the column is removed under reduced pressure using the vacuum line. Then degassed solvent is added into the column under a nitrogen atmosphere. The bubbles in the column can be removed by pumping until the solvent and the silica gel are mixed well followed by refilling the column with nitrogen. Note, the solvent level must be higher (~ 6 cm) than that of silica gel before pumping. This process usually removes all bubbles in the column but can be repeated if necessary. The solution collected from the yellow band in the column contains a small amount of $Fe(CO)_5$ that can be removed with the solvent hexane using a vacuum line at $-15°C$ or below. The yield is 10–15% of $Fe_2(CO)_6B_2H_6$ based on boron as determined by ^{11}B NMR.

Properties

$Fe_2(CO)_6B_2H_6$ is volatile, and the solvent-free, pure compound (viscous, yellow-brown, oil) can be obtained only by removing the solvent under vacuum at $-15°C$ or below. The compound is very air-sensitive (but not pyrophoric) in either the pure state or in solution. Further, it can be stored for long periods of time in an inert atmosphere (e.g., N_2) at low temperature (e.g., $< -20°C$). MS, $P^+ = {}^{56}Fe_2{}^{11}B_2{}^{12}C_6{}^{16}O_6{}^1H_6{}^+$, 307.9048 calculated, 307.9066 observed. The ^{11}B NMR spectrum in hexane at 25°C shows an apparent broad quartet at 24.4 ppm with $J_{B-H} \approx 60$ Hz. The infrared spectrum shows two weak B–H absorptions at 2526 and 2444 cm^{-1}, and CO absorptions at 2090 (s), 2046 (vs), 2026 (s), 2022 (s), 1996 (s), and 1990 (s) cm^{-1}. The variable temperature (from $-110°C$ to $+35°C$) ^1H NMR spectra in CD_2Cl_2 are shown in Fig. 1 ($-110°C$, δ 2.89, 2H, -2.44, 1H, -12.88, 1H, -15.57, 2H).[2b]

B. $Fe_3H(CO)_{10}BH_2$

$$Fe_2(CO)_9 + BH_3 \cdot SMe_2 \rightarrow Fe_3H(CO)_{10}BH_2$$

The trinuclear ferraborane, $Fe_3H(CO)_{10}BH_2$, can also be obtained from the above procedure for the preparation of $Fe_2(CO)_6B_2H_6$ by employing a 8 : 1

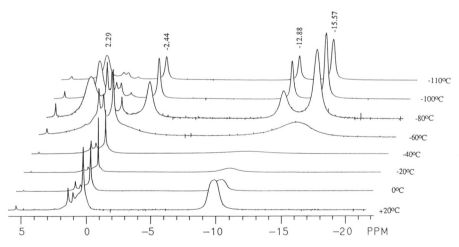

Fig. 1. Variable-temperature ^1H NMR spectra of $Fe_2(CO)_6B_2H_6$ in CD_2Cl_2.

ratio (Fe : B) of $Fe_2(CO)_9 + BH_3 \cdot SMe_2$ rather than 1 : 1. That is, a 1.5 g (4.1 mmol) sample of $Fe_2(CO)_9$ was reacted with 0.10 mL of $BH_3 \cdot SMe_2$ (1.0 mmol) according to the procedures in Section A. After the reaction solution cools to room temperature, a silica gel column (about 3 cm diameter, 8 cm long) is used to separate the solution. The product $Fe_3H(CO)_{10}BH_2$ is deprotonated on a room-temperature silica gel column and therefore stays at origin. However, it can be eluted with degassed methanol as the corresponding anion after all other compounds are washed out from the column by degassed toluene. The methanol solvent is removed under reduced pressure. Then 15 mL of hexane is added to the dry residue followed by slow addition of 0.03 mL of CF_3COOH (~ 0.39 mmol) (or 10 mL of degassed 80% aqueous phosphoric acid[4]) while stirring the solution. This protonation is allowed to proceed for 30 min and yields the neutral compound as a dark brown solution in hexane. The excess acid in the solution is removed by pumping until dry, and fresh hexane is then added. This procedure produces a yield of about 30% of $Fe_3H(CO)_{10}BH_2$ based on boron from ^{11}B NMR. $Fe_3H(CO)_{10}BH_2$ can also be obtained in somewhat higher yield from the reaction mixture as the third band (brown) from a low-temperature silica gel column[5] [$-15°C$, about 2.5 cm diameter, 12 cm long, degassed hexane; the first band, yellow, is $Fe_2(CO)_6B_2H_6$ and the second band, green, is $HFe_4H(CO)_{12}BH_2$ plus a small amount of $Fe_3(CO)_{12}$].

Properties

$Fe_3H(CO)_{10}BH_2$ is air-sensitive in solution and is pyrophoric in the solid state. The solution can be stored for long periods of time under a N_2 atmosphere at low temperature (e.g., $-20°C$). The crystal structure has not been obtained, but the compound has been characterized spectroscopically. ^{11}B NMR[2b]: hexane, 20°C, 56.5 ppm, doublet, $J_{B-H} = 110$ Hz. IR: hexane, 2495 (w) (v_{B-H});[2b] 2106 (w), 2073 (s), 2054 (vs), 2041 (sh), 2031 (m), 2022 (m), 2010 (m), 1995 (m), 1868 (m) (v_{CO}).[4] ^1H NMR: toluene-d_8, $-60°C$, δ 5.9 (br, 1H), -13.7 (br, 1H), -25.6 (s, 1H). MS: $P^+ = {}^{56}Fe_3{}^{10}B_1{}^{12}C_{10}{}^{16}O_{10}{}^1H_3{}^+$, 460.790 calculated, 460.788 observed.[4]

References

1. E. L. Andersen and T. P. Fehlner, *J. Am. Chem. Soc.*, **100**, 4606 (1978).
2. (a) G. B. Jacobsen, E. Andersen, C. E. Housecroft, F.-E. Hong, M. L. Buhl, G. J. Long, and T. P. Fehlner, *Inorg. Chem.*, **26**, 4040 (1987), (b) X. Meng Ph.D. thesis, University of Notre Dame, 1990.
3. J. C. Vites, Ph.D. thesis, University of Notre Dame, 1984.

4. J. C. Vites, C. E. Housecroft, G. B. Jacobsen, and T. P. Fehlner, *Organometallics*, **3**, 1591 (1984).
5. R. C. Buck and M. S. Brookhart, in *Experimental Organometallic Chemistry*, A. L. Wayda, and M. Y. Darensbourg (eds.), ACS Symposium Series, American Chemical Society, Washington, DC, 1987, p. 27.

61. PREPARATION OF THE METAL- AND HYDROGEN-RICH FERRABORANE Fe₃H(CO)₉BH₄

Submitted by XIANGSHENG MENG* and THOMAS P. FEHLNER*
Checked by KENTON H. WHITMIRE†

The ferraborane, $Fe_3H(CO)_9BH_4$, was first synthesized from the reaction of $Na[Fe(CO)_4C(O)CH_3]$ with $BH_3 \cdot THF‡$ at 65°C followed by acidification.[1] Later the procedure was modified to yield $\approx 5\%$ of the compound.[2] The low yield of this ferraborane limits the investigation of its potentially interesting properties. Herein we report a new method that gives a 30% yield of the ferraborane.

Procedure

$$[PPN]_2[Fe_4(CO)_{13}]\S + BH_3 \cdot THF \rightarrow Fe_3H(CO)_9BH_4$$

■ **Caution.** *The reaction must be run in a hood because of the toxicity of the borane complex and iron carbonyls.*

All the following manipulations are performed under nitrogen using Schlenk techniques or a nitrogen-atmosphere dry bag. All solvents are degassed and distilled under nitrogen before use unless mentioned otherwise. $BH_3 \cdot THF$ and CF_3CO_2H (99%) are purchased from Aldrich Chemical Company and are used as received.

After a 1-g (0.6-mmol) sample of $[PPN]_2[Fe_4(CO)_{13}]$[3] in 20 mL of THF is heated to 75°C in a double-necked, round-bottomed, 250-mL flask fitted with a water-cooled reflux condenser connected to nitrogen, a single 10-mL (10-mmol) aliquot of $BH_3 \cdot THF$ is quickly added and the solution stirred for

---

* Department of Chemistry, University of Notre Dame, Notre Dame, IN 46556.
† Department of Chemistry, Rice University, Houston TX 77251.
‡ THF = tetrahydrofuran.
§ PPN = μ-nitrido-bis(triphenylphosphorus) (H).

2 h. Before the $BH_3 \cdot THF$ is added the compound $[PPN]_2[Fe_4(CO)_{13}]$ is only slightly dissolved. As the $BH_3 \cdot THF$ is added, the solution becomes dark-red with all the starting material finally dissolving in the THF. After 2 h the THF solvent is removed under reduced pressure at 75°C until brown oil-like residue remains in the flask. The residue is then cooled to room temperature and 20 mL of hexane is added followed by the slow addition of 1.5 mL of CF_3COOH while the solution is stirred. During this step gas is evolved. The acidification is carried out for 1 h, then the solution is transferred by a transfer needle into a second flask. In order to extract more of the products, 2×10 mL of hexane (but no more acid) is sequentially added to the reaction flask and transferred to the second flask. The pure compound $Fe_3H(CO)_9BH_4$ is obtained by collecting the second band [dark red following a yellow band, $H_3Fe_3(CO)_9CCH_3$] in a cold chromatographic column[4] (− 20°C, about 3 cm diameter, 12 cm long) filled with silica gel using about 1 L of degassed hexane to elute it. To prepare the column silica gel 60, EM Science 70/230 mesh, is dried in an oven overnight before use. Air contained in the silica gel in the column is removed under reduced pressure using the vacuum line. Then solvent is added into the column under a nitrogen atmosphere. The bubbles in the column are removed by pumping until the solvent and the silica gel are mixed well (the solvent level must be higher, about 6 cm, than that of silica gel before pumping) followed by refilling the column with nitrogen. This process usually removes all bubbles in the column but can be repeated if necessary. Removing the hexane from the second fraction under reduced pressure using the vacuum line yields 105 mg of $Fe_3H(CO)_9BH_4$ (30% based on iron). Dark red crystals are obtained by placing a saturated $Fe_3H(CO)_9BH_4$ hexane solution in a refrigerator (≈ 0°C) overnight.

Properties

The compound is air-sensitive in solution and moderately air-sensitive in the solid state (but not pyrophoric in either state). The solution can be stored for long periods of time under a N_2 atmosphere at low temperature (e.g., − 20°C) and survives for days under a N_2 atmosphere at room temperature. The compound has been fully characterized. MS, $P^+ = {}^{56}Fe_3{}^{11}B_1{}^1H_5{}^+$, 183.855 observed, 183.853 calculated. ^{11}B NMR (hexane, 20°C, δ), 1.5 (m), $J_{B-H} \cong 50$ Hz; IR ν_{CO} (hexane, cm^{-1}) 2096 (m), 2061 (s), 2042 (s), 2030 (s), 2021 (s), 2013 (s), 1998 (m); ^1H NMR (toluene-d_8, 20°C, δ), 3.2 (b, d, 1 H), − 14.7 (b, asymmetric, ≈ 3 H), − 24.3 (b, ≈ 1 H) (The molecule is fluxional at 20°C and near coalescence at 300 MHz.)[2] The X-ray crystal structure as well as some other properties have also been reported.[2,5]

References

1. J. C. Vites, Ph.D. thesis, University of Notre Dame, 1984.
2. J. C. Vites and T. P. Fehlner, *J. Am. Chem. Soc.*, **106**, 4633 (1984); J. C. Vites, C. E. Housecroft, M. L. Buhl, G. J. Long, and T. P. Fehlner, *J. Am. Chem. Soc.*, **108**, 3304 (1986).
3. K. Whitmire, J. Ross, C. B. Cooper, and D. F. Shriver, *Inorg. Synth.*, **21**, 66 (1982).
4. R. C. Buck and M. S. Brookhart, in *Experimental Organometallic Chemistry*, A. L. Wayda and M. Y. Darensbourg (eds.), ACS Symposium Series, American Chemical Society, Washington, DC, 1987, p. 27.
5. X. Meng, Ph.D. thesis, University of Notre Dame, 1990.

62. TETRAMMONIUM HEXAKIS[TETRAETHYL-2,3-DI-OXOBUTANE-1,1,4,4-TETRACARBOXYLATO(2 −)-$O^{1'},O^2:O^3,O^{4'}$]TETRAMETALATE(4 −)

$M = Mg^{2+}, Mn^{2+}, Co^{2+}, Ni^{2+}, Zn^{2+}$

Submitted by ROLF W. SAALFRANK,* ARMIN STARK,* and ROMAN BURAK*
Checked by WILL E. LYNCH† and RICHARD L. LINTVEDT†

The synthesis described here represents a general procedure for preparing four-nucleus metal chelate complexes in a kind of "spontaneous self-assembly." The products are interesting as model species in the bioinorganic chemistry. For example, four manganese atoms per photosystem II reaction center are essential for activity. There water is oxydated by the photosynthetic process in plants. So four-nucleus manganese complexes will provide useful models for the oxygen-evolving complex in plants.

Method A is applicable only to the metal Mg^{2+}; the modified method B, however, allows synthesis of four-nucleus metal chelate complexes with different metals such as Mg^{2+}, Mn^{2+}, Co^{2+}, Ni^{2+}, and Zn^{2+}. Both methods have several advantages: (1) easy availability of the reagents, (2) one-pot operation, and (3) high yields and easy isolation of the products.

Procedure

■   **Caution.**   *Oxalyl chloride is toxic. This preparation should be carried out in a well-ventilated hood. All solvents should be tried thoroughly and air-free before use. The reaction must be performed under an atmosphere of nitrogen during the course of the reaction.*

A.   METHOD A (M = Mg²⁺)

A 500-mL, three-necked, round-bottomed flask is equipped with a magnetic stirring bar, a thermometer, a reflux condenser, and a 50-mL dropping funnel fitted with a nitrogen inlet adapter. The flask is charged with 1.5 g (0.060 mol) of magnesium and 50 mL of dry diethyl ether. Then 8.5 g (0.060 mol) of methyl iodide is added dropwise. The resultant solution of methyl magnesium iodide is added with freshly distilled tetrahydrofuran (100 mL) and is cooled to $-78°C$ with a Dry Ice–acetone bath. Diethyl malonate (9.6 g, 0.060 mol) in dry tetrahydrofuran (50 mL) is added over 30 min from the dropping funnel. The addition is accompanied by evolution of methane. The white reaction mixture is stirred 2 h at $-78°C$. A solution of distilled oxalyl chloride (1.9 g, 0.015 mol) in dry tetrahydrofuran (50 mL) is slowly added to the cooled suspension within 1 h while the mixture is allowed to warm to

* Institut für Organische Chemie, University of Erlangen-Nürnberg, Henkestrasse 42, 8520 Erlangen, Germany.
† Department of Chemistry, Wayne State University, Detroit, MI 48202.

− 70°C. The temperature of the reaction mixture is allowed to rise to room temperature over a period of 18 h. Then 200 mL of saturated aqueous ammonium chloride solution is added in one portion to the resulting light yellow reaction mixture. After the mixture is stirred for 10 min, the yellow organic phase is separated and washed with 50 mL of saturated aqueous sodium hydrogen carbonate solution and 50 mL of water. The aqueous layer is further extracted with two 100-mL portions of dichloromethane and washed as described above. The combined colorless organic layers are dried over anhydrous magnesium sulfate, and the drying agent is filtered. The solvents are removed with a rotary evaporator under reduced pressure. The oily residue is treated carefully with a small amount of diethyl ether and filtered. One recrystallization from warm acetone gives 5.1 g (85%) of product as colorless crystals.

Anal. Calcd. for $C_{96}H_{136}O_{60}N_4Mg_4$: C, 47.99; H, 5.71; N, 2.33; Mg, 4.05. Found: C, 47.46; H, 5.75; N, 2.32; Mg, 4.03.

B.   METHOD B ($M = Mg^{2+}$, Mn^{2+}, Co^{2+}, Ni^{2+}, Zn^{2+})

A 250-mL, three-necked, round-bottomed flask is equipped with a magnetic stirring bar, a thermometer, and a 50-mL dropping funnel fitted with a nitrogen inlet adapter. The flask is charged with 4.8 g (0.030 mol) of diethyl malonate, and 50 mL of dry tetrahydrofuran and the resultant solution is cooled to − 78°C with a Dry Ice–acetone bath. Methyl lithium (1.6 *M* in diethyl ether, 0.030 mol) is added over 10 min from the dropping funnel. The addition is accompanied by evolution of methane. The solution is stirred for 1 h at − 78°C), dry solid metal chloride ($MgCl_2$, 2.9 g; $MnCl_2$, 3.8 g; $CoCl_2$, 3.9 g; $NiCl_2$, 3.9 g; $ZnCl_2$, 4.1 g, 0.030 mol) is given into the solution and is stirred for a further 1 h. A solution of distilled oxalyl chloride (0.9 g, 0.0075 mol) in dry tetrahydrofuran (20 mL) is slowly added to the cooled suspension within 30 min. The temperature of the reaction mixture is allowed to rise to room temperature over a period of 18 h. Then 100 mL of saturated aqueous ammonium chloride solution is added in one portion to the resulting reaction mixture. After the mixture is stirred for 10 min, the organic phase is separated. The aqueous layer is extracted with two 100-mL portions of dichloromethane. The combined organic layers are dried over anhydrous sodium sulfate, and the drying agent is filtered. The solvents are removed with a rotary evaporator under reduced pressure. The solid residue is treated carefully with a small amount of diethyl ether and filtered.

Anal. Calcd. for $C_{96}H_{136}O_{60}N_4Mg_4$: C, 47.99; H, 5.71; N, 2.33; Mg, 4.05. Found: C, 47.82; H, 5.87; N, 2.28; Mg, 4.03. Colorless crystals, 2.6 g (87%).

Anal. Calcd. for $C_{96}H_{136}O_{60}N_4Mn_4$: C, 45.65; H, 5.43; N, 2.22; Mn, 8.70. Found: C, 45.34; H, 5.56; N, 2.23; Mn, 8.52. Colorless crystals, 2.7 g (86%).

Anal. Calcd. for $C_{96}H_{136}O_{60}N_4Co_4$: C, 45.36; H, 5.39; N, 2.21; Co, 9.27. Found: C, 44.96; H, 5.47; N, 2.28; Co, 9.03. Pink crystals, 2.4 g (79%).

Anal. Calcd. for $C_{96}H_{136}O_{60}N_4Ni_4$: C, 45.38; H, 5.40; N, 2.20; Ni, 9.24. Found: C, 44.88; H, 5.46; N, 2.14; Ni, 9.13. Green crystals, 3.0 g (94%).

Anal. Calcd. for $C_{96}H_{136}O_{60}N_4Zn_4$: C, 44.90; H, 5.34; N, 2.18; Zn, 10.19. Found: C, 44.51; H, 5.43; N, 2.24; Zn, 10.12. Colorless crystals, 2.7 g (84%).

Properties

All complexes have nearly the same properties. Here the properties of the magnesium complex are represented. The four-nucleus magnesium complex is an air-stable, crystalline solid that decomposes at 180°C. The infrared spectrum of this compound shows absorbances at 3150, 1690, and 1635 cm^{-1} (KBr). The 400-mHz proton NMR spectrum in $[D_6]$-acetone (Me$_4$Si internal standard) has two triplet methyl peaks at 1.18 amd 1.26 ppm, complex multiplets for the methylene protons at 3.92–4.23 ppm, and a broad singlet ammonium peak at 7.52 ppm. The 100-mHz ^{13}C NMR spectrum in $[D_6]$-acetone shows peaks as follows: 14.30 and 14.41 ppm (24 CH$_3$), 60.13 and 60.95 ppm (24 OCH$_2$), 95.62 (12=C), 168.05, 171.49, and 185.52 ppm (24 C=O, 12 C=O).

References

1. R. W. Saalfrank, A. Stark, K. Peters, and H. G. von Schnering, *Angew. Chem. Int. Ed. Engl.*, **27**, 851 (1988).
2. R. W. Saalfrank, A. Stark, M. Bremer, and H. U. Hummel, *Angew. Chem. Int. Ed. Engl.* **29**, 311 (1990).

63. MIXED-METAL-GOLD PHOSPHINE CLUSTER COMPOUNDS

Submitted by A. M. MUETING,* B. D. ALEXANDER,*
P. D. BOYLE,* A. L. CASALNUOVO,* L. N. ITO,* B. J. JOHNSON,*
and L. H. PIGNOLET *
Checked by M. LEEAPHON,†‡ K. E. MEYER,†‡ R. A. WALTON,†‡
D. M. HEINEKEY,§ and T. G. P. HARPER§

The study of mixed-metal–gold cluster compounds is a rapidly expanding area with the potential for significant impact in the field of homogeneous and heterogeneous bimetallic catalysis.[1–5] Various chemical methods are used in the preparation of mixed-metal–gold cluster compounds. Addition and substitution reactions of transition metal complexes with the $[Au(PPh_3)]^+$ moiety are generally the most useful of these synthetic methods. The form in which the $[Au(PPh_3)]^+$ moiety is introduced depends on the character of the metal complex and the type of reaction required. For example, if the metal complex is capable of abstracting a chloride, $Au(PPh_3)Cl$ can be used. If this is not possible, the *in situ* generation of $[Au(PPh_3)]^+$ by the addition of $TlNO_3$ or $Tl[PF_6]$ to $Au(PPh_3)Cl$ may provide better results. The addition of thallium salts can be avoided, however, by the use of $Au(PPh_3)NO_3$, which has a weakly coordinated counterion. Reduction reactions in the presence of $[Au(PPh_3)]^+$, which are analogous to those used for the synthesis of gold cluster compounds, can also be applied to the synthesis of large mixed-metal–gold clusters. Additionally, the reaction of PPh_3 with known clusters often produces new clusters.

Transition metal hydride complexes are important starting materials for reactions with $[Au(PPh_3)]^+$. The foundation for this method is the isolobal relationship between $[Au(PPh_3)]^+$ and H^+.[6] Although many examples have shown this preparative method to be useful, there is always a competition between addition and substitution. For example, $[AuIr(\mu\text{-}H)_2(bpy)(PPh_3)_3]^{2+}$ (bpy = 2,2′-bipyridine),[7] is synthesized from the simple addition of $Au(PPh_3)NO_3$ to $[Ir(H)_2(bpy)(PPh_3)_2]^+$. $[Ir(H)_2(PPh_3)_2(acetone)_2]$-$[BF_4]$, on the other hand, undergoes both addition and substitution. One $[Au(PPh_3)]^+$ is simply added to the iridium complex and one $[Au(PPh_3)]^+$ replaces an H^+, yielding $[Au_2Ir(H)(PPh_3)_4(NO_3)][BF_4]$.[8] Further H^+ substitution of $[Au_2Ir(H)(PPh_3)_4(NO_3)]^+$ occurs in the presence of

* Department of Chemistry, University of Minnestoa, Minneapolis, MN 55455.
† Department of Chemistry, Purdue University, West Lafayette, IN 47907.
‡ Checked syntheses A, H, J, and K.
§ Department of Chemistry, Yale University, New Haven, CT 06520.

MgMeCl and $Au(PPh_3)NO_3$, producing $[Au_3Ir(PPh_5)_5(NO_3)]^{+}$.[8] Similarly, the reaction of $Re(H)_7(PPh_3)_2$ with 6 equiv of $Au(PPh_3)NO_3$ producing $[Au_5Re(H)_4(PPh_3)_7]^{2+}$ involves both addition and substitution.[9]

In the reaction of $Au(PPh_3)NO_3$ with $[Ir_3(\mu_3\text{-}H)(\mu\text{-}H)_3(H)_3(dppe)_3]$ $[BF_4]_2$ [dppe = 1,2-ethanebis(diphenylphosphine)] only substitution occurs. The triply bridging hydrogen ligand is substituted by $Au(NO_3)$ to form $[AuIr_3(\mu\text{-}H)_3(H)_3(dppe)_3(NO_3)][BF_4]$.[10] In this reaction the stable bisphosphine $[Au(PPh_3)_2]^{+}$ is formed as a by-product, effectively removing the phosphine ligand from the coordinated gold atom. The addition of PPh_3 replaces the NO_3^{-} ligand, giving the $Au(PPh_3)$ adduct $[AuIr_3(\mu\text{-}H)_3$ $(H)_3(dppe)_3(PPh_3)]^{2+}$ [10] The reaction of PPh_3 with mixed-metal–gold phosphine clusters also leads to the formation of new clusters. For example, the reaction of $[Au_5Re(H)_4[P(p\text{-}tol)_3]_2(PPh_3)_5]^{2+}$ $[P(p\text{-}tol)_3 = tris(p\text{-}methyl$phenyl)phosphine] with an excess of PPh_3 results in the formation of the cluster $[Au_4Re(H)_4[P(p\text{-}tol)_3]_2(PPh_3)_4]^{+}$, in good yield.[11]

Nonhydride metal phosphine complexes also undergo addition reactions with $Au(PR_3)X$. For example, it has been shown that $Pt(PPh_3)_2(C_2H_4)$ adds 2 equiv of $Au(PPh_3)NO_3$ to produce $[Au_2Pt(PPh_3)_4(NO_3)][NO_3]$.[12] Reduction of the $[Au(PPh_3)]^{+}$ moiety, which has been carried out with the use of reducing agents such as $Na[BH_4]$ for the synthesis of gold clusters,[13,14] has recently been improved by the use of milder reagents such as CO and H_2.[15] This method of reduction results in much cleaner reactions, and has also proved useful for the synthesis of mixed-metal–gold clusters. For example, the gold–platinum cluster $[Au_6Pt(PPh_3)_7]^{2+}$ has been synthesized in high yield by the reaction of $[Au_2Pt(PPh_3)_4(NO_3)][NO_3]$ with H_2 in dichloromethane.[12]

For the following synthetic strategies, all manipulations are carried out under a purified N_2 atmosphere with use of standard Schlenk techniques. ^{31}P NMR spectra are run with proton decoupling and are reported in ppm relative to internal standard trimethylphosphate (TMP), with positive shifts downfield. Compound concentrations used in the conductivity experiments are $3 \times 10^{-4} M$ in CH_3CN. All solvents are dried and distilled prior to use.

A. NITRATO(TRIPHENYLPHOSPHINE)GOLD(I)

$$Au(PPh_3)Cl + AgNO_3 \rightarrow Au(PPh_3)NO_3 + AgCl$$

Procedure

The following is adapted from the procedure described by L. Malatesta[16] and is carried out without the use of an inert atmosphere.* A solution of

---

* The checkers note that the synthesis must be done in the absence of light without unnecessary heating.

Au(PPh$_3$)Cl (2.00 g, 4.04 mmol)[17] in 40 mL of dichloromethane is added dropwise to a 250-mL round-bottomed flask containing a stirred solution of AgNO$_3$ (Aldrich Chemical Co.) (1.37 g, 8.08 mmol) in 125 mL of methanol. A white precipitate of AgCl forms immediately. After this mixture is stirred at room temperature for 1 h, it is transferred to a fritted filter containing a 1-cm layer of diatomaceous earth and washed through with 50 mL of dichloromethane. The filtered solution is evaporated to dryness under reduced pressure on a rotary evaporator. Au(PPh$_3$)NO$_3$ is obtained from this residue on the addition of 75 mL of dichloromethane, filtration of this solution through a fritted filter containing a 1 cm layer of diatomaceous earth, and evaporation to dryness under vacuum. The product is dissolved in a minimum amount of dichloromethane in a 500-mL round-bottomed flask, and approximately three times this volume of absolute ethanol is added. This mixture is stirred under a slow stream of N$_2$ for about 1 h to reduce the volume of dichloromethane. White crystals form in 94% yield, which are collected on filtration, washed successively with absolute ethanol (3 × 5 mL) and diethyl ether (3 × 5 mL), and dried under vacuum.

Properties

The white crystals of Au(PPh$_3$)NO$_3$ are light- and temperature-sensitive; soluble in acetone, tetrahydrofuran, chloroform, and dichloromethane; and insoluble in alcohols and saturated hydrocarbons. Samples may be stored in the dark in a refrigerator for extended periods of time. ^{31}P NMR (dichloromethane, 25°): δ 25.0 (s). IR (KBr): v(NO$_3$) 1495, 1268 cm^{-1}.

B. CARBONYLDI-μ-HYDRIDO-TRIS(TRIPHENYL-PHOSPHINE)(TRIPHENYLPHOSPHINEGOLD) RUTHENIUM(1+) HEXAFLUOROPHOSPHATE(1−)

$$Ru(H)_2(CO)(PPh_3)_3 + Au(PPh_3)NO_3 \rightarrow [AuRu(H)_2(CO)(PPh_3)_4][NO_3]$$

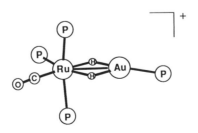

Procedure

A dichloromethane suspension of $Ru(H)_2(CO)(PPh_3)_3$[18] (5 mL, 530 mg, 0.577 mmol) and a dichloromethane solution of $Au(PPh_3)NO_3$ (7 mL, 316 mg, 0.606 mmol) are cooled to $-30°$ in separate 100-mL sidearm flasks. The $Au(PPh_3)NO_3$ solution is transferred to the flask containing the $Ru(H)_2(CO)(PPh_3)_3$ suspension by means of a cannula tube. The mixture is warmed to $0°$ and then stirred for 1 h. During this time the ruthenium complex dissolves, resulting in a very pale yellow solution. An off-white precipitate of $[AuRu(H)_2(CO)(PPh_3)_4][NO_3]$ is obtained on concentration of the solution under vacuum to approximately 4 mL followed by addition of 35 mL of diethyl ether. The solid is collected on filtration, redissolved in a minimum amount of cold dichloromethane ($-10°$), and filtered into a methanol solution (30 mL) containing 710 mg of $NH_4[PF_6]$ (Aldrich Chemical Co.). This solution is colled to $-70°$, at which time a white precipitate forms. The product is collected on a fritted filter and washed successively with 5 mL each of cold methanol ($-10°$), toluene, and diethyl ether. It is then redissolved in a minimum amount of cold dichloromethane ($-10°$) and filtered through a fritted filter containing a 1-cm layer of diatomaceous earth. On cooling this pale yellow solution to $-30°$ and adding 35 mL of diethyl ether, a white solid precipitates in 79% yield, which is then collected, washed with 5 mL of diethyl ether, and dried under vacuum. The solid is dissolved in a minimum amount of dichloromethane in a narrow Schlenk tube and layered with approximately three times this volume of diethyl ether. Slow solvent diffusion of the solvent mixture at ambient temperature produces clear white crystals of $[AuRu(H)_2(CO)(PPh_3)_4][PF_6]$.

Properties

The compound $[AuRu(H)_2(CO)(PPh_3)_4][PF_6]$ is stable in the solid state in air at room temperature but slowly decomposes while in solution. It is soluble in acetone and dichloromethane but insoluble in diethyl ether and hydrocarbons. ^{31}P NMR (acetone-d_6, 25°): d 41.8 (P_{Au}, d, $J = 20.1$ Hz, int $= 1$), 41.0 (P_{Ru}, q, $J = 20.4$ Hz, int $= 1$), 37.2 (P_{Ru}, d, $J = 20.8$ Hz, int $= 2$). ^1H NMR in hydride region (acetone-d_6, 25°): $d - 2.7$ (d of quint, int $= 1$), -5.4 (m, int $= 1$). IR (Nujol): $v(CO)$ 1954 cm^{-1}. Equivalent conductance: 84.3 cm^2 mho mol^{-1}. FABMS (*m*-nitrobenzyl alcohol matrix): m/z 1377 ($[AuRu(H)_2(CO)(PPh_3)_4]^+ = (M)^+$), 1113 ($(M - 2H - PPh_3)^+$).

Anal. Calcd. for $AuRuP_5C_{73}H_{62}OF_6$: C, 57.59; H, 4.13; P, 10.17. Found: C, 56.90; H, 4.38; P, 9.47.

C. (η⁴-1,5-CYCLOOCTADIENE)BIS(TRIPHENYLPHOSPHINE) IRIDIUM(1+) TETRAFLUOROBORATE(1−)

$$\tfrac{1}{2} [Ir(\eta^4\text{-cod})Cl]_2 + Ag[BF_4] \rightarrow [Ir(\eta^4\text{-cod})(PPh_3)_2][BF_4] + AgCl$$

(cod = 1,5-cyclooctadiene)

Procedure

The following is adapted from the synthetic strategy described by Shapley et al.[19] Commercially available $[Ir(\eta^4\text{-cod})Cl]_2$[20] (Strem Chemicals) (930 mg, 1.38 mmol) and $Ag[BF_4]$ (Aldrich Chemical Co.) (540 mg, 2.77 mmol) are placed in a 100-mL sidearm flask to which 30 mL of acetone is added. The mixture is heated to refluxing temperature for approximately 15 min. After it is cool, the orange solution that contains the AgCl suspension is filtered through a fritted filter containing a 1-cm layer of diatomaceous earth and rinsed with small portions of acetone until the orange color is no longer apparent in the diatomaceous earth. A solution of PPh_3 (Aldrich Chemical Co.) (1.45 g, 5.53 mmol) dissolved in a minimum amount of toluene is added to this orange filtrate. The solution color immediately turns red. On removal of the acetone under vacuum, a red precipitate forms. This product is collected on a fritted filter, washed with several small portions of toluene (3 × 5 mL), redissolved in a minimum amount of dichloromethane, and filtered. The dichloromethane solution is layered with approximately three times this volume of diethyl ether in a 100-mL sidearm flask. Red crystals are obtained in approximately 85% yield on slow diffusion of this solvent system.

Properties

The red crystals of $[Ir(\eta^4\text{-cod})(PPh_3)_2][BF_4]$ are stable in air; soluble in dichloromethane, acetone, and alcohols; and insoluble in diethyl ether and hydrocarbons. ^{31}P NMR (dichloromethane, 25°): δ 15.3 (s).

D. BIS(ACETONE)DIHYDRIDOBIS(TRIPHENYLPHOSPHINE) IRIDIUM(1+) TETRAFLUOROBORATE(1−)

$$[Ir(\eta^4\text{-cod})(PPh_3)_2][BF_4] + xsH_2 \rightarrow [Ir(PPh_3)_2(H)_2(acetone)_2][BF_4]$$
$$+ \text{cyclooctane}$$

Procedure

The following is adapted from the procedure described by Crabtree.[21] $[Ir(\eta^4\text{-cod})(PPh_3)_2][BF_4]$ (310 mg, 0.34 mmol) is dissolved in 6 mL of acetone and cooled to 0°. Hydrogen is bubbled through this red solution until the color fades to pale yellow (~ 45 min). A white precipitate forms in 92% yield with the addition of 50 mL of diethyl ether. The solid is collected by filtration, washed with three 10-mL portions of diethyl ether, and dried under vacuum. Colorless crystals of [Ir(PPh$_3$)$_2$(H)$_2$(acetone)$_2$][BF$_4$] are obtained by dissolving the solid in a minimum amount of dichloromethane in a narrow Schlenk tube, layering with three times this volume of diethyl ether, and placing this mixture in a cold bath (4°) for several days.*

Properties

Infrared (Nujol): v(Ir—H) 2230 and 2260 cm^{-1}, v(CO) 1650 and 1660 cm^{-1} (sharp doublet indicative of coordinated acetone). ^{31}P NMR (acetone, 25°): δ 24.9 (s). ^{1}H NMR (acetone-d_6, 25°, hydride region): δ − 27.7 (t, J_{P-H} = 15.8 Hz).

E. η-HYDRIDO(NITRATO-*O,O'*)BIS(TRIPHENYLPHOSPHINE) BIS(TRIPHENYLPHOSPHINEGOLD)IRIDIUM(1+) TETRAFLUOROBORATE(1−)

$$[\text{Ir(PPh}_3)_2(\text{H})_2(\text{acetone})_2]^+ + 2\text{Au(PPh}_3)\text{NO}_3 \xrightarrow{\text{acetone}}$$

$$[\text{Au}_2\text{Ir(H)(PPh}_3)_4(\text{NO}_3)]^+ + \text{HNO}_3$$

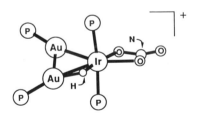

Procedure

The following synthesis is carried out without the use of an inert atmosphere [Ir(PPh$_3$)$_2$(H)$_2$(acetone)$_2$][BF$_4$] (150 mg, 0.163 mmol) is dissolved in 2 mL of acetone in a 100-mL sidearm flask and cooled to − 50°. A solution of Au(PPh$_3$)NO$_3$ (178 mg, 0.342 mmol) in 3 mL of acetone is added to this colorless [Ir(PPh$_3$)$_2$(H)$_2$(acetone)$_2$][BF$_4$] solution by means of a cannula tube. The resultant solution is warmed to − 15° and stirred for about 1 h,

---

* The checkers obtained a recrystallized yield of 71%.

during which time it gradually turns yellow. The solution is concentrated to approximately 1 mL. The rapid addition of 40 mL of cold diethyl ether (− 15°) results in the precipitation of a flaky yellow solid that slowly changes into a fine yellow powder at − 10°. The product is collected, washed with three 5-mL portions of cold diethyl ether (− 15°), and dried under vacuum. It is then dissolved in a minimum amount of cold acetone (− 15°) in a narrow Schlenk tube and layered with approximately three times this volume of diethyl ether. Slow solvent diffusion at − 20° affords large yellow crystals of $[Au_2Ir(H)(PPh_3)_4(NO_3)][BF_4]$ in 85% yield.*

Properties

Solutions of $[Au_2Ir(H)(PPh_3)_4(NO_3)][BF_4]$ decompose slowly over several days even at − 20°. Crystals of $[Au_2Ir(H)(PPh_3)_4(NO_3)][BF_4]$ also slowly decompose over a period of several weeks at 25° and are best stored at temperatures below − 10°. ^{31}P NMR (acetone-d_6, − 90°): δ 7.3 (s, int = 2), 7.9 (s, int = 1), 35.9 (s, int = 1). ^1H (acetone-d_6, − 90°): δ − 25.3 (quint, J_{P-H} = 13 Hz). An elemental analysis is not easily obtained due ot the thermal instability of the compound.

F. (NITRATO-*O,O'*)BIS(TRIPHENYLPHOSPHINE) TRIS(TRIPHENYLPHOSPHINEGOLD)IRIDIUM(1+) TETRAFLUOROBORATE(1−)

$$[Au_2Ir(H)(PPh_3)_4(NO_3)]^+ + Au(PPh_3)NO_3 \xrightarrow{\text{MgMeCl}}$$

$$[Au_2Ir(PPh_3)_5(NO_3)]^+ + HNO_3$$

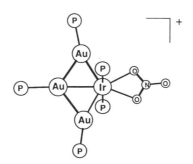

Procedure

A tetrahydrofuran solution of 2.9 M MeMgCl (Aldrich Chemical Co.) (21 μL, 0.061 mmol) is added dropwise to a suspension of crushed

---

*Checkers obtained a crystallized yield of 48%.

$[Au_2Ir(H)(PPh_3)_4(NO_3)][BF_4]$ (100 mg, 0.056 mmol) and $Au(PPh_3)NO_3$ (30 mg, 0.058 mmol) in 2 mL of tetrahydrofuran at 0°. A very deep green solution results immediately. The solvent is removed under vacuum leaving a dark green residue. A solution of $Au(PPh_3)NO_3$ (30 mg, 0.056 mmol) in 3 mL of dichloromethane is added to the green residue. The solution is stirred for 10 min at 25°, during which time the color changes to deep red-brown. The solvent is removed under vacuum, leaving a red-brown residue. The residue is redissolved in approximately 1 mL of acetone and the resultant mixture is stirred for 6 h. During this time a red-brown powder precipitates from the solution. The crude product is filtered, washed with three 5-mL portions of tetrahydrofuran to remove $Au(PPh_3)Cl$, redissolved in a minimum amount of dichloromethane, and filtered again. The addition of a large excess of diethyl ether to the red-brown filtrate affords $[Au_3Ir(PPh_3)_5(NO_3)][BF_4]$ as a red-brown powder in 60% yield.

Properties

$[Au_3Ir(PPh_3)_5(NO_3)][BF_4]$ is insoluble in acetone and tetrahydrofuran, but dissolves readily in dichloromethane to give air-stable solutions. ^{31}P NMR (dichloromethane, 25°): δ 39.2 (t, $J = 9.5$ Hz, int = 1), 10.9 (m, int = 2), -7.25 (t, $J = 7.5$ Hz, int = 2).

Anal. Calcd. for $IrAu_3P_5C_{90}H_{75}BF_4NO_3$: C, 48.18: H, 3.37. Found: C, 47.36; H, 3.41.

G. TRI-µ-HYDRIDO-TRIS(TRIPHENYLPHOSPHINE)BIS(TRI-PHENYLPHOSPHINEGOLD)OSMIUM(1+) HEXAFLUOROPHOSPHATE(1−) AND ITS RUTHENIUM ANALOG

$$Os(H)_4(PPh_3)_3 + 2Au(PPh_3)NO_3 \rightarrow [Au_2Os(H)_3(PPh_3)_5]^+ + HNO_3$$

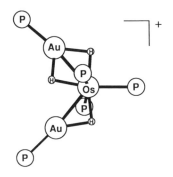

Procedure

A dichloromethane suspension of $Os(H)_4(PPh_3)_3{}^{22}$(10 mL, 623 mg, 0.635 mmol) and a dichloromethane solution of $Au(PPh_3)NO_3$ (10 mL, 662 mg, 1.27 mmol) are cooled to 0° in separate 100-mL sidearm flasks. The $Au(PPh_3)NO_3$ solution is transferred to the flask containing the $Os(H)_4(PPh_3)_3$ suspension by means of a cannula tube. The mixture is allowed to warm to room temperature and then stirred for 1 h, during which time the osmium complex dissolves and a black solution results. Concentration of the solution to 5 mL and precipitation by addition of approximately 40 mL of diethyl ether results in a dark gray product. The solid is collected by filtration, redissolved in a minimum amount of dichloromethane, and filtered into a methanol solution (30 mL) containing 660 mg of $[NH_4][PF_6]$ (Aldrich Chemical Co.). The light gray precipitate that forms is collected, washed successively with 5 mL of methanol and 10 mL of diethyl ether, and redissolved in 5 mL of dichloromethane. Filtration of this solution through a fritted filter containing a 1-cm layer of diatomaceous earth results in a clear, pale yellow solution. A white solid precipitates in 60% yield on the addition of approximately 40 mL of diethyl ether. This solid is collected by filtration, washed with diethyl ether (3 × 5 mL), dried under vacuum, and redissolved in a minimum volume of dichloromethane. Layering of this solution with three times the volume of diethyl ether in a narrow Schlenk tube and slow diffusion at ambient temperature produces clear white crystals.

Properties

Crystals of $[Au_2Os(H)_3(PPh_3)_5][PF_6]$ are stable in air at room temperature; soluble in dichloromethane and acetonitrile; and insoluble in diethyl ether, methanol, and hydrocarbons. Solution of $[Au_2Os(H)_3(PPh_3)_5][PF_6]$ are stable for a few days under a nitrogen atmosphere. ^{31}P NMR (dichloromethane-d_2, − 70°): δ 48.6 (P_{Au}, br s, int = 2), 21.8 (P_{Os}, br s, int = 1), 5.7 (P_{Os} s, int = 2). CP MAS ^{31}P{^{1}H} NMR (25°): δ 56.8 and 48.8 (P_{Au}), 19.6 (P_{Os}, br s), 5.8 (P_{Os}, br s). ^{1}H NMR (dichloromethane-d_2, − 70°, hydride region): δ − 4.1 (m, J = 27.3, 23.4, and 15.4 Hz, int = 1), − 5.5 (d of d of t, J = 57.6, 15.9, and 7.8 Hz, int = 2). Equivalent conductance: 87.1 cm^2 mho mol^{-1}. FABMS (*m*-nitrobenzyl alcohol matrix): m/z 1899 $[[(Au_2Os(H)_3(PPh_3)_5]^+ = (M)^+]$, 1633 $[(M - 3H - PPh_3)^+]$, 1557 $[(M - 2H - PPh_3 - Ph)^+]$.

Anal. Calcd. for $Au_2OsP_6C_{90}H_{78}F_6$: C, 52.90; H, 3.85; P, 9.09. Found: C; 53.31; H, 3.99; P. 8.89.

Analogous Compound

The ruthenium analog of $[Au_2Os(H)_3(PPh_3)_5][PF_6]$ is prepared in a similar manner with one notable exception: the reaction requires the use of a hydrogen atmosphere. A toluene suspension of $Ru(H)_4(PPh_3)_3$[23] (15 mL, 349 mg, 0.392 mmol) is combined with a toluene solution of $Au(PPh_3)NO_3$ (15 mL, 450 mg, 0.863 mmol) under a hydrogen atmosphere at ambient temperature. An immediate reaction takes places; it leads to complete dissolution of the ruthenium complex followed by precipitation of pale yellow product. The suspension is placed under a purified N_2 atmosphere and allowed to stir for 1 h. The solid is collected, washed with three 5-mL portions of toluene, redissolved in a minimum amount of dichloromethane, and filtered into a methanol solution containing 385 mg of $[NH_4][PF_6]$. An off-white precipitate forms in 40% yield; this is collected and washed with 5 mL of methanol followed by 10 mL of diethyl ether. The solubility and stability of $[Au_2Ru(H)_3(PPh_3)_5][PF_6]$ are similar to those of the osmium analog. ^{31}P NMR (dichloromethane-d_2, 25°): δ 56.3 (P_{Ru}, m, int = 1), 46.7 (P_{Ru}, d, $J = 24.0$ Hz, int = 2), 43.3 (P_{Au}, d, int = 2), ^1H NMR (dichloromethane-d_2, $-70°$, hydride region); δ -3.5 (br m, int = 1), -4.2 (br d of m, $J = 67.2$ Hz, int = 2). Equivalent conductance: 84.6 cm^2 mho mol^{-1}. FABMS (*m*-nitrobenzyl alcohol matrix): m/z 1809 $[Au_2Ru(H)_3(PPh_3)_5]^+ = (M)^+]$, 1545 $((M - 2H - PPh_3)^+)$.

Anal. Calcd. for $Au_2RuP_6C_{90}H_{78}F_6$: C, 55.31; H, 4.02; P, 9.52. Found: C, 55.85; H, 4.29; P, 8.76.

H. TETRAHYDRIDOBIS(TRIPHENYLPHOSPHINE)PENTA-KIS(TRIPHENYLPHOSPHINEGOLD)RHENIUM(2+) BIS[HEXAFLUOROPHOSPHATE(1−)]

$$Re(H)_7(PPh_3)_2 + 5Au(PPh_3)NO_3 \rightarrow [Au_5Re(H)_4(PPh_3)_7]^{2+} + 3HNO_3$$

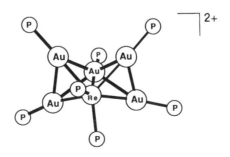

Procedure

A dichloromethane solution of $Re(H)_7(PPh_3)_2$[24] (7 mL, 103 mg, 0.143 mmol) and a dichloromethane solution of $Au(PPh_3)NO_3$ (6 mL, 447 mg, 0.857 mmol) are cooled to $-60°$ in separate 100-mL sidearm flasks. The $Au(PPh_3)NO_3$ solution is transferred to the flask containing the $Re(H)_7(PPh_3)_2$ solution by means of a cannula tube, and the reaction mixture immediately turns yellow. The solution is further cooled to $-75°$ and allowed to stir for 15 min. Cold ethanol (10 mL, $-75°$) is added to the solution, and the mixture is allowed to warm to room temperature. The dichloromethane is removed under vacuum, and an ethanol solution of $[NH_4][PF_6]$ (Aldrich Chemical Co.) (8 mL, 140 mg, 0.859 mmol) is added to the reaction mixture, causing the formation of a yellow precipitate. The mixture is stirred for 18 h, filtered through a fritted filter containing a 1-cm layer of diatomaceous earth, and washed with ethanol (3 × 5 mL). The yellow solid, which is on the frit with the diatomaceous earth, is then redissolved in a minimum amount of dichloromethane ($\sim$ 15 mL), passed through the frit, and reprecipitated with an excess of diethyl ether ($\sim$ 15 mL). The solid is collected by filtration in 94% yield and washed with diethyl ether (3 × 5 mL). Recrystallization by means of slow solvent diffusion from a dichloromethane–diethyl ether solvent mixture yields yellow crystals of $[Au_5Re(H)_4(PPh_3)_7][PF_6]_2$.

Properties

An alternative method may be used for the metathesis of the NO_3^- salt of $[Au_5Re(H)_4(PPh_3)_7]^{2+}$ to the PF_6^- salt that involves the use of 8 mL of methanol, 158 mg of $K[PF_6]$, and a reaction time of only 3 h. $[Au_5Re(H)_4(PPh_3)_7][PF_6]_2$ is stable in air in both the solid state and solution, and is soluble in dichloromethane, acetone, and warm tetrahydrofuran. ^{31}P NMR (dichloromethane, 25°): δ 49.8 (P_{Au}), 18.3 (P_{Re}, sextet, $J_{P-P} = 8.0$ Hz). ^{1}H NMR (dichloromethane-d_2, 25°): δ -2.07 (8 line mult, $J_{H-P_{Au}} = 9.1$ Hz, $J_{H-P_{Re}} = 18.6$ Hz).

Analogous Compounds

$[Au_5Re(H)_4[P(p\text{-}tol)_3]_2(PPh_3)_5][PF_6]_2$ can be prepared using a procedure analogous to that for $[Au_5Re(H)_4(PPh_3)_7][PF_6]_2$ as described above, starting with $Re(H)_7[P(p\text{-}tol)_3]_2$. ^{31}P NMR (dichloromethane, 25°): δ 49.7, 14.3. ^{1}H NMR (dichloromethane-d_2, 25°): δ 1.98 (CH_3), -2.08 (8 line mult, $J_{P-H} = 9.4$ Hz). Equivalent conductance: 186 cm² mho mol^{-1}. UV–VIS (λ(nm), log ε): 350 (sh), 3.90. FABMS (*m*-nitrobenzyl alcohol matrix): m/z 3096

$[[Au_5Re(H)_4[P(p\text{-tol})_3]_2(PPh_3)_5]^+ = (M)^+]$, 3240 $((M + PF_6 - H)^+)$, 3162 $((M + PF_6 - H - Ph)^+)$, 3018 $((M - H - Ph)^+)$, 2636 $((M - Au - PPh_3)^+)$, 2372 $((M - Au - H - 2PPh_3)^+)$.

Anal. Calcd. for $Au_5ReP_9F_{12}C_{132}H_{121}$: C, 46.8; H, 3.6. Found: C, 46.61; H, 3.38.

I. TRIS[1,2-ETHANEBIS(DIPHENYLPHOSPHINE)] HEXAHYDRIDO(NITRATO-*O,O'*-GOLD) TRIIRIDIUM(1+) TETRAFLUOROBORATE(1−)

$$[Ir_3(\mu_3\text{-H})(\mu\text{-H})_3(H)_3(dppe)_3]^{2+} + 2Au\,(PPh_3)NO_3 \rightarrow$$

$$[AuIr_3(\mu\text{-H})_3H_3(dppe)_3(NO_3)\,]^+ + [Au(PPh_3)_2\,]^+$$

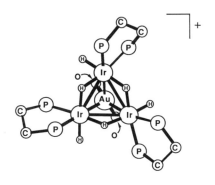

Procedure

A solution of $Au(PPh_3)NO_3$ (31 mg, 0.06 mmol) in 6 mL of acetone is added by means of a cannula tube to a yellow solution of $[Ir_3(\mu_3\text{-H})(\mu\text{-H})_3(H)_3(dppe)_3][BF_4]_2$[25] (60 mg, 0.03 mmol) dissolved in approximately 5 mL of dichloromethane in a 100-mL sidearm flask. This solution is stirred for 24 h at 25° while in the dark, during which time the solution color becomes orange-yellow. The solvent mixture is removed under vacuum, leaving an orange-yellow residue. Approximately 1 mL of acetone is added to this residue. A combination of orange-yellow crystals and orange-yellow powder precipitates from this solution on standing at −20° for 1–2 days. Recrystallization of this solid mixture by means of slow solvent diffusion from a chloroform–diethyl ether solvent system affords transparent, orange-yellow crystals of $[AuIr_3(\mu\text{-H})_3H_3(dppe)_3(NO_3)][BF_4]$ in 40–45% yield.

Properties

The transparent, orange-yellow crystals of $[AuIr_3(\mu\text{-}H)_3H_3(dppe)_3(NO_3)]$ $[BF_4]$ are temperature- and light-sensitive, soluble in chloroform and dichloromethane, and slowly become opaque when removed from the solvent mixture. ^1H NMR (chloroform-d_1, 25°): δ -8.86 (d, $J_{P-H} = 72$ Hz, int = 1), -15.04 (m, int = 1). ^{31}P NMR (dichloromethane, 25°): δ 41.2 (br s, int = 1), 29.0 (m, int = 1). IR (KBr): $v(Ir\text{—}H)$ 2158 cm^{-1} (br). FABMS (*m*-nitrobenzyl alcohol matrix): m/z 2036 $[[AuIr_3(dppe)_3(H)_6(NO_3)]^+ = (M)^+]$, 1974 $[(M - NO_3)^+]$, 1814 $[(M - NO_3 - 2Ph - 6H)^+]$, 1737 $[(M - NO_3 - 3Ph - 6H)^+]$, 1381 $[(M - NO_3 - Au - dppe + 2H)^+]$.

Analogous Compound

$[AuIr_3(\mu\text{-}H)_3(H)_3(dppe)_3(PPh_3)][BF_4, NO_3]$ can be generated *in situ* by the addition of 1 equiv of PPh_3 to $[AuIr_3(\mu\text{-}H)_3H_3(dppe)_3(NO_3)][BF_4]$ in chloroform-*d*. The originally orange-yellow solution immediately turns orange. The solvent is removed under vacuum, yielding an orange residue. ^1H NMR (chloroform-*d*): δ -6.79 (d, $J_{P-H} = 74$ Hz, int = 1), -13.4 (m, int = 1). ^{31}P NMR (chloroform): δ 78.7 (P_{Au}, q, $J_{P-P} = 38$ Hz, int = 1), 40.0 (P_{Ir}, s, int = 3), 38.5 (P_{Ir}, d, $J_{P-P} = 38$ Hz, int = 3). IR (KBr): $v(Ir\text{—}H)$ 2155 cm^{-1} (br).

J. HEXAHYDRIDOTETRAKIS(TRIPHENYLPHOS-PHINE)BIS(TRIPHENYLPHOSPHINEGOLD)DIRHENIUM(1 +) TETRAPHENYLBORATE(1 −)

$$Re_2(\mu\text{-}H)_4(H)_4(PPh_3)_4 + 2Au(PPh_3)NO_3 \rightarrow [Au_2Re_2(H)_6(PPh_3)_6]^+ + ?$$

Procedure

Cold dichloromethane (7 mL, $-20°$) is added to a 100-mL sidearm flask containing $Re_2(\mu\text{-}H)_4(H)_4(PPh_3)_4$[24] (150 mg, 0.015 mmol) and $Au(PPh_3)NO_3$ (110 mg, 0.211 mmol). This red suspension is stirred and allowed to warm slowly to room temperature. During this time the solid materials dissolve, producing a red-brown solution and then a dark green solution. After several minutes of stirring at room temperature, the dichloromethane solvent is removed under vacuum, and the residue is redissolved in a minimum amount of methanol. A precipitate forms on the addition of 3 equiv of either $Na[BPh_4]$ (Aldrich Chemical Co.) (108 mg, 0.136 mmol) or $K[PF_6]$ (Alfa Products) (58 mg, 0.315 mmol) dissolved in a minimum amount of methanol (~ 3 mL). The precipitate is collected on a fritted filter containing a 1-cm layer of diatomaceous earth, washed with a small amount of methanol (5–10 mL), redissolved in a minimum volume of dichloromethane (~ 4 mL) while still on the fritted filter, and filtered. Formation of a dark green precipitate of $[Au_2Re_2(H)_6(PPh_3)_6][X]$ (X = BPh_4 or PF_6) in 62% yield occurs on the addition of a large excess of diethyl ether. Reprecipitation can be carried out with the use of dichloromethane/diethyl ether.

Properties

$[Au_2Re_2(H)_6(PPh_3)_6]X$ is soluble in dichloromethane and acetone and slightly soluble in tetrahydrofuran, is stable in air in the solid state, and decomposes in solution on standing for several days. The following characterization data are for the PF_6^- salt of $[Au_2Re_2(H)_6(PPh_3)_6]X$. Equivalent conductance: 79.1 cm^2 mho mol^{-1}. UV–VIS (λ (nm), log ε) 612, 3.70; 458, 3.57; 335, 4.10; 270, 4.58. Magnetic data: gram susceptibility $\chi_g = -1.26 \times 10^{-8}$ cgs g^{-1}, corrected molar susceptibility $\chi_M = 1.30 \times 10^{-3}$ cgs mol^{-1}, magnetic moment $\mu = 1.77$ BM. FABMS (*m*-nitrobenzyl alcohol matrix): *m/z* 2346 $[[Au_2Re_2(H)_6(PPh_3)_6]^+ = M^+]$.

Anal. Calcd. for $Au_2Re_2C_{132}H_{116}BF_6$ (BPh_4^- salt): C, 59.48; H, 4.39 P, 6.97. Found: C, 59.19; H, 4.60 P, 6.63.

K. TETRAHYDRIDOBIS(BIS[TRIS(4-METHYLPHENYL)] PHOSPHINE)TETRAKIS(TRIPHENYLPHOSPHINEGOLD) RHENIUM(1+) HEXAFLUOROPHOSPHATE(1−)

$$[Au_5Re(H)_4[P(p\text{-tolyl})_3]_2(PPh_3)_5]^{2+} + PPh_3 \rightarrow$$
$$[Au_4Re(H)_4[P(p\text{-tolyl})_3]_2(PPh_3)_4]^+ + [Au(PPh_3)_2]^+$$

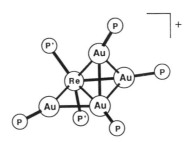

Procedure

Dichloromethane (7 mL) is added to a 100-mL sidearm flask containing $[Au_5Re(H)_4[P(p\text{-tolyl})_3]_2(PPh_3)_5][PF_6]_2$ (218 mg, 0.0644 mmol) and PPh_3

(51.6 mg, 0.197 mmol). The yellow solution is stirred for 12 h. The solvent is then removed under vacuum. Methanol (5 mL) is added to this residue, producing a suspension that is stirred and filtered through a fritted filter containing a 1-cm layer of diatomaceous earth. Methanol is passed through this filter until the yellow color in the filtrate is no longer apparent (~ 20 mL). A yellow solid precipitates on reduction of the volume of the methanol solution to approximately 1 mL. Recrystallization by means of slow solvent diffusion from a dichloromethane–pentane solvent system produces yellow crystals in 56% yield.

Properties

$[Au_4Re(H)_4[P(p\text{-tolyl})_3]_2(PPh_3)_4][PF_6]$ is stable in air both in the solid state and solution and is soluble in dichloromethane and acetone. ^{31}P NMR (dichloromethane, 25°): δ 53.2 (P_{Au}, t, $J = 6.7$ Hz), 23.7 (P_{Re}, quin, $J = 6.7$ Hz). ^1H NMR (dichloromethane-d_2, 25°): $\delta - 3.43$ (9 line mult; triplet with P_{Au} decoupled, $J_{H-P_{Re}} = 20.4$ Hz; quintet with P_{Re} decoupled, $J_{H-P_{Au}} = 8.5$ Hz). Equivalent conductance: 60 cm^2 mho mol^{-1}. Unit cell dimensions determined by X-ray crystallography at 23° with use of MoK$_\alpha(\lambda = 0.71069$ Å) radiation: $a = 17.06$ (1) Å, $b = 28.09$ (1) Å, $c = 28.04$ (1) Å; $\alpha = 88.60$ (3), $\beta = 107.74$ (4), $\gamma = 89.99$ (5)°, V = 12,795 Å^3. FABMS (*m*-nitrobenzyl alcohol matrix): m/z 2636 ($\{Au_4Re(H)_4[P(p\text{-tol})_3]_2(PPh_3)_4\}^+$ = M$^+$), 2781 [(M + PF$_6$)$^+$], 2372 [(M − H − PPh$_3$)$^+$], 2107 [(M − 4H − 2PPh$_3$)$^+$], 1845 [(M − 4H − 3PPh$_3$)$^+$], 1803 [(M − 4H − P(p-tol)$_3$)$^+$], 1729 [(M + PF$_6$ − 3H − 4PPh$_3$)$^+$], 1583 [(M − 4H − 4PPh$_3$)$^+$], 1541 [(M − 4H − 3PPh$_3$ − P(p-tol)$_3$)$^+$].

L. (NITRATO-*O,O'*)BIS(TRIPHENYLPHOSPHINE) BIS(TRIPHENYLPHOSPHINEGOLD) PLATINUM(1+) HEXAFLUOROPHOSPHATE(1−)

$$Pt(PPh_3)_2(C_2H_4) + 2Au(PPh_3)NO_3 \rightarrow [Au_2Pt(PPh_3)_4(NO_3)][NO_3] + C_2H_4$$

Procedure

$Pt(PPh_3)_2(C_2H_4)^{26}$ (185 mg, 0.248 mmol) and $Au(PPh_3)NO_3$ (270 mg, 0.518 mmol) are combined in a 100-mL sidearm flask equipped with a magnetic stirring bar, to which approximately 15 mL of tetrahydrofuran is added. Almost immediately a cream-colored precipitate forms. The slurry is then stirred for 1 h, and transferred by means of a cannula tube to a fritted filter for collection. The precipitate is washed successively with 20 mL of tetrahydrofuran and 20 mL of diethyl ether, dried under vacuum, and dissolved in a minimum amount of dichloromethane. Excess NH_4PF_6 (Aldrich Chemical Co.) (404 mg, 2.48 mmol) dissolved in a minimum amount of methanol is added to this dichloromethane solution of the NO_3^- salt and stirred for 1 h. The solution volume is reduced under vacuum to approximately 5 mL, which results in the precipitation of 0.40 g of $[Au_2Pt(PPh_3)_4$ $(NO_3)][PF_6]$ (92% yield)*.

Properties

$[Au_2Pt(PPh_3)_4(NO_3)][PF_6]$ is air-, light-, and moisture-stable; soluble in acetone and dichloromethane; and insoluble in tetrahydrofuran, diethyl ether, and saturated hydrocarbons. The analytical data reported below are for the $[PF_6]^-$ salt of $[Au_2Pt(PPh_3)_4(NO_3)]^+$. IR (KBr pellet): v (bound NO_3) 1490, 1480, 1270 cm^{-1} (s), v (P—F) 840 cm^{-1} (s). ^{31}P NMR (dichloromethane, 22°): δ 28.8 (s with ^{195}Pt satellites, $^2J_{195Pt-P} = 804$ Hz, int = 1), 22.6 (s with ^{195}Pt satellites, $J_{195Pt-P} = 2502$ Hz, int = 1). ^{31}P NMR (CH$_3$CN, 22°): δ 34.0 (s with ^{195}Pt satellites, $^2J_{195Pt-P} = 718$ Hz, int = 1), 21.6 (s with ^{195}Pt satellites, $J_{195Pt-P} = 2390$ Hz, int = 1). Equivalent conductance: 184 cm^2 mho mol^{-1}.

Anal. Calcd. for $PtAu_2C_{72}H_{60}F_6NO_3P_5$: C, 46.87; H, 3.28: P, 8.39. Found: C, 46.52 H, 3.00; P, 8.08. Unit cell dimensions determination by X-ray crystallography at 23° with use of MoK$_\alpha$($\lambda = 0.71069$ Å) radiation (crystal grown from dichloromethane/diethyl ether by slow solvent diffusion and mounted on glass rod with epoxy): triclinic, $a = 12.295$ (9) Å, $b = 22.54$ (3) Å, $c = 13.303$ (8) Å, $\alpha = 89.86$ (8), $\beta = 113.21$ (6), $\gamma = 93.03$ (9)°, V = 3389 Å^3. FABMS (*m*-nitrobenzyl alcohol matrix): m/z 1700 $[[Au_2Pt(PPh_3)_4 (NO_3)]^+$ = (M)$^+$], 1638 $[(M - NO_3)^+]$, 1561 $[(M - NO_3 - Ph)^+]$, 1437 $[(M - PPh_3)^+]$, 1298 $[(M - NO_3 - PPh_3 - Ph)^+]$, 1221 $[(M - NO_3 - PPh_3 - 2Ph)^+]$.

* The checkers used a modified procedure that involved using Pt(PPh$_3$)$_2$(C$_2$H$_4$) *in situ* without further purification. To a suspension of 400 mg of Pt(PPh$_3$)$_2$Cl$_2$ in THF (saturated with ethylene and under an ethylene atmosphere) (50 mL) was added a sufficient quantity of 0.135 *M* sodium napthalide to produce a homogeneous yellow solution. To this THF solution was added 551 mg of Au(PPh$_3$)NO$_3$. The recommended procedure suggests that a solid will be produced at this point. Filtration of this solution results in the harvest of 72 mg of gray solid, which when metathesized with 200 mg [NH$_4$][PF$_6$] produces 167 mg (18% based on starting Pt) of a pale yellow solid. If the THF is removed from the golden suspension (rather than filtering the solid) and the subsequent red-orange oil is taken up in CH$_2$Cl$_2$ (4 mL) and precipitated with the addition of diethyl ether (50 mL), a beige solid is obtained (532 mg, 60% based on Pt). This material can be metathesized with [NH$_4$][PF$_6$] according to the recommended procedure in good yield (95%) or used in the subsequent reaction.

M. (TRIPHENYLPHOSPHINE)HEXAKIS-(TRIPHENYLPHOSPHINEGOLD)PLATINUM(2+) BIS[TETRAPHENYLBORATE(1−)]

$$[Au_2Pt(PPh_3)_4(NO_3)]^+ \xrightarrow{\ H_2\ } [Au_6Pt(PPh_3)_7]^{2+} + ?$$

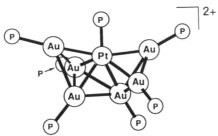

Procedure

$[Au_2Pt(PPh_3)_4(NO_3)][NO_3]$ (100 mg, 5.67×10^{-2} mmol) is dissolved in 10 mL of dichloromethane in a 100-mL sidearm flask equipped with a magnetic stirring bar and placed under a hydrogen atmosphere. The solution is stirred for about 1 h, during which time the color changes from yellow to dark orange-brown. Diethyl ether (60 mL) is added to precipitate a brown material. This product is collected on a fritted filter, washed with 40 mL of diethyl ether, and dried under vacuum. The solid is then redissolved in a minimum amount of dichloromethane, and a solution of $NaBPh_4$ (Aldrich Chemical Co.) (45 mg, 0.13 mmol) in 10 mL of methanol is added. The volume of solution is reduced under vacuum until a brown solid begins to deposit. At this point the solution is allowed to stir overnight. The solid is collected on a fritted filter in 62% yield (based on Au), washed successively with 10 mL of methanol and 20 mL of diethyl ether, and dried under vacuum. Dark brown crystals of the unmetathesized complex $[Au_6Pt(PPh_3)_7][NO_3]_2$ can be obtained by dissolving 50 mg of the NO_3^- salt in approximately 2 mL of methanol, adding small portions of diethyl ether, with mixing after each addition, until the solution becomes slightly cloudy (~ 15 mL of diethyl ether is required), and allowing the solution to stand overnight at room temperature.

Properties

$[Au_6Pt(PPh_3)_7][NO_3]_2$ is soluble in alcohols, dichloromethane, chloroform, and acetone; insoluble in saturated hydrocarbons and diethyl ether; and air-, light-, and moisture-stable in both the solid state and solution. The $[PF_6^-]$ salt is less soluble in alcohols. ^{31}P NMR (dichloromethane, 22°): δ 50.2 (d with ^{195}Pt satellites, $J_{P-P} = 30$ Hz, $^2J_{195Pt-P} = 413$ Hz), 62.3 (m, $J_{P-P} = 30$ Hz). ^{1}H NMR (acetone-d_6): δ 6.5–8.0 (m, aromatic H), no peaks in 0 to -30 ppm region. Equivalent conductance: 125 cm^2 mho mol^{-1}. FABMS (*m*-nitrobenzyl alcohol matrix): m/z 3212 $[[Au_6Pt(PPh_3)_7]^+ = (M)^+]$, 3532 $[(M + BPh_4)^+]$, 2950 $[(M - PPh_3)^+]$, 2688 $[(M - 2PPh_3)^+]$, 2491 $[(M - Au - 2PPh_3)^+]$, 2194 $[(M - 3PPh_3 - 3Ph)^+]$.

Analogous Compound

$[AuIr(dppe)_2(PPh_3)][BF_4]_2$ is also made from a nonhydride metal phosphine complex; however, $H[BF_4]$ is required, which suggests that a hydride intermediate is involved. $Ir(dppe)_2[BF_4]$ (100 mg, 0.093 mmol) and $Au(PPh_3)NO_3$ (53 mg, 0.100 mmol) are dissolved in 4 mL of acetone. $H[BF_4]$ solution (48–50%, 3 μL) is added to this mixture and stirred. The initially red-orange solution immediately turns to a light orange. The rapid addition of diethyl ether results in the precipitation of an orange powder. Recrystallization of the orange powder from dichloromethane–diethyl ether solution affords orange needles of $[AuIr(dppe)_2(PPh_3)][BF_4]_2$ in 55% yield. ^{31}P NMR (dichloromethane, 25°): δ 41.4 (d, $J_{P-P} = 6$ Hz).

Anal. Calcd. for $IrAuP_5C_{70}H_{63}B_2F_8 \cdot 0.4\ CH_2Cl_2$: C, 51.06; H, 3.89. Found: C, 50.49; H, 4.02.

N. DI-μ-HYDRIDO-BIS(TRIPHENYLPHOSPHINE) TETRAKIS(TRIPHENYLPHOSPHINEGOLD) IRIDIUM(1+)TETRAPHENYLBORATE (1−)

$$[Au_3Ir(PPh_3)_5(NO_3)]^+ \xrightarrow{\ H_2\ } [Au_4Ir(H)_2(PPh_3)_6]^+ + ?$$

Procedure

A red-brown solution of 75 mg of $[Au_3Ir(PPh_3)_5(NO_3)][BF_4]$ dissolved in 3 mL of dichloromethane is stirred vigorously under a hydrogen atmosphere at 23° for 45 min. Within 4 min the initial red-brown color of the solution fades to a golden yellow and then gradually changes to red-orange over the next 15 min. A yellow-orange powder precipitates from this stirring solution on the addition of 25 mL of diethyl ether. This powder is collected, re-dissolved in a minimum amount of tetrahydrofuran, and filtered through a fritted filter containing a 1-cm layer of diatomaceous earth. The tetrahydro-furan is removed under vacuum and the remaining yellow-orange residue is redissolved in a minimum amount of dichloromethane in a narrow Schlenk tube and layered with approximately three times this volume of diethyl ether. Recrystallization by diffusion from this solvent system precipitates a combi-nation of very thin golden fibers and large red-orange plates. Both materials give identical ^{1}H and ^{31}P NMR spectra. Yields typically range from 55–60% based on Ir.

Properties

Crystals of $[Au_4Ir(H)_2(PPh_3)_6][BF_4]$ are soluble in dichloromethane, acet-one, and tetrahydrofuran, and lose solvent and fracture on removal from the solvent mixture. ^{31}P NMR (acetone, 23°): δ 46.5 (t, $J_{P-P} = 33$ Hz, int = 2), ~ 33.5 (br s, int = 1). ^{1}H NMR (acetone-d_6, 23°): δ -5.71 (m). IR (KBr): ν(B—F) 1050 cm^{-1} (s).

Anal. Calcd. for $IrAu_4C_{108}H_{92}P_6BF_4$: C, 49.08; H, 3.51; F, 2.88. Found: C, 48.58; H, 3.44: F, 3.08.

References

1. P. G. Jones, *Gold Bull.*, **19**, 46 (1986) and references cited therein.
2. P. G. Jones, *Gold Bull.*, **16**, 114 (1983) and references cited therein.
3. P. Braunstein, J. Rosé, *Gold Bull.*, **18**, 17 (1985).
4. K. P. Hall and D. M. P. Mingos, *Prog. Inorg. Chem.*, **32**, 237 (1984) and references cited therein.
5. A. M. Mueting, W. Bos, B. D. Alexander, P. D. Boyle, J. A. Casalnuovo, S. Balaban, L. N. Ito, S. M. Johnson, and L. H. Pignolet, "Recent Advances in Di- and Polynuclear Chemistry," *New J. Chem.*, **12**, 505 (1988).
6. J. W. Lauher and K. Wald, *J. Am. Chem. Soc.*, **103**, 7648 (1981).
7. B. D. Alexander, B. J. Johnson, S. M. Johnson, A. L. Casalnuovo, and L. H. Pignolet, *J. Am. Chem. Soc.*, **108**, 4409 (1986).
8. A. L. Casalnuovo, T. Laska, P. V. Nilsson, J. Olofson, L. H. Pignolet, W. Bos, J. J. Bour, and J. J. Steggerda, *Inorg. Chem.*, **24**, 182 (1985).

9. P. D. Boyle, B. J. Johnson, A. Buehler, and L. H. Pignolet, *Inorg. Chem.*, **25**, 5 (1986).
10. A. L. Casalnuovo, L. H. Pignolet, J. W. A. van der Velden, J. J. Bour, and J. J. Steggerda, *J. Am. Chem. Soc.*, **105**, 5957 (1983).
11. B. D. Alexander, P. D. Boyle, B. J. Johnson, S. M. Johnson, J. A. Casalnuovo, A. M. Mueting, and L. H. Pignolet, *Inorg. Chem.*, **26**, 2547 (1987).
12. (a) P. D. Boyle, B. J. Johnson, B. D. Alexander, J. A. Casalnuovo, P. R. Gannon, S. M. Johnson, E. A. Larka, A. M. Mueting, and L. H. Pignolet, *Inorg. Chem.*, **26**, 1346 (1987); (b) L. N. Ito, J. D. Sweet, A. M. Mueting, L. H. Pignolet, M. F. J. Schoondergang, and J. J. Steggerda, *Inorg. Chem.*, **28**, 3696 (1989).
13. J. J. Steggerda, J. J. Bour, and J. W. A. van der Velden, *Rec. trav. chim. Pays-Bas*, **101**, 164 (1982).
14. B. D. Alexander, B. J. Johnson, S. M. Johnson, P. D. Boyle, N. C. Kann, A. M. Mueting, and L. H. Pignolet, *Inorg. Chem.*, **26**, 3506 (1987).
15. W. Bos, J. J. Bour, J. J. Steggerda, and L. H. Pignolet, *Inorg. Chem.*, **24**, 4298 (1985).
16. L. Malatesta, L. Naldini, G. Simonetta, and F. Cariati, *Coord Chem. Rev.*, **1**, 255 (1966).
17. P. Braunstein, H. Lehner, and D. Matt, *Inorg. Synth.*, **27**, 218 (1990).
18. N. Ahmad, J. J. Levison, S. D. Robinson, and M. F. Uttley, *Inorg. Synth.*, **15**, 48 (1974).
19. J. R. Shapley, R. R. Schrock, and J. A. Osborn, *J. Am. Chem. Soc.*, **91**, 2816 (1969).
20. J. L. Herde, J. C. Lambert, and C. V. Senoff, *Inorg. Synth.*, **15**, 18 (1974).
21. R. H. Crabtree, G. G. Hlatky, C. P. Parnell, B. E. Segmuller, and R. J. Uriarte, *Inorg. Chem.*, **23**, 354 (1984).
22. N. Ahmad, J. J. Levison, S. D. Robinson, and M. F. Uttley, *Inorg. Synth.*, **15**, 56 (1974).
23. R. O. Harris, N. K. Hota, L. Sadavoy, and J. M. C. Yuen, *J. Organomet. Chem.*, **54**, 259 (1973).
24. C. J, Cameron, G. A. Moehring, and R. A. Walton, *Inorg. Synth.*, **27**, 14 (1990).
25. H. H. Wang, A. M. Mueting, J. A. Casalnuovo, S. Yan, J. K.-H. Barthelmes, and L. H. Pignolet, *Inorg. Synth.*, **27**, 22 (1990).
26. R. A. Head, *Inorg. Synth.*, **24**, 213 (1986).

INDEX OF CONTRIBUTORS

Prepared by THOMAS E. SLOAN*

*Chemical Abstracts Service, Columbus, OH.

SUBJECT INDEX

Prepared by THOMAS E. SLOAN*

Names used in this Subject Index for Volumes 26–30 are based upon IUPAC *Nomenclature of Inorganic Chemistry*, Second Edition (1970), Butterworths, London; IUPAC *Nomenclature of Organic Chemistry*, Sections A, B, C, D, E, F, and H (1979), Pergamon Press, Oxford, U.K.; and the Chemical Abstracts Service *Chemical Substance Name Selection Manual* (1978), Columbus, Ohio. For compounds whose nomenclature is not adequately treated in the above references, American Chemical Society journal editorial practices are followed as applicable.

Inverted forms of the chemical names (parent index headings) are used for most entries in the alphabetically ordered index. Organic names are listed at the "parent" based on Rule C-10, *Nomenclature of Organic Chemistry*, 1979 Edition. Coordination compounds, salts and ions are listed once at each metal or central atom "parent" index heading. Simple salts and binary compounds are entered in the usual uninverted way, e.g., *Sulfur oxide* (S_8), *Uranium(IV) chloride* (UCl_4).

All ligands receive a separate subject entry, e.g., *2,4-Pentanedione*, iron complex. The headings *Ammines, Carbonyl complexes, Hydride complexes,* and *Nitrosyl complexes* are used for the NH_3, CO, H, and NO ligands.

*Chemical Abstracts Service, Columbus, OH.

309

FORMULA INDEX

Prepared by THOMAS E. SLOAN*

The Formula Index, as well as the Subject Index, is a Cumulative Index for Volumes 26–30. The Index is organized to allow the most efficient location of specific compounds and groups of compounds related by central metal ion or ligand grouping.

The formulas entered in the Formula Index are for the total composition of the entered compound, e.g., F_6NaU for sodium hexafluorouranate (V). The formulas consist solely of atomic symbols (abbreviations for atomic groupings are not used) and arranged in alphabetical order with carbon and hydrogen always given last, e.g., $Br_3CoN_4C_4H_{16}$. To enhance the utility of the Formula Index, all formulas are permuted on the symbols for all metal atoms, e.g., in the order, e.g., $FeO_{13}Ru_{13}C_{13}H_{13}$ is also listed at $Ru_3FeO_{13}C_{13}H_{13}$. Ligand groupings are also listed separately in the same order, e.g., $N_2C_2H_8$, 1,2-ethanediamine, cobalt complexes. Thus individual compounds are found at their total formula in the alphabetical listing; compounds of any metal may be scanned at the alphabetical position of the metal symbol; and compounds of a specific ligand are listed at the formula of the ligand, e.g., NC for cyano complexes.

Water of hydration, when so identified, is not added into the formulas of the reported compounds, e.g., $Cl_{0.30}N_4PtRb_2C_4 \cdot 3H_2O$.

$AlCl_4P_3C_{36}H_{30}$, Aluminate(1−), tetrachloro-, 1,1,1,3,3,3-hexaphenyltriphosphenium, 27:254

$Al_2Cl_8FeN_6C_{12}H_{18}$, Aluminate(1−), tetrachloro-, hexakis(acetonitrile)iron(II) (2:1), 29:116

$AsBrF_6S_7$, Arsenate(1−)hexafluoro-, cyclo-heptasulfur(1+), bromo-, 27:336

$AsC_{18}H_{15}$, Arsine, triphenyl-, cobalt complex, 29:180
   iron complex, 28:171
   nickel complex, 28:103
   ruthenium complex, 29:162

AsF_6IS_7, Arsenate(1−), hexafluoro-, iodo-cyclo-heptasulfur(1+), 27:333

AsF_6O_2, Dioxygenyl hexafluoroarsenate(1−), 29:8

$AsFeO_4C_{22}H_{15}$, Iron(0), tetracarbonyl(triphenylarsine)-, 26:61, 28:171

$As_2C_{10}H_{16}$, Arsine, 1,2-phenylenebis(dimethyl-, gold complex, 26:89
   nickel complex, 28:103

$As_2Cl_3NRuSC_{36}H_{30}$, Ruthenium(II), tri-

chloro(thionitrosyl)bis(triphenylarsine)-, 29:162

$As_2H_4O_{70}Rb_4W_{21}.34H_2O$, Tungstate(4−), aquadihydroxohenhexacontaoxobis[trioxoarsenato(III)]henicosa-, tetrarubidium, tetratricontahydrate, 27:113

$As_2H_8O_{70}W_{21}.xH_2O$, Tungsten, aquahexahydroxoheptapentacontaoxobis[trioxoarsenato(III)]henicosa-, hydrate, 27:112

$As_4Au_2F_{10}C_{32}H_{32}$, Gold(I), bis[1,2-phenylenebis(dimethylarsine)]-, bis(pentafluorophenyl)aurate(I), 26:89

$As_4Co_2H_{100}N_{24}O_{142}W_{40}.19H_2O$, Ammoniodicobaltotetracontatungstotetraarsenate(23−), tricosaammonium, nonadecahydrate, 27:119

$As_4Na_{28}O_{140}W_{40}.60H_2O$, Sodiotetracontatungstotetraarsenate(27−), heptacosasodium, hexacontahydrate, 27:118

$As_4NiC_{20}H_{32}$, Nickel(0), bis[1,2-phenylenebis(dimethylarsine)]-, 28:103

*Chemical Abstracts Service, Columbus, OH.